AF255733

Beyond the Boundary

Exploring the science and culture of interstellar spaceflight

Edited by Kelvin F Long

Initiative for Interstellar Studies
www.i4is.org

Published by the Initiative for Interstellar Studies ©2015

i4is.org
info@i4is.org

ISBN: 978-0-9935109-08

Cover designed by Keith Cooper.
Background – The Christmas Tree Cluster and Cone Nebula. Credit:ESO.
Inset 1 – Exoplanet. Credit: ESA/NASA/G Tinetti (UCL)/M Kornmesser (ESA/Hubble).
Inset 2 – Photon starship. Credit: David A Hardy.
Inset 3 – Voyager 1. Credit: Adrian Mann.
Back cover – star cluster NGC 3590. Credit: ESO/G Beccari.

We dedicate this book to Dr Robert L Forward, a man who exemplified everything that was good about humanity, and who worked towards the fulfillment of the interstellar vision with leadership, courage, inspiration, intellectual pioneering and joy.

May we make him proud by our efforts.

Beyond the Boundary

ACKNOWLEDGEMENTS

This book would not have been possible if not for the outstanding efforts of many people. First, we would like to thank the contributors, including Jon Lomberg for his foreword and for believing in our vision as being consistent with that of Carl Sagan, a hero to many of us. We would also like to say a big thank you to our other contributors: Ian Crawford, Giovanni Vulpetti, Stephen Ashworth, Rachel Armstrong, Bill Cress, Richard Osborne, Jonathan Brooks, Angelo Genovese, Adam Crowl, Remo Garattini, Tiffany Frierson, Divya Shankar, David Fields, Jeremy Clark, Keith Cooper, Robert Kennedy, Kenneth Roy, Andreas Hein, Martin Ciupa, David Hardy and Alex Storer.

A special thank you to two people who have also contributed significantly toward the book's development – Keith Cooper for the production of the chapters and Adrian Mann for the use of his artwork; he remains the best darned graphical engineer in the world.

We would also like to thank Chris Welch and Gregory Matloff for their continued guidance as the leadership team of the I4IS Advisory Committee. We would like to thank the continued support and mentoring by members of the British Interplanetary Society, in particular its most recent President, Alistair Scott, its Executive Secretary Suszann Parry and the Editor of its popular publication *Spaceflight*, David Baker. We also thank the Managing Director of Reaction Engines Ltd, Alan Bond, who is an inspiration and whose continued support is appreciated.

Many others in the Initiative for Interstellar Studies have played a role in our success to date, even if they did not directly contribute towards this volume. We would like to thank them, because this is their success too. Several others outside of the organisation have also helped at some of our various events and this is our opportunity to mention them as well. So a big thank you to Robert Swinney, George Abbey Jnr, Sarah Margree, Nicholas Eftimiades, Freeman Dyson, Claudio Maccone, Roman Kezerashvili, Michael Minovitch, Austin Tate, Raghavan Gopalaswami, Martyn Fogg, Pamela Menges, Tobias Lugoloobi, Randy Chung, Tom Shirley, John Quel, Damian Evans, Michael Million, Tom Pothecary, Maria Solomon, Jeff Lee, Tim Gurshin, Kieran Twaites, Andrew Alexander, Sissi Enestam, James Harpur, Samiksha Mestry, Gillian Norman, Ian Norman, John Davies, Lindsay Wakeman, Karen Hart, Gemma Long, Amelia Scarlett Long, Terry Regan, Michael Dawson, Paul Hildebrandt, Paul Campbell, James Benford, James French, Andreas Tziolas, Richard Godwin, Mark Hempsell, Terry Kammash, Colin McInnes, Tibor Pacher, Austin Tate, Joe Ritter, Gerry Webb, Friedwardt Winterberg, Frank da Silva, Mark Raimondeau, Stephen Baxter, Alastair Reynolds, Gregory Benford, Louis Savy and Esther Armstrong.

We also thank other organisations with which we have been collaborating on specific activities, including DMTLab, Sci-Fi London, Simon Bowden Architects, Tamke-Allan Observatory, Commercial Space Technologies Ltd, the International Space University, Stellar Engines Ltd and Reaction Engines Ltd.

We did not use a publishing house for this book, but are publishing the book via Lulu.com, and so we thank Lulu for providing such a great service.

Finally, we thank all the other interstellar organisations out there for their continued efforts to bring about an interstellar capable society. As well as the British Interplanetary Society, we would like to give a mention in dispatches to Tennessee Valley Interstellar Workshop, Icarus Interstellar and the Tau Zero Foundation. May our shared dreams and our shared humanity help us to forge a path towards the stars.

CONTENTS

Dedication...iii
Acknowledgements..v
Contents...vii

Foreword...1
Jon Lomberg

Main introduction...9
Kelvin F Long

SECTION I: THE CHALLENGE OF INTERSTELLAR FLIGHT
Introduction...13
Kelvin F Long

1. Scientific and Societal Benefits of Interstellar Exploration.....................14
Ian Crawford

2. The Conceptual Basis for Interstellar Flight...43
Giovanni Vulpetti

3. When Will Voyager 1 Leave the Solar System?......................................63
Stephen Ashworth

SECTION II: BUILDING A BRIDGE TO THE STARS
Introduction...77
Kelvin F Long

4. Building A Starship From the Lessons of History.............................79
Jonathan Brooks

5. Starship Cities...87
Rachel Armstrong

6. Building Heavy Lift Rockets: How United Launch Alliance Do It.............105
Bill Cress

7. Future Launch Vehicles...125
Richard Osborne

8. The Road Before the Stars: Interstellar Precursor Missions...............145
Kelvin F Long

SECTION III: INTERSTELLAR TECHNOLOGIES & SOLUTIONS
Introduction...177
Kelvin F Long

9. Advanced Electric Propulsion for Interstellar Precursor Missions.................179
Angelo Genovese

10. Power for the Stars: Boot-strapping Our Way to Alpha Centauri...............215
Adam Crowl

11. Warp Drives and Faster-Than-Light Travel...................................227
Tiffany Frierson

12. Interstellar Travel and Traversable Wormholes............................239
Remo Garattini

Contents

13. Interstellar Communication: RF and Optical Establishment......................253

Divya Shankar

14. Starship Communication Via Quantum Entanglement.............................265

David E Fields

15. An Introduction to Artificial Intelligence As Applied to Space Exploration......275

Jeremy Clark

SECTION IV: INTERSTELLAR DESTINATIONS, COLONISATION AND THE SEARCH FOR LIFE

Introduction...299

Kelvin F Long

16. Destinations: The Planets Next Door...301

Keith Cooper

17. What Do We Do Once We Get There?.......................................327

Kenneth Roy, David E Fields and Robert G Kennedy

18. The Greatest Challenge: Manned Interstellar Travel....................349

Andreas M Hein

19. The Drake Equation, Fermi's Paradox and Active SETI.............377

Martin Ciupa

SECTION V: COMMUNICATION OF THE VISION

Introduction...395

Kelvin F Long

20. Travelling to the Stars on a Paintbrush..397
David A Hardy

21. Musica Universalis: The Interstellar Relationshiop Between Space, Music and
Popular Culture..405
Alex Storer

About the Initiative for Interstellar Studies...................................421
Meet the Authors..423
Glossary...433
Index..435

FOREWORD

JON LOMBERG

Science fiction (sf) has always served as a test bed for the future. Long before humans set foot on the Moon, the technical challenges and social implications were thrashed out in detail in stories and novels. The ships, the spacesuits, the launch and landing sites, the hazards of the space environment, the qualities and skills required from the crew – all were presented in endless variations from the most plausible to the most fantastic tales. Furthermore, both sorts of fiction inspired generations of scientists, engineers and astronauts to live out the dream.

Interestingly, sometimes the most ridiculous fantasy can have the deepest impact. While researching the history of Mars science fiction for the Planetary Society's *Visions of Mars* DVD sf collection (now on Mars aboard the Phoenix lander) I found to my surprise that some of the most serious people had been inspired by the silliest fiction. Rocket pioneer Robert Goddard was strongly influenced by Garrett P Serviss' now-forgotten *Edison's Conquest of Mars* (1898), a completely unauthorised sequel to H G Wells' *The War of the Worlds*. It so impressed Goddard that he had an epiphany in which he saw his life's work as creating a method of interplanetary travel. Carl Sagan became Mars-enchanted by Edgar Rice Burroughs' swashbuckling John Carter of Barsoom novels. Meanwhile, for many of us, the first Martian we ever met was the Warner Brothers' cartoon character Marvin Martian. It is not the quality of the science that determines the impact of the fiction.

The same can be said for sf's role in envisioning interstellar travel. The first starships appeared in the early twentieth century, when artists illustrating stories written for pulp magazine or newspaper comic strips invented the genre of sf art. These earliest spaceships were unencumbered by any engineering concerns. Just imagining the existence of such ships was hard enough, never mind how they looked or worked. The spaceship was a new concept; who could say how it would look?

Along with still images, from the beginning of movies there were filmmakers creating their own visions of starships. In what has become a film icon, George Melies' *A Trip to the Moon* (1902) has a scene where the Man in the Moon has his eye impacted by a vehicle from Earth. In the 1930s the sf pulps found a cinematic counterpart in Flash Gordon and other adventure serials, often based on popular comic strips. A more serious attempt was made in *Things to Come* (1936), a British film version of the H G Wells novel where a Europe destroyed by war takes a century to rebuild itself into a modern, scientific, perfect society, symbolised by the launch of their first vehicle into space. "It's the Universe, or nothing," is the bold conclusion of that great film and it can still send a shiver up your spine – and is, I suspect, the core belief of many space enthusiasts.

After World War II rockets were no longer science fiction and a new mania appeared, centered on UFOs – alleged flying saucers from other planets. Many movies imprinted their image of spaceships or flying saucers on the popular imagination: *The Thing from Another World* (1951), *The Day the Earth Stood Still* (1951), *Earth Versus the Flying Saucers* (1956) and *Forbidden Planet* (1956). When television entered the culture, spaceships appeared on that new medium as well with *Tom Corbett, Space Cadet* (1950–55) and other space adventures for children. The 1950s ended with *The Twilight Zone*, which brought science fiction to adults in prime time television, featuring many stories about spaceships of one sort or another.

In the 1960s, *2001: A Space Odyssey* and *Star Trek* set the standards for adult oriented sf in films and television and the look of their ships became the iconic images of space vehicles. The ships in *2001* were as strictly engineered as if they were real spacecraft. Writer Arthur C Clarke's incredible knowledge about the topic, combined with Director Stanley Kubrick's obsession for perfection in every detail, put that film into a class of its own, setting a high standard for everything since, especially the look of spaceships.

As for the original *Star Trek*, well I still have a fondness for the first Enterprise, tacky as the sets and models were compared to later incarnations of Starfleet. They did not have Kubrick's budget either. The original series was ahead of its time in many ways, including a female black officer in the crew for example. It was also ahead of itself in the graphics possible on television.

In the mid-1970s computers began to remake both media. At first, cameras could be programmed to make complicated moves repeatedly, allowing for more convincing compositing of physical models, backgrounds, and actors, with *Star Wars* (1977) being the wildly successful result. This movie was followed soon by *Close Encounters of the Third Kind* (1977), which showed how polished and intricately crafted spaceship models could be. The mothership in *Close Encounters* was beautiful as a Faberge egg – a long way from the sparking and smoking thermos bottle spaceships of Flash Gordon. For a few years, special effects in films became a hybrid of techniques, combining traditional model and cel animation with liquid-into-liquid tank and lighting effects and, increasingly, digital graphics. *The Last Starfighter* (1984) was, I believe, the first film in which the ships were completely digital, using no models at all (not counting 1982's *Tron*, which was set inside a computer). Meanwhile, the likes of *Space: 1999* (1975–77) and *Battlestar Galactica* (1978) brought this revolution to television.

In the decades that followed the need for models and painted backgrounds diminished as digital images became more complex and realistic. Significantly, *Star Trek* was reincarnated with a new level of detail and plausibility.

Star Trek: The Next Generation (1987) and the subsequent series in Roddenberry's universe – *Deep Space Nine*, *Star Trek: Voyager* and *Enterprise* were unusually self-conscious and intentional about their powerful societal role in expressing an optimistic vision of a positive future. The lavish sets and thoughtfully designed technology took the future very seriously, creating an internally consistent (if fake) future physics. Dilithium crystals may not exist in reality, but aboard any Starfleet vessel they worked the same way and required the same precautions. The series' producers took their invented physics seriously enough to spend a lot of time and money on the architecture of the ships, weapons, and astronomical backgrounds. Technical manuals were published and studied by fans who wanted to know how to get to the ship's bar in Ten Forward, as well as the configuration of weapons on a Romulan warbird. The intensity of the devotion of the fans – Trekkies (or Trekkers) becoming the embodiment of all sf fandom – was not lost on NASA, who took every opportunity to leverage the popularity of the series into support for the American space programme. You just could not ignore a television show so popular that it had fans learning fake alien languages (at least one opera has been written and performed entirely in Klingon, while in more than one family, Klingon is the language used within the home). Unlike the *Star Wars* universe, which was pure adventure fantasy "long, long ago and far, far away," the *Star Trek* universe seemed like one that might actually be near and close, maybe just a century or two away.

Star Trek helped evolve a standard look for starships employing a common set of visual conventions – stars smearing during warp drive, lots of little lighted windows giving the ship its scale, a complex maze of panels and pipes, vents and valves studding the exterior surfaces (that one started in *2001: A Space Odyssey*). Ships like this flew through many movies; the number of sleek and slick starships became too many to count. Yet like the branching twigs on an evolutionary tree, they all trace back to common ancestors. A consensus interstellar future has acquired a characteristic look agreed to by nearly all fans, that shapes our concepts even in serious discussions of real starships.

At some point in the last third of the twentieth century the cinematic and video images replaced book and magazine illustration as the dominant influence on popular culture and the public imagination. If you ask anybody to think of a spaceship, it is far more likely that they will summon up the Enterprise or the Millennium Falcon rather than anything described by Robert Heinlein or Larry Niven. Both writers have masterfully described their starships far more plausibly than anything on film or television, but it is not their ships we dream of.

Carl Sagan,
on the set of
Cosmos in his
'starship of the
imagination'.

Carl Sagan's *Cosmos* television series (1980) presented a different approach to the problem of getting around quickly in this vast Universe. Early in the production of this 13 episode series, for which I was Chief Artist, there was a problem imagining the sort of vehicle in which Carl could travel to cosmic places he was describing. Executive Producer Adrian Malone absolutely wanted a place where Carl was talking from, a "spaceship of the imagination". If Carl were merely a voiceover narrator, the animation would lose the sense of the series subtitle 'A Personal Voyage'. He had to be in some sort of vehicle.

However, Carl did not want a ship of the type seen in *Star Trek* or *Star Wars*. Even if he had preferred a ship Han Solo would pilot, I would have argued against it. We were in production at KCET-TV in Hollywood at the same time as both *Star Trek: The Motion Picture* and *The Empire Strikes Back* were in production. We had seen some effects-industry previews of the work on those films and I knew we could not compete against animation budgets like Paramount's or Lucasfilm's. We needed a different look entirely, so I pulled out a painting that I had done a year earlier. Called 'Starseeds', it shows a metaphorical blend of astronomy and biology. (Carl Sagan got me started on this theme when he had commissioned me to do a painting called 'The Backbone of Night', which also ended up in *Cosmos*. I became fascinated by the poetic possibilities of combining

'Starseeds', the artistic inspiration for the spacecraft on Carl Sagan's *Cosmos*.
© Jon Lomberg.

cosmic and biological imagery and did a whole series of paintings as a result.)
Carl immediately said that this was the approach he wanted and thus was born
the dandelion metaphor repeatedly used in the series, as Carl's spaceship of the
imagination wafts across the *Cosmos*. From inside, it was a somewhat cavernous
room with a large window (designed by KCET's Art Director John Retsek) From
the outside, it was a bit of fluff drifting through the Galaxy.

In this case the spaceship was more than just a point of view. It made a
statement in which a sense of wonder and imagination replaced dilithium crystals
and antimatter as the motive force. Perhaps it also anticipated the emerging
convergence of biology and engineering, in which truly advanced machines –
perhaps even starships – will be grown rather than built.

Of course, until we build actual starships, nobody knows what they will
really look like. Today's digital spectaculars will no doubt seem as quaint and

dated as Victorian images of flying machines seem to us.

Still, there is a subset of spaceship enthusiasts who are quite serious about the feasibility of real starship designs. Lately there seems to have been a resurgence of interest in the question of how far we are as a culture from building starships. Science fiction continues to collaborate with real science and engineering in envisioning the future of astronautics. In the short term, small robot explorers using laser or microwave sail propulsion might be feasible. Ships with people in them will take a lot longer. Nuclear propulsion vehicles and gigantic worldships each have their champions. Ably illustrated and clearly presented, starship schemes abound in the lively sub-culture in which they grow. This book, *Beyond the Boundary*, is proof that the topic is alive and well and still awesome to think about.

Today's dreams serve to inspire tomorrow. They motivate us and provide some confidence that there is a future awaiting humanity other than the myriad dystopias favored by current sf. Amid the nightmare futures of malevolent robots, all-powerful computers, a poisoned planet, genetically engineered monsters, violent people living violent lives in the ruins of civilisation, is there an image of the future as positive as a starship? It might be a hundred or a thousand years until people journey to the stars, but the very fact that someday we might do it encourages the hope that someday we will do it.

Introduction

KELVIN F LONG

EXECUTIVE DIRECTOR,
INITIATIVE FOR INTERSTELLAR STUDIES (I4IS)

When you go out into the garden at night and look up at the stars, look towards the constellation Camelopardalis, the Giraffe. In that constellation is a star called Gliese 445, a red dwarf star close in the sky to the Pole Star, Polaris, and is 17.6 light years from the Sun. It's not an inherently particularly special star as far as we are aware, having a mass at most one third that of the Sun, but it is special for one important reason. In around 40,000 year's time, a spacecraft will pass within 1.6 light years of it. That spacecraft is from an alien civilisation that lives on a world that hangs like a pale blue dot in the darkness of the void. That world is called Earth and that spacecraft is called Voyager 1.

Voyager 1 is one of two probes from the exceptionally successful Voyager programme. Its brother, Voyager 2, is also on a journey to the stars and is headed into the direction of the star Ross 248 in the constellation of Andromeda, which it will pass by in around 40,000 years. The Voyager probes were launched in 1977 to explore the outer Solar System in a series of fly-bys and using gravitational slingshots to propel them from one planet to the next (a technique devised by I4IS' own Dr Michael Minovitch, who at the time was at the Jet Propulsion Laboratory). NASA had the vision to include an 'interstellar mission' at the end of Voyager's exploration of the planets. Although it seems to be a matter of some debate over whether the Voyagers have escaped from the solar heliosphere yet or not, what cannot be denied is that they are already out there. And so, by default, are we – in some sense, at least.

Gone is the time when interstellar travel was considered mere science fiction, and instead it is now considered a realistic part of our plausible future. Our efforts to build machines that go beyond the known boundaries are an important part of our cultural heritage and there is every reason to think we will continue to do so. From the flight dreams of Leonardo da Vinci in the 1500s, to the Wright Flyer of 1903, to the breaking of the sound barrier in 1947, to the escape of the atmosphere by a manned capsule in 1961, to the boots on the lunar soil in 1969, it is within our nature and our spirit to strive to soar ever higher, to find and conquer the next frontier. Our robotic probes are all over the Solar System and it is just a matter of time before we send them further and faster. Interstellar flight represents the ultimate end point of our astronautical endeavours and the future migration of the human species.

In 2012, many of the contributors to this book set about creating an organisation that would work towards this goal, and we founded the 'Initiative for Interstellar Studies' (I4IS), an organisation that has the aspirations of creating a global institute through its various educational, research and business entrepreneurship activities. This book represents the fruits of our labours over the last year or so and I am very proud to both be leading this team and also acting as Editor for this volume, which I hope will be one of many more to come as we continue to reach for the stars in every way that we can.

We did not start this book with a structured plan; we simply put out a call to our team and asked them to write about what mattered to them; what they cared about and what they were interested in. What we received in return was such a broad collection of submissions that they makes for a fascinating read across the spectrum of interstellar issues, problems and solutions. The chapters range from the philosophical to the technological to the strategic to the poetic and the artistic. And that's another thing we aspired to: we included the artists. It is partly through artwork that we can pull on the heart strings of human kind and we seek to give the artists the platform they deserve, to inspire us and help us to see what our shared vision for interstellar humanity might look like, feel like or even sound like.

Although the chapters that have been chosen to appear in this book were chosen with care, we should say at the outset that the views expressed do not necessarily represent the official policies or positions of I4IS, but those of the author who wrote that chapter. We believe there is value in plurality of views and we seek to provide a platform for divergent and creative thinking.

The book has been structured into five key sections. Part I is to provide a gentle introduction to the subject matter so that you, the reader, can have the context upon which to assess the difficult challenge that awaits us, but also the tremendous optimism we share for how we can approach those problems.

Part II is designed to show the journey that we must go on before we can accomplish true interstellar flight, from dealing with problems down here on Earth to addressing crucial issues of sustainability and development, and eventually launching those first interstellar precursor missions in order to demonstrate the technology to reach the stars.

Part III is about presenting many of the potential solutions and thoughts on how we might solve the difficult engineering and physics challenges, from space propulsion to communicating over large distances to the adoption of artificial intelligence. We don't claim that this section covers every option or problem that is to be addressed, but it is our hope that we can provide you with a taster of some of the issues.

Part IV leaps right out there amongst the stars as we examine the many objects that we have to explore with our robotic ambassadors. And of course, what we do once we reach those distance points of light? Do we observe them? Colonise them? Are we going to meet intelligent life from other cultures? If so, how will we respond to that? All of this depends on manned interstellar flight and the adoption of world ships, themselves small worlds that cruise in the ocean of night between the stars.

Finally, part V is about some of the methods we have been using to communicate the vision of the stars through artistic efforts. In particular, the paint brush and music. Have you ever been moved by a picture or a piece of music? This shows that there are other ways to communicate our ideas other than the adoption of complex mathematical language.

We at the Initiative for Interstellar Studies are very proud to publish this book and we hope that you get both enjoyment and education from reading it. If you are inspired by this tome, please do contact us and give us your feedback, and you are welcome to join us in our efforts to reach the stars. We are a growing organisation with very ambitious plans that go beyond the individual and which represent the hopes and dreams of the entire population of Earth. If this book goes some way towards helping you to believe in this seemingly impossible vision, then we have succeeded. If you decide to work towards it in your own unique way, through writing, music, art, research, education, entrepreneurship, then the chances of us succeeding in establishing an interstellar capable society over the next century will be vastly improved. Welcome, to the voyage of the starship as we seek to go *Beyond the Boundary*.

SECTION I

THE CHALLENGE OF INTERSTELLAR FLIGHT

INTRODUCTION BY KELVIN F LONG

Before we jump into the details of technology, it is important that we firstly become immersed into the exciting subject of interstellar studies. Here we have three authors who will do just that and help to bring you up to speed.

Ian Crawford, a Professor of Space Science, makes the case for star-flight and presents us with the various places that we might choose to go as well as the various motivations for going. This includes scientific reasons, such as going for astrophysical or astrobiological interest, but also for the long-term survival of the human race.

Giovanni Vulpetti, a physicist, also lays out the arguments for interstellar flight, but gives a more gives the question of why we should go a more philosophical grounding. He also develops a novel 'Conscious Life Expansion Principle', which is for the reader to adopt and scrutinise, but at the very least get much enjoyment from discussing it.

Finally, researcher **Stephen Ashworth** revisits our old friends the Voyager probes and asks when they will leave the Solar System. His thought provoking question relates to the uncertainty over how we define the edge of the Solar System. But what he also does is eloquently map out the dynamical challenge of sending spacecraft beyond our world, giving us an insight into the scale of the problem and what we have achieved to date.

CHAPTER 1

SCIENTIFIC AND SOCIETAL
BENEFITS OF INTERSTELLAR EXPLORATION

IAN A CRAWFORD

The growing realisation that planets are common companions of stars [1–2] has reinvigorated astronautical studies of how they might be explored using interstellar space probes (for reviews see references [3-7], and also other chapters in this book). The history of Solar System exploration to-date shows us that spacecraft are required for the detailed study of planets, and it seems clear that we will eventually require spacecraft to make in situ studies of other planetary systems as well. The desirability of such direct investigation will become even more apparent if future astronomical observations should reveal spectral evidence for life on an apparently Earth-like planet orbiting a nearby star. Definitive proof of the existence of such life, and studies of its underlying biochemistry, cellular structure, ecological diversity and evolutionary history will require in situ investigations to be made [8]. This will require the transportation of sophisticated scientific instruments across interstellar space.

Moreover, in addition to the scientific reasons for engaging in a programme of interstellar exploration, there also exist powerful societal and cultural motivations. Most important will be the stimulus to art, literature and philosophy, and the general enrichment of our world view, which inevitably results from expanding the horizons of human experience [9,10]. In the longer term, interstellar colonisation will lead to increased opportunities for the spread and diversification of life and culture through the Galaxy and greatly increase the survival chances of homo-sapiens and our evolutionary successors.

STARSHIP SCIENCE

There can be little doubt that science, especially in the fields of astronomy, planetary science and astrobiology, will be a major beneficiary of the development of an interstellar spaceflight capability. In its long history astronomy has made tremendous advances through studying the light that reaches us from the cosmos, but there is a limit to the amount of information that can be squeezed out of the analysis of starlight and other cosmic radiation. Already we can identify areas where additional knowledge will only be gained by making in situ observations of distant astronomical objects. As I noted in an earlier review of interstellar spaceflight [3], a sense of the scientific potential may be glimpsed by considering "the advantages of taking thermometers, magnetometers, mass-spectrometers, gravimeters, seismometers, microscopes and all the other paraphernalia of experimental science to objects that can today only be observed telescopically."

The scientific objectives of interstellar probes have been described previously by Webb [11] and Crawford [8], and can be divided into the following broad categories: (i) studies conducted en route (e.g. of the local interstellar medium, and other physical and astrophysical studies that could make use of the vehicle as an observing platform); (ii) astrophysical studies of the target star itself (or stars if a multiple system is selected); (iii) planetary science studies of any planets in the target system, as well as moons, asteroids and comets; and (iv) astrobiological/exobiological studies of any habitable (or inhabited) planets or moons that may be found in the target planetary system. Each of these areas has different requirements for the overall architecture of an interstellar mission and for the scientific payload to be carried. We will now briefly consider each in turn.

INTERSTELLAR MEDIUM STUDIES

By definition, any interstellar vehicle will have to traverse the interstellar medium between the Solar System and its target star. As any target star for an early interstellar mission is certain to be within a few light years of the Sun, it follows that only the local interstellar medium (LISM) is relevant here. I have provided a detailed review of the structure of the LISM, with interstellar travel specifically in mind, in a separate article [12] to which the interested reader is referred. Briefly, the Sun is currently located close to the boundary of a small (spatial extent ≤ 10 light years), low density (n_H> ~ 0.1-0.2 cm^{-3}, where n_H is the density of hydrogen nuclei), warm (T ~ 7,500 K) and partially ionised interstellar cloud known as the Local Interstellar Cloud (LIC). Whether the Sun lies just within, or just outside, the LIC is currently a matter of debate. The LIC is only one of several broadly similar interstellar clouds within 15 light years of the Sun; for example, Redfield and Linsky [13] identified seven such clouds within this volume. These are immersed in the very empty (n_H in the range 0.005 to 0.04 cm^{-3}) and probably very hot (T ~ 10^4 to 10^6 K) Local Bubble in the interstellar medium that extends for about 300 light years from the Sun in the galactic plane before denser interstellar clouds are encountered.

These properties have been estimated by a range of astrophysical techniques and are still quite uncertain (see [12] and references therein). Key measurements that could be made from an interstellar probe and which would add enormously to our understanding of interstellar processes, would include in situ determinations of density, temperature, gas-phase composition, ionisation state, dust density and composition, interstellar radiation field and magnetic field strength, all as a function of distance between the Sun and the target star system. Such in situ measurements, even though obtained on a very local scale in the galactic context, would be invaluable for validating ('ground truth') information based on astronomical techniques that will, of necessity, continue to be used to determine interstellar medium properties at larger distances (both within our Galaxy and beyond). These measurements will also be invaluable for the planning of all future interstellar space missions. The first mission will be a pathfinder in this respect and will enable all subsequent missions to be designed with a much firmer knowledge of the properties of the material through which they will have to travel.

We note that none of these measurements impose stringent constraints on the architecture of an interstellar mission. From the perspective of interstellar medium studies a simple undecelerated interstellar probe would be sufficient to obtain the necessary measurements.

STELLAR STUDIES

We know far more about the Sun than any other star, simply by virtue of the fact that it is so close to us. Interstellar spaceflight would enable us to obtain comparable information about stars of other spectral types. Such observations are likely to lead to significant advances in stellar astrophysics, although their extent will depend, at least in part, on the time window that is available in which to make the measurements and this will have implications for the mission architecture.

There are really three reasons for our enhanced knowledge of the Sun compared to other stars: (i) vastly increased spatial resolution, which permits the observation of small scale features on the photosphere (e.g. sunspots and associated phenomena), chromosphere and corona; (ii) greatly increased brightness, which permits very high-signal-to-noise observations (that among other things facilitates the use of helioseismology to probe the Sun's interior structure); and (iii) a long time base of observations (hundreds of years of recorded human observations and millions of years of relevant geological records on the Earth and other planets).

Although we might expect interstellar space travel to help principally with the first two of these, we have to recognise that long before rapid interstellar spaceflight becomes feasible, astronomical instrumentation is likely to have advanced to the point where many nearby stars will be resolvable from observations conducted from the Solar System. Indeed, we have already reached the point where the radii of nearby low mass stars can be measured directly using ground-based optical interferometry and the next generation of space-based interferometers may be able to resolve surface features [14]. Similarly, the advent of very large ground- and space-based telescopes will go some way to address signal-to-noise limitations caused by the relative faintness of other stars compared to the Sun. Nevertheless, it will always be true that the spatial resolution and signal-to-noise of observations conducted from a vantage point in the vicinity of a target star will be vastly higher than comparable observations attempted from the vicinity of the Earth. So, while we should avoid exaggerating the benefits to observational stellar astronomy from interstellar missions to the closest stars, we can nevertheless be sure that such advantages do exist.

While undoubtedly scientifically valuable, stellar observations conducted from an interstellar fly-by mission would suffer from disadvantages arising from the short time span available for the highest resolution observations. Much greater benefits would result if it proved possible to decelerate at the target star system. It would then be possible to ring the star with satellites to acquire long term observations of the whole stellar surface and to obtain time-resolved, high-resolution, multi-wavelength observations of the corona and stellar wind. As such observations are of demonstrable importance for understanding of our Sun, it follows that they would also be desirable for studies of other stars, but this will require the interstellar carrier spacecraft to decelerate essentially to rest in the target star system. In addition, all stars are surrounded by circumstellar matter to varying degrees and in situ studies of this would also be of scientific interest. Undoubtedly of greatest interest would be studies of protoplanetary discs from which planets may have recently formed, or still be forming. Measurements of the density, temperature, magnetic field and, crucially, dust particle size as a function of radial distance from the star and distance from the disc mid-plane would greatly add to our understanding of planet formation processes. However, the nearest known example of a circumstellar disc of this type is around the star epsilon Eridani at a distance of 10.5 light years (see Table One later in this chapter) and, although a possible candidate for an early interstellar mission, its relatively large distance means it is unlikely to be a high priority for the first such missions.

PLANETARY SCIENCE

Over 1,800 planets are now known to orbit other stars [1,2], with new discoveries being made every month. The Kepler Space Telescope has already identified over 4,000 additional candidates (most of which will be confirmed as planets), and there will be many more to come as the full Kepler dataset is analysed [15]. Indeed, a conservative view of the statistics, discussed later in this chapter, implies that most stars in the Galaxy will be accompanied by planetary systems.

Future astronomical observations are certain to improve our knowledge of planetary systems around nearby stars. These discoveries are likely to be followed in the coming decades by observations conducted with increasingly sophisticated space-based telescopes able to directly image planets orbiting nearby stars (say out to 30 light years) and to obtain spectroscopic measurements of their atmospheres [16]. Indeed, it is salutary to reflect that, within the coming decades, astronomical observations will very likely have raised our knowledge of planetary systems around nearby stars to a level comparable to that obtained for the planets in our own Solar System prior to the Space Age. That is to say, we will know the number of planets in each system (down to some minimum mass that will probably be significantly less than that of Earth), together with their orbital parameters, masses and densities, presence or absence of an atmosphere, atmospheric composition, presence of large natural satellites, etc. All this can probably be learned without having to leave the Solar System.

That said, the history of the exploration of our Solar System shows that obtaining significantly more knowledge of extra-solar planetary systems will require in situ observations by spacecraft. We can be sure of this because, over the last half century, spacecraft have completely revolutionised the study of the planets of the Solar System, providing information that could never have been obtained telescopically from the surface of the Earth or its immediate vicinity. To highlight just three out of hundreds of possible examples, consider the structure of the lunar interior as probed by the Apollo seismic experiments, the fine scale (i.e. millimetres to centimetres) resolution of mineralogical and sedimentary structures at the landing sites of the Mars Exploration Rovers (with their implications for the volcanic and hydrological histories of that planet) and the discovery of lakes of liquid methane (and indeed an entire methane hydrological cycle) under the orange smog of Titan's atmosphere by the Cassini–Huygens mission. It follows that if we wish to obtain comparable knowledge of the planets orbiting other stars then we will have to go there and look.

The analogy with the exploration of our own Solar System has implications for the architecture of an interstellar mission designed with planetary science in mind. There is a hierarchy of architectural options for planetary missions, in order of increasing complexity and energy requirements, but also in increasing scientific return: (i) fly-by missions; (ii) orbital missions; (iii) hard landers; (iv) soft landers (with or without rover-facilitated mobility); and (v) sample return missions. The same general ordering will apply in the study of extra-solar planetary systems, although the relative jumps in difficulty between them are not the same in the two cases.

An undecelerated fly-by will be the easiest to implement and, for this reason, was adopted in the pioneering Daedalus study [17]. However, the exploration of the Solar System shows that, while appropriate for the initial reconnaissance of a planetary body, fly-bys are very limited in terms of the knowledge they are able to collect (and sometimes this information can be misleading, as in the case of the Mariner 4 fly-by of Mars in 1965 that revealed a lunar-like landscape and gave little intimation of the geological diversity discovered by later missions). The limitations of fly-bys in an interstellar mission will be exacerbated by the high speeds involved – the Daedalus study proposed to conduct planetary investigations from multiple sub-probes flying close to target planets at 12 percent of the speed of light. This would permit less than a second of time available for detailed observations at distances comparable to the radii of planetary-sized bodies, although perhaps several hours of useful observations might be obtained on the approach to and departure from the planet in question.

Much more scientific information would be obtained if it proved possible to decelerate an interstellar vehicle (or at least any sub-probes designed to conduct planetary observations) from its interstellar cruise velocity. The benefits will be immediately obvious by comparing the results of the initial fly-by reconnaissance of Mars by Mariners 4, 6 and 7 with those of the early orbital missions (i.e. Mariner 9 and Vikings 1 and 2) that discovered, amongst other things, the giant Tharsis volcanoes, the Valles Marineris canyon system and numerous dried-up river valleys indicating a warmer, wetter Martian past. Of course, even more detailed information has resulted from the handful of soft landers and rovers that have successfully reached the surface.

Although in terms of Solar System exploration there is a big jump in energy requirements between orbital missions and soft landers, this would not be a major consideration in terms of an interstellar mission – the energy differential between orbital insertion and a soft landing is trivial in comparison to that of decelerating a probe from a significant fraction of the speed of light. As for Solar System missions, landers would permit a range of geochemical, geophysical and astrobiological investigations that are simply not possible from an orbiting spacecraft. Thus, despite the added complexity involved, the potential scientific benefits are such that the designers of any interstellar mission capable of decelerating at its destination should consider including sub-probes that are capable of actually landing on the surfaces of suitable planets.

The most ambitious Solar System missions involve sample return, which allow detailed investigation of planetary materials in terrestrial laboratories. However, for any reasonable extrapolation of foreseeable technology, sample return is essentially impossible from an extra-solar planetary system on any reasonable timescale. It follows that the kinds of sophisticated geochemical (and biological) analyses that today require samples to be returned to Earth will have to be automated for in situ robotic operation within the target planetary system. Fortunately, by the time we will be in a position to build interstellar probes, the capabilities of autonomous laboratory analyses should have advanced considerably beyond present capabilities.

Astrobiology/Exobiology

Astrobiology is the science relating to the search for life elsewhere in the Universe, especially the astronomical and planetary environments that may nurture it. By adding to our knowledge of other stellar and planetary environments, the in situ scientific investigations outlined above would be of considerable astrobiological value even if no indigenous life is present in the target system. Nevertheless, it is clear that the greatest scientific interest would be in the discovery and characterisation of any life-forms that may be present. If such extraterrestrial organisms are found, their study will presumably become the subject of a new sub-discipline of biology where, by definition, the study of living things properly belongs [18].

As noted above, before rapid interstellar space travel becomes possible we will almost certainly have identified which of the nearest stars are accompanied by planetary systems. Indeed, we are likely to know the basic architecture of these systems in some detail and Solar System-based instruments will have the capability of detecting any molecular biosignatures that may be present in the atmospheres and/or on the surfaces of these planets [16]. We can therefore be confident that astronomical observations will be able to establish a hierarchy of priorities among any planets that may be detected around the nearest stars: (i) planets where bona fide bio-signatures are detected; (ii) planets that appear habitable (e.g. for which there is spectral evidence for water and carbon dioxide, but no explicit evidence of life being present); and (iii) planets that appear to have uninhabitable surfaces (either because of atmospheric compositions deemed non-conducive to life or because they lack a detectable atmosphere), but which might nevertheless support a subsurface biosphere. Thus, when planning an interstellar mission with astrobiology/exobiology in mind, we are likely to have a priority list of target systems prepared well in advance.

As for the planetary science cases discussed on pages 18–21 and for the same reasons, it is not immediately obvious that simple fly-by missions could add significantly to information likely to be obtained by the astronomical observations from the Solar System. There will be some advantages: even travelling at a significant fraction of the speed of light, probes (or sub-probes) targeted to fly close to planets could presumably perform much more detailed analyses of their atmospheric compositions than would be possible astronomically from the Earth. Nevertheless, it seems clear that only an interstellar probe that decelerated into its target star system would be able to deploy the kind of instrumentation that biologists would need to begin an investigation of an alien biosphere in any detail.

We can get an idea of the kind of instruments that would be required by considering those that have either been used (e.g. the Viking biology package [19] and the Phoenix high-resolution microscope [20]), or are planned to be used (e.g. the Urey organic molecule analyser [21] and the Life Marker Chip [22]), in the search for life on Mars. Doubtless much more sophisticated analytical tools will be available by the time of the first interstellar mission. However, it seems clear that deployment of instruments such as these would require the soft-landing of suitably instrumented sub-probes on a planetary surface – such analyses cannot be done while flying through the target system at ten percent of the speed of light!

Cultural and Societal Motivations for Interstellar Exploration

In addition to the scientific reasons for wanting to travel to the stars there are also a number of compelling cultural and societal reasons. As I have argued previously [23], many of these societal benefits would result from any large-scale programme of space exploration, beginning in our own Solar System. However, exploration and colonisation on interstellar scales will greatly increase the potential benefits, to which we now turn.

Survival

Currently humanity exists on a single small planet adrift in what is at best an uncaring and at worst a dangerous Universe. Our civilisation and perhaps even our existence as a species is therefore vulnerable to a range of natural hazards that could affect the habitability of our planet (examples include asteroid and comet impacts, large volcanic eruptions and unanticipated changes in solar activity). Moreover, we have to recognise that we are also at risk from ourselves, through the accidental or deliberate misuse of our own technology (obvious examples include nuclear or bacteriological warfare or terrorism, and irrecoverable environmental degradation). While there are in principle technological means to mitigate the former risks and political means to mitigate the latter, the fact remains that humanity will remain vulnerable to extinction while we remain a single-planet species. As Shepherd [24] put it, in what was arguably the first ever detailed discussion of the technical possibility of interstellar spaceflight,

"Humanity dispersed over many worlds would appear to be more secure than humanity crowded on one single planet."

Although it is true that many of the existential threats facing humanity (e.g. asteroid impacts and enhanced terrestrial volcanic activity) would be greatly alleviated by establishing human colonies elsewhere in the Solar System, others (e.g. unanticipated solar or galactic events) could in principle render the entire Solar System uninhabitable. Of course, on the longest (albeit multi-billion-year) timescales we know that the inexorable evolution of the Sun towards becoming a red giant star will eventually completely sterilise first the Earth, then Mars and eventually even the moons of the outer Solar System [25]. Ultimately, therefore, the survival of humanity (and, on longer timescales, our evolutionary successor species) and indeed of terrestrially-evolved life itself will depend on interstellar colonisation.

Diversification of Life, Culture and Intelligence in the Galaxy

In addition to 'mere' survival, the expansion of humanity (and ultimately post-humanity) into space opens up opportunities for the diversification of culture, what John Stuart Mill termed "different experiments of living" [26], that would otherwise not occur. As long ago as 1948 this was recognised as a potential benefit of space colonisation by the English philosopher Olaf Stapledon when, in a lecture to the British Interplanetary Society [27], he expressed the view that "The goal for the Solar System would seem to be that it should become an interplanetary community of very diverse worlds each inhabited by its appropriate race of intelligent beings, its characteristic 'humanity'… Through the pooling of this wealth of experience, through this 'commonwealth of worlds' new levels of mental and spiritual development should become possible, levels at present quite inconceivable to man."

Although, as envisaged by Stapledon, opportunities for diversification of culture will presumably result from colonising the planets and moons of the Solar System, these will remain very limited compared to the possibilities resulting from interstellar colonisation. Readers interested in exploring some of the possibilities are referred to chapters in the books *Interstellar Migration and the Human Experience* [28] and *Starship Century* [29] and, of course, to countless science fiction stories too numerous to mention here.

As Stapledon himself realised (see discussion in [30]), the scope for human (and post-human) colonisation and diversification throughout the Galaxy depends crucially on the presence or absence of other intelligent species. We do not yet know how common, or otherwise, extraterrestrial intelligence may be, but the so-called Fermi Paradox (i.e. the observation that the Earth has not itself been colonised by other technological civilisations [31-33]) suggests that other civilisations may be very rare or even non-existent. If our Galaxy, or at least our part of it, really is devoid of other intelligent civilisations, it follows that the future of intelligence in the Galaxy will depend on us. It may then be desirable for humanity (or post-humanity) to start moving out through the Galaxy colonising uninhabited planets because this would enhance the diversity and creative potential of intelligent life in the Universe. We must however recognise, as Stapledon also foresaw, that these alien planetary environments are unlikely to be such as to support human life directly and that terraforming the planets (i.e. rendering them more Earth-like) and/or genetic engineering the colonists (to adapt them to the local environments), is likely to be necessary.

There is one further important point to make. Not only do we not know how common extraterrestrial intelligence is in the Galaxy, we do not as yet even know how common life itself is. It may be that the transition from non-life to life is so unlikely that it only happens very rarely and for all we know it may have happened only once (see discussion by Paul Davies [33]). If our astronomical searches and eventual follow-up interstellar probes do reveal life to be absent even on apparently habitable planets then the question will arise as to whether we should seed them with life derived from Earth. As discussed by Francis Crick [34], even if it proves too difficult to send human beings to the stars, we can certainly envisage spreading Earth-life to other planetary systems, perhaps in the form of micro-organisms specifically genetically engineered to survive the journey and thrive on a particular target planet. If one accepts that life, with its vast potential for growth and diversity, is preferable to non-life then this could be seen as a desirable activity (although great care would need to be taken to ensure that the target planets really were uninhabited – damaging an indigenous biosphere by introducing life from Earth would be unconscionable). Conceivably, therefore, the future evolution of all life in the Galaxy and not just human (and post-human) intelligent life, may depend on interstellar exploration and colonisation activities initiated by humanity within the next few centuries.

AVOIDING INTELLECTUAL STAGNATION AND 'THE END OF HISTORY'

In 1989 the American political philosopher Francis Fukuyama published a remarkable essay entitled *The End of History?* in an obscure American journal [35]. The essence of Fukuyama's argument (subsequently expanded in his book *The End of History and the Last Man* [36]) was that humanity might be approaching the end of a long ideological evolution toward a stable form of political organisation (specifically that the whole world will soon be organised on liberal democratic principles). Although subsequent events (perhaps most worryingly the increasing influence of fundamentalist religious ideologies in some parts of the world) may suggest that Fukuyama was overly optimistic, a general trend towards democratic and liberal values has been apparent in world history for several centuries and seems likely to continue. However, despite the essentially optimistic nature of the argument, Fukuyama himself was ambivalent towards this outcome because he believed that an end to human ideological competition would also mean an end to human achievement and creativity. As he put it [35]: "The end of history will be a very sad time. The struggle for recognition, the willingness to risk one's life for a purely abstract goal, the worldwide ideological struggle that called forth daring, courage, imagination and idealism, will be replaced by economic calculation, the endless solving of technical problems, environmental concerns and the satisfaction of sophisticated consumer demands. In the post-historical period there will be neither art nor philosophy, just the perpetual care-taking of the museum of human history."

In contrast to this rather depressing vision of the future, what we really want to do is build a human civilisation that is both stable and dynamic. That is, a civilisation that is at peace with itself, but which is nevertheless an exciting place in which to live and, crucially, one whose history remains open. As I argued in an initial response to Fukuyama's ideas [9], an ambitious programme of space exploration is ideally and, perhaps uniquely, suited to satisfying these socially desirable objectives.

Fukuyama was especially concerned that a homogenised future world would mean an effective end to human creativity because many traditional sources of artistic and intellectual stimuli would have dried up. I think it must be true that new sources of intellectual stimuli will be required if human (and eventually post-human) culture is not to stagnate, because ultimately all our science, art, and philosophy is built on what John Locke [37] called 'simple ideas' – that is ideas based on sense perception and reflection on these perceptions. It follows that we cannot imagine genuinely new things, but instead need to discover them. This is the ultimate cultural benefit of all exploration.

In the present context, it is clear that space exploration presents a vast new field of activity with literally infinite potential for discovery and intellectual stimuli of multiple kinds – certainly a far richer range of stimuli than we could ever hope to experience by remaining on our home planet. As was the case for the survival and diversification issues discussed above, humanity will begin to experience these intellectual and cultural benefits by exploring and colonising our own Solar System, but it is on the far larger stage of interstellar exploration that they will really come into play.

Possible destinations: Stars and Planets within 15 Light Years

Given the sheer technical difficulties of achieving interstellar spaceflight on a human timescale (see discussion in references [3-7], and other chapters in this book), for the foreseeable future only the very nearest stars are likely to be candidates for exploration. Here I consider a radius of 15 light years from the Sun to enclose the volume likely to be of interest for the early phases of an interstellar exploration programme (i.e. a distance that could be covered in 100 years at an ambitious cruising speed of 15 percent that of light).

Within 15 light years of the Sun there are approximately 58 stars, in 39 separate stellar systems. These are listed in Table One. The number is approximate for several reasons. Firstly, at the outer boundary the errors on the distances can amount to a few tenths of a light year, which could mean that some stars notionally just beyond 15 light years might actually be closer (and vice versa). Secondly, not all stars within this volume may yet have been discovered, although this is only likely for the very dimmest red or brown dwarfs; that this is a very real possibility was reinforced by the 2013 discovery of the nearby brown dwarf binary WISE J104915.57-531906.1 at a distance of only 6.6 light years [38] – suddenly we found that the Sun has a new third-closest star system! Thirdly, perhaps surprisingly, there are still slight discrepancies between the various astronomical catalogues of nearby stars.

Probably the most authoritative recent compilation of nearby stars and the one on which the number of 58 stars is based (with the addition of WISE J104915.57-531906.1), is the Research Consortium on Nearby Stars (RECONS) list of the 100 nearest star systems [39]. Of these 58 stars, there is one star of spectral type A (Sirius); one F star (Procyon); 2 G stars (alpha Centauri A and tau Ceti); five K stars; 41 M stars (red dwarfs); 3 white dwarfs; and five probable brown dwarfs (Table One). A visual impression of the spatial distribution of these nearby stars (restricted to a distance of 12.5 light years for clarity) is given in Figure One on the next page.

Excellent summaries of known extra-solar planets can be found in the Extra-Solar Planet Encyclopaedia maintained by Jean Schneider at Strasbourg Observatory [2] and the NASA Exoplanet Archive [41]. Of the stars listed in Table One, four are currently known or suspected to have accompanying planets. These are alpha Centauri B (a member of the closest star system to the Sun at a distance of only 4.36 light years, although this planet has yet to be confirmed);

epsilon Eridani (a single K2 star at a distance of 10.5 light years); tau Ceti (a single G8 star at a distance of 11.9 light years, although in this case also the existence of planets is at present somewhat controversial); and GJ 674 (a M3 red dwarf at a distance of 14.8 light years).

In the following sections I briefly describe what is known about these nearby planetary systems, before moving on to a discussion of the implications of the known statistical properties of more distant planetary systems for the actual frequency of planets within 15 light years from the Sun.

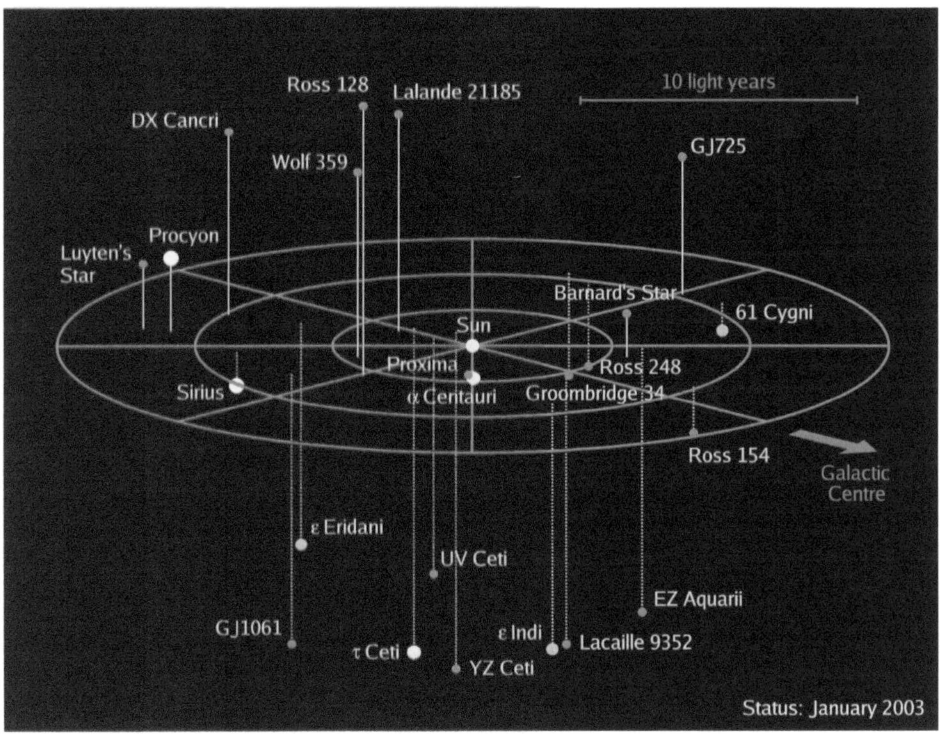

Figure One: a 3D map of all known stellar systems in the solar neighbourhood within a radius of 12.5 light-years as of January 2003. The colour is indicative of the temperature and the spectral class – white stars are (main sequence) A and F dwarfs; yellow stars like the Sun are G dwarfs; orange stars are K dwarfs; and red stars are M-dwarfs, by far the most common type of star in the solar neighbourhood. The blue axes are oriented along the galactic coordinate system, and the radii of the rings are 5, 10, and 15 light-years, respectively. Image: ESO/R–D Scholz et al (AIP).

ALPHA CENTAURI

In one of the most exciting astronomical discoveries of recent years, researchers using instruments at the European Southern Observatory in Chile reported in November 2012 the detection of an Earth-mass planet orbiting alpha Centauri B [42]. The planet was reported to have a minimum mass of 1.13 Earth-masses (the exact mass depends on the unknown inclination of the orbit), with an orbital radius of 0.04 astronomical units (AU; where 1 AU is the radius of Earth's orbit) and an orbital period of just 3.24 days. This is far too close to the star for the planet to be habitable (being one tenth of Mercury's distance from the Sun!) and, depending on its unknown surface composition, its surface temperature is probably about 1,470 K (1,200 degrees Celsius). Nevertheless, the detection of one planet orbiting alpha Centauri B would augur well for the presence of others in potentially more habitable orbits, although these will be more difficult to detect using current methods.

If confirmed by future work, this discovery is of enormous significance for starship planners because it proves that the most accessible star system to the Sun contains at least one planet and probably more. However, it has to be said that the detection of this planet is right on the limit of current observational techniques and, as a result, the detection cannot yet be considered secure. Indeed, in mid-2013 a re-analysis of the data failed to confirm its existence [43]. Further observations are therefore urgently required to confirm or refute the existence of this planet orbiting alpha Centauri B.

Even if this particular planet is not confirmed, however, it is entirely possible that other, as yet undetected, planets exist within the alpha Centauri system. Super-earths in the habitable zones of both alpha Centauri A and B should be just about detectable with current techniques, although confirming (or disproving) the existence of Earth and sub-Earth-mass planets in habitable orbits will be very challenging technically and is unlikely to be possible in the near future. Nevertheless, the statistics on the frequency of planetary systems, suggest that such planets are quite likely to be present.

TABLE ONE

Star systems within 15 light years of the Sun [38,39]. GJ is each star's number in the Gliese-Jahreiß catalogue [40]; l, b are each star's galactic longitude and latitude, respectively, where l=0 points towards the galactic centre and b is the angle above or below the galactic plane.

GJ	Popular Name	Spectral Type	Distance (ly)	l (deg)	b (deg)
1 551	Proxima Cen	M5.5V	4.2	313.9	−01.9
559	alpha Cen A	G2V	4.4	315.7	−00.7
"	alpha Cen B	K0V	"	"	"
2 699	Barnard's Star	M4V	6.0	031.0	+14.1
3 –	WISE J104915 A	L7.5	6.6	285.2	+05.3
–	WISE J104915 B	T0.5	"	"	"
4 406	Wolf 359	M6V	7.8	244.1	+56.1
5 411	Lalande 21185	M2V	8.3	185.1	+65.4
6 244	alpha CMa A (Sirius)	A1V	8.6	227.2	−08.9
"	alpha CMa B	DA2	"	"	"
7 65	Luyten 726-8 A	M5.5V	8.7	175.5	−75.7
"	Luyten 726-8 B	M6V	"	"	"
8 729	Ross 154	M3.5V	9.7	011.3	−10.3
9 905	Ross 248	M5.5V	10.3	110.0	−16.9
10 144	epsilon Eridani	K2V	10.5	195.8	−48.1
11 887	Lacaille 9352	M1.5V	10.7	005.1	−66.0
12 447	Ross 128	M4V	10.9	270.1	+59.6
13 866	EZ Aqr A	M5V	11.3	047.1	−57.0
"	EZ Aqr B	M?	"	"	"
"	EZ Aqr C	M?	"	"	"
14 280	alpha CMi A(Procyon)	F5IV-V	11.4	213.7	+13.0
"	alpha CMi B	DA	"	"	"
15 820	61 Cyg A	K5V	11.4	082.3	−05.8
"	61 Cyg B	K7V	"	"	"
16 725	Struve 2398 A	M3V	11.5	089.3	+24.2
"	Struve 2398 B	M3.5V	"	"	"
17 15	(A) GX And	M1.5V	11.6	116.7	−18.4
"	(B) GQ And	M3.5V	"	"	"

TABLE ONE CONTINUED

GJ		Popular Name	Spectral Type	Distance (ly)	l (deg)	b (deg)
18	845	epsilon Ind A	K5Ve	11.8	336.2	−48.0
"		epsilon Ind B	T1	"	"	"
"		epsilon Ind C	T6	"	"	"
19	1111	DX Can	M6.5V	11.8	197.0	+32.4
20	71	tau Ceti	G8V	11.9	173.1	−73.4
21	1061	–	M5.5V	12.0	251.9	−52.9
22	54.1	YZ Ceti	M4.5V	12.1	149.7	−78.8
23	273	Luyten's Star	M3.5V	12.4	212.3	+10.4
24	–	Teegarden's Star	M7V	12.5	160.3	−37.0
25	–	SCR1845-6357 A	M8.5V	12.6	331.5	−23.5
	–	SCR1845-6357 B	T	"	"	"
26	191	Kapteyn 's Star	M1.5V	12.8	250.5	−36.0
27	825	AX Mic	M0V	12.9	003.9	−44.3
28	860	Kruger 60 A	M3V	13.1	104.7	+00.0
"		Kruger 60 B	M4V			
29	–	DEN J1048-3956	M8.5V	13.2	278.7	+17.1
30	234	Ross 614 A	M4.5V	13.3	212.9	−06.2
"		Ross 614 B	M8V	"	"	"
31	628	Wolf 1061	M3.0V	13.8	003.4	+23.7
32	35	Van Maanen's Star	DZ7	14.1	121.9	−57.5
33	1	–	M3V	14.2	343.6	−75.9
34	473	Wolf 424 A	M5.5V	14.3	288.8	+71.4
"		Wolf 424 B	M7V	"	"	"
35	83.1	TZ Ari	M4.5	14.5	147.7	−46.5
36	687	–	M3V	14.8	098.6	+32.0
37	3622	LHS 292	M6.5V	14.8	261.0	+41.3
38	674	–	M3V	14.8	343.0	−06.8
39	1245	V1581 Cyg A	M5.5V	14.8	078.9	+08.5
"		V1581 Cyg B	M6.0V	"	"	"
"		V1581 Cyg C	M?	"	"	"

Similarly, in the case of Proxima Centauri (alpha Centauri C), there are as yet no actual detections of orbiting planets, but statistically super-earth-mass planets are known to be common around such red dwarf stars [44]. It is true that planets more massive than five Earth-masses orbiting Proxima within the habitable zone would have been detected by now (as would any planets more massive than about 20 Earth masses in more distant orbits [44]), but Earth- or Mars-mass planets could easily be present but are currently undetectable.

Clearly much more observational work is required to determine the presence or absence of planets in the alpha Centauri system. Given the system's proximity to the Sun and that it is already scientifically very interesting because it contains three stars of different spectral types, and because the path to it passes through a particularly diverse part of the local interstellar medium [12], the confirmation that planets are present would likely ensure that alpha Centauri remains at the top of the priority list for humanity's first interstellar mission.

EPSILON ERIDANI

The planet orbiting epsilon Eri is a giant planet, with a mass about 1.5 times that of Jupiter [45]. It has a highly eccentric orbit, which brings it as close to its star as 1.0 AU (i.e. the same distance as the Earth is from the Sun) and out as far as 5.8 AU (i.e. just beyond the orbit of Jupiter in our Solar System), with a period of 6.8 years. Although this would span the habitable zone in the Solar System (i.e. the range of distances from a star on which liquid water would be stable on a planetary surface given certain assumptions about atmospheric composition), this orbit lies wholly outside the likely habitable zone for a K2 star like epsilon Eridani. Also, being a gas giant, this planet itself it not a likely candidate for life and its eccentric orbit would not help in this respect (although it is possible that the planet may have astrobiologically interesting moons, perhaps similar to Jupiter's moon Europa, which could in principle support sub-surface life).

There is an unconfirmed detection of another planet in the epsilon Eridani system, of intermediate mass (a tenth of Jupiter's mass) in a very distant (40 AU) orbit [46]. It is possible that the system contains lower mass, more Earth-like, planets that might be more interesting targets for investigation, especially closer to the star than the giant planet that is known to exist. Epsilon Eridani is also known to be surrounded by a disc of dust [47] that may be derived from collisions between small planetesimals (i.e. asteroids and/or comets) and which is an indirect argument for smaller planets also being present. Only further research will tell how many planets actually reside in the epsilon Eridani system and whether any are of astrobiological interest. The existence of at least one planet, along with the dust disc (itself of great astrophysical interest), would make epsilon Eridani a high scientific priority candidate for interstellar exploration if it were not for its distance of 10.5 light years. Although within the 15 light year radius considered here, this is still a very challenging distance for the first attempt at an interstellar voyage.

TAU CETI

Tau Ceti is notable as the closest single G-type star to the Sun and has long been of interest to astronomers as a possible Solar System analogue. However, until recently there was no indication of it having a planetary system. This changed in early 2013 when an international group of astronomers led by Mikko Tuomi at the University of Hertfordshire published a paper [48] arguing that the star's radial velocity variations were consistent with the presence of up to five planets with minimum masses between 2.0 and 6.6 Earth-masses, with orbital distances between 0.1 and 1.4 AU. Although this tight spacing of multiple super-earth-mass planets within about one astronomical unit of a solar-type star might seem unusual, it is actually quite consistent with the statistical properties of planetary systems around more distant stars as determined by NASA's Kepler Space Telescope (discussed in Section 4.5 below). Moreover, tau Ceti is also orbited by a disc of dust [49] that supports, but does not prove, the presence of a planetary system.

Clearly this would be a very exciting result if confirmed, not least because the two outermost planets (a 4.3 Earth-mass world orbiting at 0.55 AU, and a 6.6 Earth-mass planet orbiting at 1.35 AU) probably bracket the habitable zone.

Although the habitability of super-earth-mass planets is unknown, it is possible that they might be accompanied by lower mass habitable moons. Unfortunately, however, it has to be said that the statistical significance of these planet detections is quite low and, for the reasons discussed above in the context of the postulated planet orbiting alpha Centauri B, it is far too soon to consider these detections to be confirmed. Further observations are therefore urgently required. It is also true that, as for epsilon Eridani, the distance of 11.9 light years would make tau Ceti a very challenging exploration target even if its planetary system is confirmed.

GJ 674

At a distance of 14.8 light years GJ 674 is on the limit of the distance range considered here. With a mass of about 11 Earth masses the known planet orbits its star every 4.7 days, in a moderately elliptical orbit at a mean distance of 0.04 AU [50]. This is a similar orbital distance as the possible planet orbiting alpha Centauri B and, even though GJ 674 is a much cooler star, this is probably too close to be habitable. Nevertheless, as one planet exists around this star it is possible that others will be discovered in more habitable orbits as observations continue. Only time will tell, but in any case the distance of this star probably renders it of marginal interest for direct investigation.

STATISTICAL PROPERTIES OF EXOPLANETS: IMPLICATIONS FOR THE PREVALENCE OF PLANETS WITHIN 15 LIGHT YEARS

Over the last few years our knowledge of the statistical prevalence of exoplanets has improved considerably as a result of increasingly sensitive measurements by ground-based telescopes [44, 51] and results from Kepler [15, 52]. For an excellent up-to-date summary of this fast-moving field, interested readers are referred to a recent review article in the journal *Science* [1]. Briefly, these may be summarised as follows:

• The vast majority (at least 75 percent) of solar type stars (i.e. stars of spectral types F, G, K) have a planet of some kind with orbital periods less than

10 years (surveys are not yet complete for longer orbital periods).

• Over half (about 60 percent) of solar type stars harbour at least one planet with a period up to 100 days (recall that in our Solar System Mercury has an orbital period of 88 days) and that multiple super-earth-mass planets are common in these short period orbits. However, most of these planets will probably be too hot to be habitable.

• Similar statistics apply for red dwarf stars (i.e. stars of spectral type M), where again about 60 percent are found to have super-earth-mass planets with orbital periods up to 100 days [44]. Although longer period (and lower mass) planets probably also exist, it is these shorter period ones that are the most interesting because, owing to the lower luminosities of red dwarf stars, many of these are theoretically within the habitable zone.

• Estimates for the fraction of M-dwarf (red dwarf) stars with Earth to super-earth-sized planets in the habitable zone range from 15 percent to over 60 percent [53]. Statistics for the fraction of solar-type (F/G/K) stars with planets in the habitable zone are not yet complete owing to greater difficulty in detecting these.

High though these estimates already are for the prevalence of planets, the actual fraction of stars with planets is likely to be considerably higher still. This is because both the ground-based radial velocity and the Kepler transit surveys are still incomplete. For example, neither method can yet reliably detect Earth (and still less sub-Earth) mass planets in habitable (approximately one astronomical unit) orbits around solar type stars. Nor can the radial velocity method yet detect giant planets (and still less low-mass planets) with orbital periods longer than about 20 years (i.e. planets in Saturn-like or more distant orbits), yet such planets presumably do exist. Clearly if we already know that about 75 percent of stars have planets and that this is a lower-limit, then it seems certain that, once all the statistics are in, we will find that essentially all stars will have planets of one kind or another. Indeed, the near ubiquity of planets is supported by the independent technique of gravitational lensing observations of distant stars, which imply that each star in the galactic disc is, on average, orbited by at least 1.6 planets with orbital radii in the range 0.5 to 10 AU [54].

Clearly, if almost every star in the Galaxy has a planetary system it follows that almost every star within 15 light years of the Sun (i.e. essentially all those listed in Table One) will also have planets, even if we have not detected most of them yet. Therefore there will be no shortage of exploration targets for interstellar spacecraft once the capability is developed.

CONCLUSIONS

In this chapter I have argued that, once we have developed the technical capability to engage in it, interstellar exploration will advance human knowledge in multiple areas. In particular, it will advance our knowledge of the interstellar medium, stellar astrophysics, and planetary science beyond what can be plausibly achieved by astronomical instruments based in the Solar System. Its most exciting scientific potential, however, lies in the area of astrobiology – if future astronomical observations indicate the presence, or even the possible presence, of life on planets orbiting nearby stars then the incentive to develop the means to make in situ observations of these alien biospheres may become irresistible.

Moreover, the benefits of interstellar exploration will extend far beyond science. Indeed, a wide array of potential cultural and societal benefits can also be identified. These include the long-term survival of humanity (and our post-human successor species), increasing the opportunities for spreading and diversifying life and intelligence through the Galaxy and providing a literally never-ending source of cultural and intellectual stimulation (for both the subset of humanity left behind and the interstellar explorers and colonisers themselves).

As noted earlier in this chapter, there is probably no shortage of planets within 15 light years of the Sun that would lend themselves to scientific investigation by humanity's early attempts at interstellar exploration. It appears that many of these planets will lie within the habitable zones of their host stars, which will make them of great astrobiological interest. Although it is important to realise that most of these will be orbiting red dwarf stars (Table One) and, while they may support indigenous micro-organisms (or genetically engineered terrestrial micro-organisms if a decision is made to introduce them), conditions on most of these 'habitable' planets are unlikely to be appropriate for human colonisation.

Even if Earth- (or Mars-) mass planets are located orbiting within the 'habitable zones' of nearby solar-type stars (such as alpha Centauri A/B or tau Ceti) it still seems most unlikely that they would be amenable to colonisation by Homo sapiens without either considerable modification of the planetary environments (see e.g. [55]) or of human physiology, or both (a probable necessity recognised by Olaf Stapledon in 1948 [27]). That said, it may be that as humanity moves out into the Galaxy we and our successors may cease to be interested in planets as places to live, relying instead on the stellar energy and cometary/ asteroidal raw material resources of planetary systems we encounter. It seems almost certain that most of the stars within 15 light years of the Sun and, indeed far beyond, could support human (and eventually post-human) civilisations on that basis indefinitely.

ACKNOWLEDGEMENTS

I thank Kelvin Long for the invitation to contribute to this book and for his tireless efforts in promoting the cause of interstellar exploration. Sections Two and Four of this chapter are based on updated versions of two papers previously published in the *Journal of the British Interplanetary Society* (i.e. *The Astronomical, Astrobiological and Planetary Science Case for Interstellar Spaceflight*, *JBIS*, 62, 415-421, 2009; and *Project Icarus: Astronomical Considerations Relating to the Choice of Target Star*, *JBIS*, 63, 419-425, 2010) and I thank the British Interplanetary Society for permission to reproduce parts of those papers here.

REFERENCES

[1] Howard, A W; 'Observed Properties of Extrasolar Planets', *Science* 340, 572–576, (2013).

[2] Schneider, J; 'The Extrasolar Planets Encyclopaedia' (exoplanet.eu)

[3] Crawford, I A; 'Interstellar Travel: a Review for Astronomers', *Quarterly Journal of the Royal Astronomical Society*, 31, 377–400, (1990).

[4] Matloff, G L; *Deep-Space Probes*, (Springer–Praxis, 2012).

[5] Gilster, P; *Centauri Dreams: Imagining and Planning Interstellar Exploration*, (Copernicus Books, 2004).

[6] Long, K F; *Deep Space Propulsion: A Roadmap to Interstellar Flight*, (Springer, 2012).

[7] Crowl, A; 'Starship Pioneers', *Starship Century: Toward the Grandest Horizon*, eds. J Benford and G Benford, pp 169–191, (Microwave Sciences/Lucky Bat Books, 2013)

[8] Crawford, I A; 'The Astronomical, Astrobiological and Planetary Science Case for Interstellar Spaceflight', *Journal of the British Interplanetary Society*, 62, 415–421, (2009).

[9] Crawford, I A; 'Space, World Government, and The End of History,' *Journal of the British Interplanetary Society*, 46, 415-420, (1993).

[10] McLaughlin, W I (ed.); 'The Potential of Space Exploration for the Fine Arts,' *Journal of the British Interplanetary Society*, 46, 421-430, (1993).

[11] Webb, G M; 'Project Daedalus: Some Principles for the Design of a Payload for a Stellar Flyby Mission,' Project Daedalus: Final Report, *Journal of the British Interplanetary Society*, S149-S161, (1978).

[12] Crawford, I A; 'Project Icarus: A Review of Local Interstellar Medium Properties of Relevance for Space Missions to the Nearest Stars', *Acta Astronautica*, 68, 691–699, (2011).

[13] Redfield, S and Linsky, J L; 'The Structure of the Local Interstellar Medium', *The Astrophysical Journal*, 673, 283–314, (2008).

[14] Carpenter, K G, Schrijver, C J and Karovska, M; 'The Stellar Imager (SI) Project: A Deep Space UV/Optical Interferometer (UVOI) to Observe the Universe at 0.1 Milli-arcsec Angular Resolution', *Astrophysics and Space Science*, 320, 217–223, (2009).

[15] Kepler Mission Website (http://kepler.nasa.gov)

[16] Cockell, C S, et al; 'Darwin – A Mission to Detect and Search for Life on Extrasolar Planets', *Astrobiology*, 9, 1–22, (2009).

[17] Bond, A et al; 'Project Daedalus: Final Report', *Journal of the British Interplanetary Society* (Supplement, 1978).

[18] Mayr, E; *This is Biology*, Harvard, University Press, Cambridge, MA, (1998).

[19] Brown, F S et al; 'The Biology Instrument for the Viking Mars Mission', *Review of Scientific Instruments*, 49, 139–182, (1978).

[20] Staufer, U, et al; 'The PHOENIX Microscopy Experiments', Fourth International Conference on Mars Polar Science and Exploration, Davos, Switzerland, abstract No. 8097, (2006).

[21] Aubrey, A D et al; 'The Urey Instrument: An Advanced In Situ Organic and Oxidant Detector for Mars Exploration,' *Astrobiology*, 8, 583–595, (2008).

[22] Sims, M R et al; 'A Life Marker Chip for the Specific Molecular Identification of Life Experiment,' ESA SP–543, 139–146, (2004).

[23] Crawford, I A; 'Towards an Integrated Scientific and Social Case for Human Space Exploration,' *Earth Moon and Planets*, 94, 245–266, (2004).

[24] Shepherd, L R; 'Interstellar Flight,' *Journal of the British Interplanetary Society*, 11, 149–167, (1952).

[25] Schröder, K-P and Smith, R C; 'Distant Future of the Sun and Earth Revisited,' *Monthly Notices of the Royal Astronomical Society*, 386, 155–163, (2008).

[26] Mill, J S; *On Liberty*, (1859).

[27] Stapledon, O; 'Interplanetary Man?', *Journal of the British Interplanetary Society*, 7, 213–233, (1948).

[28] Finney, B R and Jones, E M (eds.), *Interstellar Migration and the Human Experience*, (University of California Press, 1985)

[29] Benford, J and Benford, G (eds.), *Starship Century: Toward the Grandest Horizon*, (Microwave Sciences/Lucky Bat Books, 2013)

[30] Crawford, I A; 'Stapledon's Interplanetary Man: A Commonwealth of Worlds and the Ultimate Purpose of Space Colonisation,' *Journal of the British Interplanetary Society*, 65, 13-19, (2012).

 [31] Crawford, I A; 'Where are They? Maybe We Are Alone In The Galaxy After All,' *Scientific American*, 283(1), 28–33, (2000).

[32] Webb, S; *Where is Everybody? Fifty Solutions to the Fermi Paradox*, (Copernicus Books, 2002).

[33] Davies, P; *The Eerie Silence: Are We Alone in the Universe?* (Allen Lane, 2010).

[34] Crick, F; *Life Itself: Its Origin and Nature*, (Simon and Schuster, 1981).

[35] Fukuyama, F; 'The End of History?', *The National Interest*, No. 16, pp. 3–18 (Summer, 1989).

[36] Fukuyama, F; *The End of History and the Last Man*, (Hamish Hamilton, 1992).

[37] Locke, J; *An Essay Concerning Human Understanding*, (1689).

[38] Luhman, K L; 'Discovery of a Binary Brown Dwarf at 2 pc from the Sun', *The Astrophysical Journal Letters*, 767, L1, (2013).

[39] Research Consortium on Nearby Stars (RECONS) [www.chara.gsu.edu/ RECONS/].

[40] Gliese, W and Jahreiß, H; 'Nearby Star Data Published 1969–1978,' *Astron. Astrophys. Suppl.*, 38, 423–448, (1979).

[41] NASA Exoplanet Archive [exoplanetarchive.ipac.caltech.edu]

[42] Dumusque, X et al; 'An Earth-mass planet orbiting alpha Centauri B', *Nature*, 491, 207–211, (2012).

[43] Hatzes, A P; 'The Radial Velocity Detection of Earth-mass Planets in the Presence of Activity Noise: The Case of alpha Centauri Bb', *The Astrophysical Journal*, 770, article id. 133, (2013).

[44] Bonfils, X., et al; 'The HARPS Search for Southern Extra-Solar Planets XXXI: The M-Dwarf Sample', *Astronomy and Astrophysics*, 549, article id A109, (2013).

[45] Benedict, G F et al; 'The Extra-Solar Planet epsilon Eridani b: Orbit and Mass,' *Astronomical Journal*, 132, 2,206–2,218, (2006).

[46] Quillen, A C and Thorndike, S; 'Structure in the epsilon Eridani Dusty Disk Caused by Mean Motion Resonances with a 0.3 Eccentricity Planet at Periastron,' *The Astrophysical Journal*, 578, L149–L152, (2002).

[47] Backman, et al; 'Epsilon Eridani's Planetary Debris Disc: Structure and Dynamics Based on Spitzer and Caltech Submillimeter Observatory Observations,' *The Astrophysical Journal*, 690, 1,522–1,538 (2009).

[48] Tuomi, M; 'Signals Embedded in the Radial Velocity Noise: Periodic Variations in the tau Ceti Velocities,' *Astronomy and Astrophysics*, 551, article id A79, (2013).

[49] Greaves, J S et al; 'The Debris Disc around tau Ceti: A Massive Analogue to the Kuiper Belt,' *Monthly Notices of the Royal Astronomical Society*, 351, L54–L58, (2004).

[50] Bonfils, X., et al; 'The HARPS Search for Southern Extra-Solar Planets X: A Msin i = 11 ME Planet Around the Nearby Spotted M Dwarf GJ 674,' *Astronomy and Astrophysics*, 474, 293–299, (2007).

[51] Mayor, M et al; 'The HARPS Search for Southern Extra-Solar Planets XXXIV: Occurrence, Mass Distribution and Orbital Properties of Super-Earths and Neptune-Mass Planets,' submitted to *Astronomy and Astrophysics*, arXiv:1109.2497, (2011).

[52] Batalha, N M; 'Planetary Candidates Observed by Kepler III: Analysis of the First 16 Months of Data,' *The Astrophysical Journal Supplement*, 204, article id 24, (2013).

[53] Gaidos, E; 'Candidate Planets in the Habitable Zones of Kepler Stars,' *The Astrophysical Journal*, 770, article id. 90, (2013).

[54] Cassan, A et al; 'One Or More Bound Planets Per Milky Way Star From Micro-lensing Observations,' *Nature*, 481, 167-169, (2012).

[55] Fogg, M J; *Terraforming: Engineering Planetary Environments*, (Society of Automotive Engineers, 1995).

Chapter 2

The Conceptual Basis for Interstellar Flight

Giovanni Vulpetti

This chapter is devoted to the characterisation of interstellar flight and its relation to physical laws. Many topics are still highly debatable. What is explained here could be taken from a standpoint that is different, but not detached, from the one regarding the technical realisation of an interstellar vehicle, either unmanned or manned. Intuitively, interstellar flight should be a huge multi-disciplinary global enterprise that humankind may, or may not, decide to perform to reach another stellar system via fully automated probes and/or starships. Is such an endeavour possible? This is perhaps the most frequent question asked by members of the public. Well, this chapter attempts to answer this question by covering some points that – seemingly – are not connected to interstellar flight at all.

Introduction

Let us begin this chapter with a (very) short historical introduction about international research into interstellar flight. It is the opinion of this author that information about this research in the second half of the twentieth century could help the reader gain a better grasp of the modern revival of the quest for interstellar flight. The reader should be aware that the bibliography regarding interstellar flight is vast. It is literally impossible to mention all contributors; nevertheless, a few papers and books will be cited since the author deems them as particularly relevant to the chapter's topics, the meaning of which (perhaps erroneously) is often taken for granted.

The basic target for initiating interstellar flight, at least for humankind, is to reach a nearby star other than the Sun. The minimum requirement for this achievement is to deliver some probe(s) into a close encounter with the chosen star at a range of distances from it, e.g. 100–200 times the radius of its photosphere. If a planet orbits the target star, the minimum objective could be to design the flight so that at least one probe may orbit the star close to the planet and send scientific information back to a primary receiving station in the form of a spacecraft that, in addition to other probes with significant scientific payloads, is devoted to star–Sun communication (SSC). In other words, the probes orbiting the target star locally communicate with the SSC station and this, in turn, delivers the data to our Solar System. If the nearby star is a binary system, then its rank as the target star would be very high. For instance, the triple system of alpha Centauri, with its appealing distance from the Sun and the recently reported planet, would be of the highest rank for beginning human-related interstellar flight.

With such a minimal background about interstellar travel, let us go back to the 1950s. The seminal papers by Les Shepherd and Eugen Sänger [1–3] have opened up the era of scientific investigation of interstellar flight. Special relativity was applied to very advanced concepts of nuclear fission and electron–positron annihilation rocket-ships in order to achieve cruise speeds, far from the Solar System, that are a significant fraction of the speed of light. In [1] Shepherd established a basic result about the optimal mass propulsion ratio for matter–antimatter-based propulsion (from the mathematical viewpoint, some optimisation problems applicable to relativistic rockets were extended and generalised strictly only in the late 1970s and 1980s [4, 5]). Some investigators pictured very complicated devices in order to handle nuclear or antimatter power onboard. A famous example is Project Orion, which was conceived and then discarded in the late 1950s and 1960s.

In the 1960s, research was focused on propulsion systems; rocket vehicles for very high-speed deep space missions were thought of as having many large parallel and series stages. However, parametric mission studies were the majority. The space ramjet concept [6] was introduced. Though popular in science fiction, many investigators subsequently showed that the concept of the ramjet is unfeasible. Thus, many variants of the original concept were proposed; they are well explained/ referenced in chapter eight of Greg Matloff's book *Deep Space Probes* [7].

The 1970s saw the birth of a strong and systematic investigation, especially in the USA and Europe, on spaceship models, including power and propulsion systems, propellant types, payload systems, navigation and guidance systems, and telecommunications over stellar distances. Many relativistic-flight missions to nearby stars were conceived; trajectories were optimised by advanced methods in control theory. Nuclear fission and fusion engines were given the main propulsion role. Technology and mission analysis were emphasised in order to carry out an acceptable starship configuration. The celebrated Project Daedalus by the British Interplanetary Society [8] represented a remarkable joint effort that has prompted new areas of investigation and highlighted without doubt the need for studying interstellar flight from multidisciplinary viewpoints.

The 1980s represented the years of enthusiastic and major international (but independent) efforts for proposing 'solutions' to interstellar travel. In particular:

(a) Researchers became aware that a rocket starship, even if powered by nuclear fission and/or thermonuclear fusion, was turning out to be too massive (nanoscience and nanotechnology were in their infancy). Thus, many non-rocket propulsion methods were proposed. For example, laser-driven and microwave-driven vehicle concepts appeared as promising designs to overcome the conceptual rocket limitations. Space sailing received particular attention for one-way flights; however, round-trip interstellar voyages were also proposed. The various sailing modes suggested/analysed in those years are well summarised and referenced in chapter seven of Matloff's book. The rocket, however, had a comeback because during the 1980s there was also significant research into matter–antimatter annihilation for potential space utilisation. Clever concepts for exploiting the fundamental properties of particle–antiparticle annihilation were set up in Europe and the USA. Although not strictly envisaged, engines of rocket-based starships could exhibit two of the basic properties for achieving fast, long-range missions: high specific energy and sufficiently high specific power. However, the so-called antiproton and positron technologies are still too complicated and expensive today for producing the enormous amounts of antimatter that would need to be stored onboard an interstellar vehicle.

(b) More so than in the previous decades, most investigators searched for complex systems, trajectories and missions to star systems by using known physics and by extrapolating technology. As a result, new concepts for spaceships, ways for exploration and methods for gathering information at the various targets were closely analysed. It was also realised that SETI plays an increasing role in space exploration and the subject expanded to include many different viewpoints, ranging from biology and philosophy to ethics and sociology.

(c) At that time, people from the space community proposed interstellar precursor missions. Besides the high cost of an interstellar mission, many investigators pointed out that it is meaningless to try a jump by many orders of magnitude if one is not able to accomplish a much smaller scale mission. Instead, speculative investigation could, and had to, continue in parallel for the general comprehension of interstellar flight. This is a fundamental concept that is still valid today.

Finally, in the 1990s:

(i) During the early nineties, some funding policy changes and a certain disenchantment in public opinion for spaceflight slowed down general research in interstellar flight, although a number of valuable collections of ideas proceeded in certain national institutions for basic and applied research. The new general approach to space considered low-cost, high-information return projects. In July 1997, NASA re-launched their initiative to evaluate the feasibility of building a roadmap towards the first interstellar probing mission. There were many targets of high scientific importance in the heliosphere and beyond, in the range of hundreds or thousands of astronomical units (AUs). Thus, one of the projected challenges was to design a practical propulsion system that would have permitted sending fast probes beyond the outer planets. The Kuiper Belt, the heliopause, the unperturbed interstellar medium and the solar gravitational lens all still represent high-interest targets.

(ii) Many researchers emphasised three important points. First, that our current knowledge of the physical laws does not seem sufficient to realise interstellar flight. Second, even if our knowledge of physics were to give us some hope for a few-light-year flight lasting a human working-lifetime of three or four decades (e.g. by nuclear fusion or antimatter propulsion), its full cost may be quite outside the capabilities of one space agency. Third, multidisciplinary areas have to progress simultaneously in order to fully understand systems for space exploration. Consequently, research into interstellar flight had the undisguised hope of finding some breakthrough in science and technology.

Perhaps, in a historical perspective, those years could be considered as a sort of resetting from a number of viewpoints: scientific, technological, economical and its relation to project management. In such a general framework, speculative research continued. A number of researchers specialising in basic physics in the USA, Japan and Europe tried to apply advanced concepts and methods of relativity and quantum field theory to investigate new ways for carrying out space travel and/ or re-evaluating old ideas. In addition, nanotechnology and robotics provided new insights and perspectives. We will return to such important points later.

In the recent years of the twenty-first century, the enthusiasm of many young investigators resembles that of young researchers during the 1950s–1970s, who practically founded scientific interstellar studies without the remarkable support of the Internet. There is a risk, though. It appears evident from the advanced topics published in prestigious astronautical and space-oriented journals that an increasing number of young scientists and engineers have been re-discovering what has been widely studied in the 1950s–1990s. Perhaps the policy of considering to be quite new almost any items that cannot be found over the Internet (simply because the old documents were never ported into digital form) is not always correct. Furthermore, the very large amount of calculations made during those times should not be ignored; young people would otherwise repeat already-solved conceptual pitfalls and inconclusive rough models. On the flipside it is also true that re-discovering something under the light of modern views and capabilities in science and technology could result in further progress. Thus, probably, an equilibrium between the past and the current Internet-dominated times should hopefully be the best way for producing new results.

At the time of writing Voyagers 1 and 2 have both long passed the termination shock and Voyager 1 has crossed the heliopause – the outer boundary of the heliosheath and the heliosphere as a whole – to enter the interstellar medium, according to NASA. The termination shock is a plasma discontinuity in the heliosphere where the solar wind flows from supersonic to subsonic. The heliosheath is the vast region of the external heliosphere where the solar wind is slowed down by the opposing interstellar-wind pressure. After many planetary fly-bys, Voyager 1 achieved a cruise speed of 3.5AU per year. Such a low value (even though it is the highest cruise speed achieved by an artefact from Earth) introduces us to the problem of time in performing a general deep-space mission: design and construction time, flight time to the target and the scientific data collection time at the target.

SOME REASONS FOR INTERSTELLAR FLIGHT

Suppose for a moment that a vehicle like Voyager 1, albeit with a payload and telecommunication systems far more advanced, but with its cruise speed of 3.5AU per year, is designed to reach Proxima Centauri; this will take place after 770 centuries. Question one: might such a vehicle be considered an interstellar probe? Question two: might such a vehicle prove that interstellar travel is feasible? The second question contains more food for thought than the first one. In order to give such questions reasonable answers (i.e. aside from pure advertisements by space agencies), let us recall that the current history of humankind is estimated to be around 60 centuries, during most of which we have been without the awareness that Earth is a planet belonging to a stellar system in a vast galaxy. In addition, we would only need to wait less than a century to build a probe considerably faster than Voyager 1. Thus, the answer to question one is 'no'.

With regard to the second question, let us go back to the beginning of the twentieth century. At a time when there were so few heavier-than-air vehicles taking to the skies, it was hard to visualise how aeronautics would benefit human activities. Therefore, the capability to design and carry out any number of missions with more and more complex interstellar vehicles is necessary to prove and utilise interstellar flight on behalf of all humankind. These short considerations tell us that flight time – relative to the history of our civilisation – and our activities in very deep space (which add to our natural curiosity) are basic ingredients for realising interstellar flight. Thus we can now briefly mention some of the reasons why interstellar flight should be encouraged and carried out progressively over the course of many decades or more.

Just for simplicity, let us take for granted that we, i.e. humans, can be considered as conscious and intelligent life. As such, we can find many specific reasons for deciding to explore extra-solar planets and expand our civilisation into other stellar systems (in the same spiral galactic arm of our star!). Let us consider a list as follows:

1) A virtually unlimited extension of scientific knowledge;

2) An intelligent search for and utilisation of new resources;

3) Higher quality of life;

4) Non-destructively getting past the limits to growth;

5) Human aggrandisement;

6) Search for and communication with intelligent extraterrestrial life, should any exist;

7) The physics and metaphysics connection.

Each of these points contains many important aspects, which have been dealt with extensively in literature, especially in the *Journal of the British Interplanetary Society*. The above list does not imply any ranking (at least here); it may be subjective at an individual level, in national priorities and dependent upon international co-operation. Motivational boundaries could be fuzzy and affected by the particular period the world's society is going through at any given time. The above reasons are not all characteristic of this and subsequent centuries; in the far past, the 'local versions' of points two and three induced ancient populations to migrate towards unknown lands and in several cases they crossed (unknown) oceans by means of simple and efficacious water-craft. Those expansions or migrations involved and spanned many generations, and were not risk-free.

On a Solar System scale, motivations one to five could find their applications via advanced spacecraft, namely space-robots and/or crewed spaceships. These motivations also become considerably stronger with time and when the time comes for items six and seven to possibly pave the way towards astonishing discoveries, then interstellar flight should appear to be unavoidable in the continuation of human evolution.

Now it is evident that the intuitive definition of interstellar travel with which we began this chapter appears to be insufficient for characterising the very nature of interstellar flight. The line of reasoning we have used in this section will be revisited later for examining unprecedented situations in human history.

BASIC ISSUES FROM CURRENT PHYSICS APPLIED TO INTERSTELLAR FLIGHT

Suppose for a moment that humankind is able to produce large amounts of antimatter, or can build a very powerful space laser (or microwave) station, or set up a very smart ramjet system, and – highly improbable – construct a starship utilising all of these propulsion systems together. In principle, by using the theory developed in [9], the combination of these basic modes of getting thrust is equivalent to a rocket with very high effective exhaust speed and high thrust. Many tonnes of anti-hydrogen and a 50 terawatt photon beam may allow a payload of the order of one tonne to reach and orbit alpha Centauri in 15 years (considering special relativity). Even a rendezvous version of Daedalus-like vehicles, based solely on rocket propulsion and avoiding antimatter and photon beams (but mining helium-3 in the Solar System), may reach the Centauri system in a matter of decades. They may carry a few advanced sail-craft as a payload for exploring alpha Centauri for many years, before escape this binary star and heading towards Proxima Centauri (e.g. by extending and applying the theory of fast solar sailing [10] to a photon sail-craft illuminated by two stars).

For very advanced rockets, the problem of optimising the payload has been particularly highlighted in past decades; it starts as a conceptual problem but is strongly related to the whole mission, including the propulsion system. For instance, special trajectories to two stars a few light years apart, like those ones studied in [11], may meet the likely conflicting requirements stemming from the various scientific communities involved in interstellar missions.

The problem of flight time for manned starships and the highly challenging goal of colonising star systems has led us to envisage world-ships; a recent revisitation of this old and still appealing concept can be found in Andreas Hein's paper, 'Worldships: Architecture and Feasibility Revisited', published in the *Journal of the British Interplanetary Society* [12]. There are enormous non-astronautical problems that must be tackled by any century-long missions carrying generations of crew members to the stars.

The large number of discipline-dependent problems that need to be solved has induced many investigators to explore unconventional solutions to interstellar flight. One may group these solutions into two wide sets we name the 'space drives' (SDs) and the 'space–time alteration drives' (STADs).

Space drives denote propulsion concepts aimed at utilising some fundamental property or properties of 3D physical space. Usually, such features come from general relativity and quantum field theory. Engines do not cause momentum reaction for attaining thrust, but utilise some on-board field for exciting the space around the ship. The most detailed explanation of this propulsive mode can be found in [13–16], while a summary of the key points can be found in [17]. The key to this form of propulsion lies in simultaneously accelerating the ship structure and all objects inside the vehicle; a very special feature indeed! One drawback of space drives is that the ultra-high, local-space excitation field represents not only a technological challenge, but a conceptual problem of physics.

Space–time alteration drives rely on inducing local changes in the properties of space–time dimensional tissue. The engine's role would be to insert and keep the spaceship in the artificially-altered region of space–time. Departing from classic physics, quantum-field theory allows states of negative energy density, often described, inappropriately, as exotic matter (this type of mass-energy should not be confused with antimatter). Perhaps the most well-known of the STAD types is the warp drive. Related concepts can be found in [18, 19]; however, the proper idea with clear calculations first appeared in 1994 [20]. Many papers by other researchers followed. Based essentially on general relativity, STADs seem to be the ideal solution to interstellar flight: not only may the ship move with the speed of light or higher with respect to distant observers (e.g. on Earth), but also the altered metric tensor is controlled such as to allow the onboard time intervals to match the corresponding intervals on Earth. Thus, this theoretical basis should guarantee that it is only a question of finding a way of manipulating the negative energy density efficiently and on a large scale to be able to build this sort of starship and the moving bubble hosting it. Is it really possible? Unfortunately, the answer is not conclusive and much more involved than one might suspect at first glance. There are at least two schools of thought.

One school relies on quantum inequalities that have to be satisfied in the neighbourhood of a generic point along the world line of an observer (e.g. [21]). A clear review for non-technical readers can be found in [22]. Substantially, if observers experience some negative energy density in a space–time region, then such energy cannot persist too long.

The other school states that some effects of negative energy density have been already been observed in laboratory conditions via the squeezed states of light (e.g. [23]) and the Casimir effect. In contrast to the quantum intequalities-based school of thought, some of these investigators assert that the validity of the various energy conditions and quantum inequalities are questionable [24].

A very good survey and explanation of warp drive (and other things) can be found in a recent book, *Time Travel and Warp Drives: A Scientific Guide to Shortcuts Through Time and Space* [25]. Explaining in detail the different views (with equations) and some philosophical implications is beyond the purpose of this chapter. Nevertheless, a few remarks are in order.

Research into SDs and STADs has characterised research into deep space propulsion over the last years of the twentieth century and the first years of the twenty-first century. After about five decades of intense investigation, people became aware that any sophisticated propulsion based on action–reaction (even in relativistic form) cannot suffice for realising fast interstellar flight. Here, fast flight refers to times comparable or (hopefully) lower than the mean human lifetime. Thus, one has to resort to the frontiers of fundamental physics. What has been done so far with supposed wormholes, warp drives and claimed time machines (this chapter would prefer to avoid time machine-like or fantasy items – with full conviction, the author agrees with the Hawking's chronology protection conjecture) re-proposes one essential point: as the history of science has proved many times, pushing a physical theory to its extreme does not guarantee real results or discoveries, i.e. reliable and verifiable via systematic experiments. In this specific case, trying to unify general relativity with quantum field theory by the use of 'exotic solutions' for interstellar flight appears to be a hopeless endeavour merely because these two scientific milestones of the twentieth century are still incompatible with each other, despite the huge efforts to unify them.

We should not forget that propulsion (in any way it may be conceived) plays a central role in interstellar flight, but it is a necessary, not a sufficient, condition. Astonishing experiments have been conducted over the decades investigating photons, artificial crystals, nano-photonics, quantum computing and aspects of advanced robotics that may find applications in interstellar flight.

The above considerations lead us to summarise the very large difficulties in setting up a propulsion theory (with features similar to the travelling properties as we know them nowadays on our planet) via the following assumption.

Conjecture: the propulsion features relevant to fast interstellar travel – meant as the regular transport of people and machines to or from other star systems within the mean human lifetime – might be fully solved in the framework of an experimentally-tested theory unifying or superseding relativistic physics and quantum physics. Either form of knowledge might then appear as a particular case of the more general theory when a new universal constant is set to either values of two admissible limits.

Of course, this conjecture does not entail that we have to wait for a breakthrough in physics to begin to implement interstellar flight. It is reasonable to assume that interstellar flight will become a reality gradually. In this view, the twenty-first century will be host to the systematic exploration of significant targets well beyond the heliosphere via many interstellar precursor missions. From a propulsion viewpoint, nuclear electric propulsion and solar-photon sailing are the best candidates for successful interstellar precursor missions, especially the latter as that has the ability to overcome the intrinsic limits of rocket propulsion.

CHARACTERISATION OF INTERSTELLAR FLIGHT

World-ships may be a way for realising interstellar flight without requiring new physics since they are entire artificial worlds that move at relatively low speeds (0.01–0.05 of the speed of light). Perhaps subsequent generations on Earth will not be extremely interested in what took place decades ago (for example, given the every day difficulties that many in the younger generation face, they do not show any particular empathy with the Apollo achievements as the 1960s generation did). A world-ship hosts a part of the original culture of the civilisation that approved its design and construction. Although we won't deal with the world-ship concept's own significant problems (they are widely dealt with in literature, such as in [12] and [17]), what concerns us are the motivations for such an immense enterprise, which would certainly involve all aspects characterising a civilisation over the course of centuries and millennia. From our list on page 55, points 1, 4, 5 and 7 are necessary; however, they could not be sufficient on their own in a financial crisis-dominated world with no peace. Reasons for interstellar flight are not, and cannot be, like those that drove the Apollo project. It is the profound belief of this author that interstellar flight can be one of the positive fruits stemming from a true global peace, not vice-versa, even though some discoveries can be utilised for peaceful aims. Of course, similar to past and current decades, parts of the scientific and technological communities will continue to research interstellar travel and communication.

The reader should now be aware that to characterise interstellar flight involves many aspects of human activity and society. To attempt such a characterisation, let us use some concepts from mathematics and begin by defining a set that we shall call 'space missions'. The elements of this set may be written as:

1. Ground-to-orbit launchers;
2. Artificial satellites around a planet bearing conscious and intelligent life;
3. Planetary and interplanetary probes;
4. Robotic and crewed exploration of a star system (the Solar System, in our case);
5. Utilisation of a star system's (the Solar System, in our case) resources;
6. Interstellar precursor missions reaching out to a small fraction of a light-year from a planet bearing conscious and intelligent life, such as Earth in our case;
7. One-way fast interstellar flights to nearby planets possible hosting conscious and intelligent life;
8. Slow expansion: robotic replication and diffusion;
9. Slow expansion: the world-ships;
10. Fast return interstellar flights to other star systems.

Here the terms fast and slow have been explained previously. One will immediately recognise that the set of 'space missions' reflects the young history of astronautics on our planet, but it is generalised to some extent. Broadly speaking, the above list has been compiled in increasing order of knowledge and system complexity. More importantly, contrary to the past, elements six through to ten need very strong international cooperation. It is extremely doubtful that any one space agency could be successful in setting up and managing a complete interstellar mission, not just because of the economics, but because such an enterprise should reflect the aims and aspirations of all humankind.

The 'space missions' set contains objects that already characterise a progressive climb towards interstellar travel. The reader will have noted that the only quantitative figure on the list appears in element six, namely the mission's regard for exploration beyond the stellar-sphere hosting the planet bearing intelligent life: the heliosphere in our case. The heliosphere is not a sphere, but a very large elongated volume, the 'surface' of which (i.e. the heliopause) stems from the interaction between the interstellar wind and the solar wind, whose relative velocity varies from point-to-point, even temporally, in a Sun-centred inertial frame of reference. A small fraction of a light year, e.g. 0.01–0.05 light years, means that the exploration radius is large with respect to the stellar-sphere and small compared to the mean distance between the stars. The above range holds for the Solar System and humankind; however, because of known astrophysical and gravitational reasons, this order of magnitude might also be correct in the galactic arm through which the Sun moves. Thus, the first phase of interstellar exploration should be driven by non-biological constraints.

In contrast, the characteristic flight times for star-probes depend on the people operating or crewing the mission. With regard to humankind (the only conscious and intelligent life we know at the moment), in the 1990s some investigators in Europe and the United States, including the author, independently introduced and highlighted the concept of 'human job time', subsequently slightly modified as the 'mean human job time', or about 35–40 years. Therefore, a young researcher/engineer participating in an interstellar precursor mission should get scientific data return almost at the end of her or his career. This criterion should hold even for politicians to encourage them to approve space missions to be completed during their lifetimes. The International Academy of Astronautics (IAA) uses this concept to assess the main feasibility of interstellar precursor missions [26]. This IAA report recommends two propulsion modes to space agencies for carrying out such exploration missions in the near-to-medium term: nuclear-electric propulsion and solar-photon sail propulsion.

From what we discussed hitherto, interstellar flight could be viewed as a giant leap in deep-space exploration and human expansion subsequent to the interstellar precursor missions in the earlier age of astronautics. Astronautics should no longer be considered as an isolated set of scientific and technological endeavours. Instead it should evolve into a global mentality consisting of science and technology for robotic exploration and human expansion into space. Astronautics will get benefits from international human welfare and global ecology under a real peace (i.e. not only the absence of military conflicts) among countries.

Is there a leading line connecting all the elements of the space missions set? In fact, each element (which in turn is a set of interrelated 'objects') represents an advance on the previous one; the first one begun during a war can be considered a revolution in the human capabilities of building flying machines. The last section of this chapter is devoted to trying to answer this question.

THE RELATIONSHIP BETWEEN INTERSTELLAR FLIGHT AND THE UNIVERSAL LAWS

Michael Michaud [27] describes grand strategies for our species; accordingly, crewed interstellar flights and the search for extraterrestrial life could proceed in parallel. Starting from a biological behaviour, whereupon a species responds to inputs coming from the environment, Michaud substantially states that the activities of the human species have to be stretched to include space. Nevertheless, here we have been searching for a general principle that does not need any environmental push from conscious and intelligent life for its application, although environmental pressures may not be neglected or underestimated.

Many principles have been proposed for linking universal laws, conscious life and mind together. Paul Davies [28] has discussed many ideas regarding the Universe that followed the (probably) best known one, namely the weak anthropic principle (WAP) [29]. Among the general motives, we remember Intelligent Design, the Life Principle, the Unique-Universe, the Multiverse and the Self-Explaining Universe. Discussing the implications of and the objections to such principles is beyond the aims of this chapter. Nevertheless, what matters is that none of these principles includes interstellar flight or the need for it.

For this basic reason, we turn our attention to something different; it has been dealt with in [18] via two forms of the same principle. Here, we will discuss a modified form stemming from what we discussed in the previous sections.

In mathematics, objects and rules seem to be abstract items for non-technical people. However, they 'live' together, in that the objects are entities satisfying the rules, or axioms, and at the same time the rules exist inasmuch as they establish basic interrelationships between objects. Thus, we make the hypothesis that 'space missions' is a set of objects satisfying the following rule:

Universal laws are conscious-life oriented. They enable a conscious species, inhabiting a planet, to eventually realise systems for progressive space exploration, including interstellar flight. A civilisation may then be motivated to either continue to live on its birth planet only or to spread to other star systems.

We name this construction the 'Conscious Life Expansion Principle', or CLEP for short. We have to point out that CLEP is different from the principles mentioned above. Those ones, discussed in [28], are current-cosmology-affected principles. For instance, WAP does not require that observers in the Universe will eventually be exploring space by space-borne artefacts, or will send self-replicating robots throughout a galaxy, or will colonise stellar systems by slow world-ships. In other words, they simply do not consider that the set 'space missions' could represent one of the basic expressions of conscious life.

Let us comment on a few essential features of CLEP. First, it re-highlights the role of conscious life in the Universe, while interstellar flight is given the role of a universal mode of conscious and intelligent life. Life is not considered an improbable event throughout galaxies, but an unavoidable consequence of the universal laws. In the contrary case, there would be no need for conceiving of laws permitting interstellar flight or for giving interstellar expansion huge importance and relevance in galactic history. Second, a civilisation eventually expresses its free will (in some particular period) about whether to limit its space evolution to its Solar System or steer itself towards a long-term expansion towards stars. Third, many civilisations might exhibit some well-substantiated motivations ultimately pushing them to contact each other, via radio or some other form of communication, such as neutrinos. A civilisation might develop sufficient knowledge and technological mastery over neutrinos (and anti-neutrinos), e.g. about the oscillations of their mass and flavour, to build coded beams of them. For instance, if a way were found for coding the three basic transcendental numbers, then (very) intense beams might be launched into the interstellar space and perhaps detected by another civilisation. All this may be done with radio waves, however, the detection of neutrino beams would be a clear signature of an extraterrestrial civilisation's presence and their level of technology.

It is highly improbable that galaxy-oriented conscious lives may alter the evolution of the Universe even locally: any conscious and intelligent life cannot change the fundamental laws. CLEP implies this fact.

Applying CLEP to our Solar System, robots may, at an unknown stellar target, carry out dangerous and complex tasks more reliably and faster than their human counterparts. Nevertheless, CLEP excludes that robots act for human beings in every respect. A full interstellar expansion is ultimately for conscious life, not for artefacts no matter how sophisticated they will be made by their human designers.

Finally, CLEP does agree with migration via world-ships. Even if one civilisation were present in a galaxy, then it could spread into nearby star systems. Suppose that at least one small society (e.g. from Earth) eventually finds a suitable planet on which to live after wandering for centuries through space and that the arriving pilgrim generation decides to stay there. It might then happen that the whole pilgrim generation's new society devolves back to stages before their ancestors departed from the birth planet. Presumably, in the new history of the descendants, records of what had taken place during interstellar travel will be retained so they can remember where they came from (i.e. the Earth in this example). As time goes by, the knowledge and the spirit of their society may evolve again towards more interstellar adventure because CLEP allows that.

CONCLUSIONS

We began this chapter with a short synopsis of interstellar studies performed during the 1950s to the 1990s years, in which the basic disciplines of interstellar flight were founded and developed to a large extent by using the physical and engineering knowledge of that time. That golden period saw the birth of various nuclear and antimatter engine concepts, non-rocket propulsion investigations and preliminary studies on interstellar precursor missions. In subsequent decades, an increasing number of researchers resorted to fundamental physics to show that, at least in principle, fast interstellar flight is possible. However, such investigations do not appear conclusive; quantum physics seems to put strong constraints on the claimed negative energy density of matter and its quantity needed to alter space-time. Yet nothing may be taken for granted about interstellar flight in the framework of the current Standard Model.

Motivations for interstellar flight have been recalled to emphasise that the expansion of humankind into the Solar System and subsequently towards other star systems should be unavoidable after overcoming many severe global crises of different origins that plague us even into the second decade of the twenty-first century.

We have arranged the progressive steps towards space as a mathematical set we named 'space missions'. The last four elements of this set specifically address interstellar flight, but they are not disjointed from the previous elements (i.e. the planetary and Solar System eras); instead, a sort of natural evolution appears.

Does this set of human progress seem like a standalone gathering of activities, or does it obey a way of working that would apply to any civilisation in the Galaxy? We have speculated on an answer through a general hypothesis named the Conscious Life Expansion Principle. It is separated from other principles regarding cosmology and physical laws, which do not require space-flight in order for them to hold. In contrast, among many meanings, CLEP asserts that astronautics – and interstellar flight in particular – is a fundamental part of the evolution of conscious life, towards which the universal laws are firmly oriented.

REFERENCES

[1] Shepherd, L R; 'Interstellar Flight', *Journal of the British Interplanetary Society*, 11, pp149–167 (1952).

[2] Sänger, E; 'Zur Theorie der Photonraketen', *Ingenieur-Archiv*, 21, pp213–226, (1953), in German.

[3] Shepherd, L R; 'Interstellar Flight', *Realities of Space Travel*, editor L J Carter, Putnam, London (1957).

[4] Vulpetti, G; 'Starship Flight Optimisation: Time Plus Energy Minimisation Criterion', *Journal of the British Interplanetary Society*, 31, pp403–410 (1978).

[5] Vulpetti, G; 'Maximum Terminal Velocity of Relativistic Rockets', *Acta Astronautica*, 12, No 2, pp81–90 (1985).

[6] Bussard, RW; 'Galactic Matter and Interstellar Flight', *Acta Astronautica*, 6 (1960).

[7] Matloff, G L; *Deep Space Probes*, 2nd edition, Springer–Praxis (2005).

[8] Bond, A et al; 'Project Daedalus Study Group: Project Daedalus, The Final Report on the BIS Starship Study', *Journal of the British Interplanetary Society Interstellar Studies*, Supplement (1978).

[9] Vulpetti, G; 'Multiple Propulsion Concepts for Interstellar Flight: General Theory and Basic Results', *Journal of the British Interplanetary Society*, 43, pp537–550 (1990).

[10] Vulpetti, G; 'Fast Solar Sailing: Astrodynamics of Special Sailcraft Trajectories', *Space Technology Library*, 30, Springer (2012).

[11] Vulpetti, G; 'Relativistic Astrodynamics: The Problem of Payload Optimisation in a Two-Star Exploration Flight with an Intermediate Powered Swing-by', *Journal of the British Interplanetary Society*, 37, pp515–521 (1984).

[12] Hein, A M et al; 'World-Ships: Architecture and Feasibility Revisited', *Journal of the British Interplanetary Society*, 65, pp119–133 (2012).

[13] Minami, Y; 'Possibility of Space Drive Propulsion', IAA-94-IAA.4.1.658, 45th IAF Congress, Jerusalem, 9–14 October (1994).

[14] Minami, Y; 'Spacefaring to the Farthest Shores: Theory and Technology of a Space Drive Propulsion System', *Journal of the British Interplanetary Society*, 50, pp263–276 (1997).

[15] Musha, T and Y Minami; *Field Propulsion Systems for Space Travel*, Bentham eBooks (2011).

[16] Hayasaka, H and Y Minami; 'Repulsive Force Generation Due to Topological Effect of Circulating Magnetic Fluids', STAIF-99 Congress, Albuquerque (New Mexico), 31 January–4 February 1999.

[17] Vulpetti, G; 'Problems and Perspectives in Interstellar Exploration', *Journal of*

the British Interplanetary Society, 52, 9/10, (1999).

[18] Morris, M S and K S Thorne; 'Wormholes in Space–Time and Their Use For Interstellar Travel: A Tool for Teaching General Relativity', *American Journal of Physics*, 56, pp395–412 (1988).

[19] Morris, M S et al; 'Wormholes, Time Machines and the Weak Energy Conditions', *Physical Review Letters*, 61, pp 1,446–9 (1988).

[20] Alcubierre, M; 'The Warp Drive: Hyper-Fast Travel Within General Relativity', *Classical and Quantum Gravity*, 11 (5), ppL73–L77 (May 1994).

[21] Pfenning, M J; 'Quantum Inequality Restrictions on Negative Energy Densities in Curved Space-Times', PhD dissertation, Department of Physics and Astronomy at Tufts University, Medford, Massachusetts (May 1998).

[22] Ford, L H and T A Roman; 'Negative Energy, Wormholes and Warp Drive', *Scientific American*, 13, pp84–91 (2003).

[23] Fox, M; *Quantum Optics: An Introduction*, Oxford University Press (2007).

[24] Davis, E W and H E Puthoff; 'Experimental Concepts for Generating Negative Energy in the Laboratory', STAIF 2006.

[25] Everett, A and T Roman; *Time Travel and Warp Drives: A Scientific Guide to Shortcuts Through Time and Space*, University of Chicago Press (2013).

[26] Bruno, C; 'Key Technologies to Enable Near-Term Interstellar Scientific Precursor Missions', IAA Study and Eighth IAA Symposium on the Future of Space Exploration: Towards the Stars, Torino, Italy, 3–5 July 2013.

[27] Michaud, M A G; 'A Grand Strategy for the Species', 33rd IAF Congress, Rome (1981).

[28] Davies, P; *The Goldilocks Enigma*, The Penguin Press (2006).

[29] Barrow, J D and F J Tipler; *The Anthropic Cosmological Principle*, Oxford University Press (1990).

CHAPTER 3

WHEN WILL VOYAGER 1 LEAVE THE SOLAR SYSTEM?

STEPHEN ASHWORTH

NASA's Voyager 1 space probe, launched in 1977, is still functioning well after more than three decades of flight and is expected to continue returning scientific data until the early 2020s. Its reports on changing particle flows and fields at the outer edge of the Sun's magnetic domain have led to months of uncertainty over claims that it was on the verge of leaving, or may already have left, the Solar System and entered interstellar space.

Rumours abounded in March 2013 that Voyager 1 had entered interstellar space, based on a new model of the edge of the Sun's magnetic domain by scientists independent of the Voyager mission [1]. This was quickly refuted on 20 March 2013, on the homepage of the mission operated by the Jet Propulsion Laboratory (JPL) in Pasadena, California. "It is the consensus of the Voyager science team that Voyager 1 has not yet left the Solar System or reached interstellar space," said the mission's Project Scientist, Dr Ed Stone [2]. "A change in the direction of the magnetic field is the last critical indicator of reaching interstellar space and that change of direction has not yet been observed."

However, by 12 September JPL was ready with the big announcement: "NASA's Voyager 1 spacecraft officially is the first human-made object to venture into interstellar space." [3, 4]

In a follow-up news release on the same day, JPL drew a distinction between entering interstellar space and leaving the Solar System. However, should the latter concept cover only the realm of the major planets out to

Neptune, or should it also include the much more distant Oort Cloud of small, icy bodies? In the latter case it would be possible both to find oneself in interstellar space and at the same time remain in the Solar System. JPL acknowledged that the nomenclature could be confusing.

Voyager 1 is currently humanity's most distant probe, 128 astronomical units from the Sun at the time of writing (the distances of Voyagers 1 and 2 from both Sun and Earth are given on their homepage, accurate to the nearest kilometre and updated at half-second intervals). It is a notable feature of their leisurely progress that for part of the year their distances from Earth are actually decreasing, because for part of its orbit Earth, moving faster in the same direction, is catching up with them. Their distances from the Sun, however, as well as from Earth year on year, are of course always increasing.

The domain of the Sun is the region of space dominated by our star's influence, mediated principally through its gravitational and magnetic fields. Here, too, the Sun is brighter at all wavelengths than any other object in the sky. Does this domain have a clear boundary which a spacecraft can cross? Does it make sense to draw a line in the sky and define the space on one side as interplanetary and on the other as interstellar?

There does exist a reasonably well-defined magnetic boundary and it is this that continues to occupy the attention of the Voyager scientists. The magnetic interaction between the Sun and the interstellar medium defines a region of space surrounding the Sun's heliosphere (the magnetic 'bubble' blown by the solar wind) called the heliosheath, whose inner boundary is called the termination shock, where the solar wind slows to sub-sonic velocities, and outer boundary is the heliopause. Voyagers 1 and 2 are both close to that outer boundary: as of July 2014, Voyager 2 was still located within the heliosheath, while Voyager 1 has now crossed the heliopause out into interstellar space.

The shape of the heliosheath is not symmetrical; as it moves into the interstellar medium, the heliosheath streams away in a tail-like shape. The thickness of the heliosheath also varies with the Sun's magnetic activity versus the pressure of the interstellar medium pressing on it from outside. For instance, both

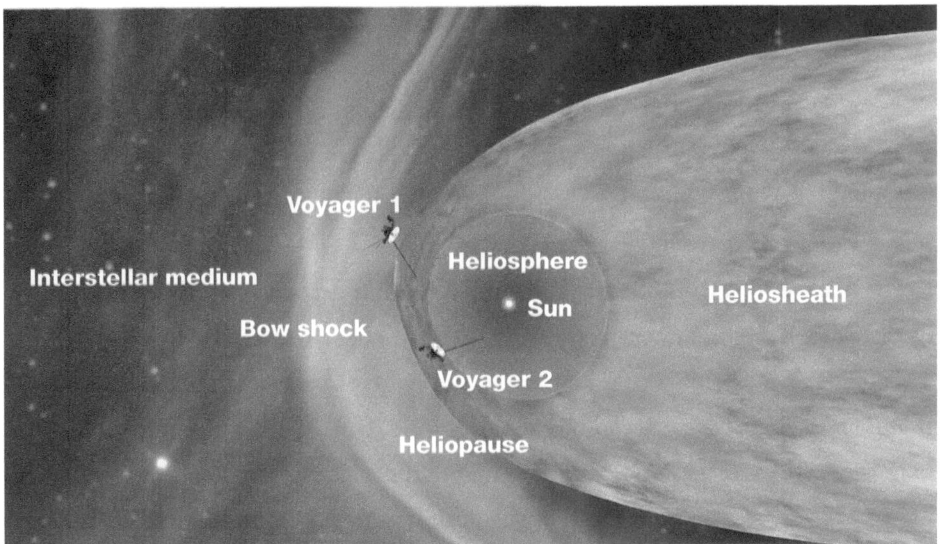

A depiction of the Sun's heliosphere and the relative positions of the Voyager spacecraft (not to scale). Within the inner bubble around the Sun the velocity of the solar wind is supersonic. Upon hitting the termination shock – the inner boundary of the heliosheath – the solar wind suddenly slows to subsonic. The heliosheath is streamlined into a long 'magneto-tail' by pressure from the interstellar medium as the Sun moves through space. On the leading edge of the heliosheath is a bow shock as interstellar gas piles up – this has been shown not to be as pronounced as expected. Voyager 1 has left the heliosphere, while Voyager 2 is still inside the heliosheath, as of July 2014. Image: NASA/JPL–Caltech.

Voyager probes crossed the termination shock multiple times as it moved inwards and outwards with the strength of the solar wind. Voyager 1 also crossed the heliopause at close to the boundary's nearest point to the Sun. Had the spacecraft been travelling in the opposite direction it would have needed to fly some ten times further away from the Sun, down the magnetic 'tail' before making that crossing. Even in any one particular direction, the distance to the heliopause differs over the solar cycle as the Sun's magnetic strength waxes and wanes.

The gravitational dominance of the Sun extends well beyond its magnetic influence. The small world 90377 Sedna, discovered in 2003, is currently close to its perihelion of 76 AU (which it will reach in 2076), but over the course of the next 5,700 years it will retreat to an aphelion of around 937 AU, which is more than seven times further from the Sun than Voyager 1 is at present and so presumably outside the heliosphere in that particular direction.

A few other small bodies are known with similarly large aphelia: 2012 VP113 reaches 449 AU; minor planet (87269) 2000 OO_{67} has aphelion at 1,068 AU, while (308933) 2006 SQ_{372} goes out as far as 1,570 AU. There must be many more waiting to be discovered.

Are we to say that Sedna and similar worldlets spend part of their orbit in the Solar System, but the rest of the time are outside the Solar System, in interstellar space? This would be bizarre: these are Solar System bodies, therefore the Solar System must extend out at least to 1,570 AU. What about the hypothesised Oort Cloud? Although no bodies in the Oort Cloud have yet been directly observed, it is believed that vast numbers of icy asteroids (that form long-period comets when occasionally perturbed into falling towards the inner Solar System) exist out to a distance of up to a light year. Do the unperturbed majority spend their entire lives orbiting the Sun in interstellar space?

These are, however, small bodies; most smaller than Pluto. Perhaps the outer boundary of the Solar System should be marked by the orbit of its outermost major planet, excluding the numerous dwarf planets of the Kuiper Belt and beyond?

For the present the outermost known major planet remains Neptune. The largest body currently known beyond Neptune is the dwarf planet Eris, which is both smaller and lighter than Neptune's largest moon Triton (which is in turn the smallest of the Solar System's seven planet-sized moons). The distance of Eris from the Sun varies from 38 to 98 AU, thus well within the heliopause. It remains possible that one or more larger planets could be discovered in a stable circular orbit well beyond the heliopause, and even more so that such worlds may one day be found orbiting other stars.

The orbit of Neptune defines a pancake 30 AU in radius but only 1.9 AU thick (thanks to the inclination of that orbit at 1.77 degrees to the ecliptic), which can be regarded as the domain of the major planets.

We thus find three possible definitions for the boundary of the Solar System:

(1) Magnetic, where the solar wind gives way to an interstellar flow of charged particles: somewhere near 125 AU out in the direction in which Voyager 1 is travelling, but up to ten times greater in other directions and in all directions likely to fluctuate in time depending on the varying strength of the solar magnetic field.

(2) Gravitational, at the orbit of the outermost major planet: that of Neptune at 30 AU in the direction of the ecliptic plane but only 0.93 AU towards the poles of the ecliptic; values that are stable over time but could change if a trans-Neptunian terrestrial or giant planet was discovered.

(3) Gravitational, at the aphelion of the outermost body that may continue to orbit the Sun for the lifetime of the Solar System, which, depending on close encounters with other stars, may be in the region of 0.1 to 1.0 light years (6,000 to 60,000 AU) away. Identifying this body is, however, impractical: for a very long time to come it will always be possible to find an even fainter planetoid orbiting at an even greater distance than the current record holder.

Clearly none of these definitions by themselves is satisfactory and they are mutually inconsistent by orders of magnitude. Taking the criterion of the Sun being the brightest object in the sky is not very helpful either, as this would abolish the whole concept of interstellar space and divide space up among the interplanetary regions of different stars, mostly empty of planets apart from the occasional interstellar nomads.

A helpful solution to this impasse does exist and it was recently raised in the context of this debate when Paul Gilster's well-known Centauri Dreams blog published an article on the Voyagers in 2013. This article attracted the following comment from a reader, James Jason Wentworth: "In his 1979 book *Planetary Encounters*, Robert M Powers usefully defined an intermediate region of space between the planetary region of the Solar System and interstellar space, which he called 'ultraplanetary space'. Its outer boundary is 0.1 light years (6,320 astronomical units) from the Sun." [5]

The reason for defining an outer boundary at 0.1 light years is that in the 1970s this was accepted as a reasonable estimate for the radius of the outer edge of the Oort Cloud [6] (present-day estimates are considerably greater). In the *Daedalus Report*, published in 1978, Alan Bond adopted the same value of 0.1 light years and assumed that the mass of the cloud of planets, asteroids and smaller debris orbiting a star was proportional to the mass of that star. Thus the radius of a star's planetary system would be 0.1 light years times the cube root of the ratio of the mass of that star to that of the Sun [7].

Given a mass of 0.15 solar masses for Barnard's Star, the target of the Daedalus study, this would give its system a radius of 0.05 light years. Bond defined the period of encounter for the Daedalus fly-by probe as being the time that it spent within 0.05 light years of that star (which, at its design speed of 12 percent of the speed of light and assuming a close fly-by of the star itself, worked out at about 300 days). This was useful as an engineering approximation, but was impressionistic rather than physical, a round-number approximation to the maximum distance at which it was thought orbiting bodies might be found that were not likely to be disturbed by stellar close encounters over the life of a planetary system.

The inner edge of ultraplanetary space would be clearly defined by (in the case of the Sun) the orbit of Neptune. Defining that inner bound by the heliopause would also be logically possible, but impractical, as this location cannot be determined except by a cloud of spacecraft continuously monitoring magnetic fields and particle fluxes in situ in every direction. However, the outer edge of ultraplanetary space would be fuzzy in the extreme.

Powers' concept of ultraplanetary space has in fact a lot in common with interstellar space. A spacecraft traversing it is exceedingly remote from any planetary body, receives minimal solar radiation and, after it has crossed the heliopause, its immediate space environment is interstellar in nature, not interplanetary – the solar wind has been replaced by a galactic regime of particles and magnetic fields.

A possible compromise solution would be to use the terms 'ultraplanetary space' for the region beyond Neptune but still within the heliopause and 'near interstellar space' for the region beyond the heliopause out to the most distant Oort Cloud members. The latter term thus expresses both the physical resemblance to interstellar space properly understood, but at the same time its sufficient nearness to the Sun for it to still contain asteroids and planets bound in orbit around the Sun. The position of the boundary between these two regions in any given direction would, however, be unknown until a spacecraft had actually made the crossing.

There may well exist other planetary systems in which a distant giant planet or brown dwarf (still too small to create a magnetosphere comparable with that of a star) orbits its primary beyond that star's heliopause, in which case such a system would not have any ultraplanetary space and its interplanetary space would overlap with its near interstellar space.

The realm of our Solar System and of the planetary systems of other stars merges gradually into that of the stars, with no clear boundary line where one could set up a border control and a passport checkpoint for humans leaving or aliens entering the Solar System.

REDEFINING THE BOUNDARY

So how can we say that our spacecraft have or have not made the crossing from interplanetary to interstellar space? In his classic book *Pale Blue Dot*, Carl Sagan answered the question by pointing out that the Voyager spacecraft are on escape trajectories from the Solar System, having long ago broken the gravitational bonds that had tied them to the Sun [8].

The difficulty in understanding Sagan's point of view comes when people still think in terms of what they are used to on planet Earth. Here on the ground, location is all-important, as every estate agent knows. Every country has a precise boundary, often physically marked with a wall or a barbed-wire fence, a river or a coastline, while crossing the border is a big deal that politicians expend much rhetoric on, while police forces watch the queues at airport arrivals with eagle eyes.

In space it is different, because nothing stays still, everything is moving under the influence of gravity. The location of a body by itself is of little account because it is constantly changing. A space 'station' is never stationary, but always in motion. The role of fixed points on the ground is played in the Solar System by fixed orbits: the path followed by a body around the Sun or a planet when it is not influenced by rocket propulsion or by the gravity of another orbiting body.

Left to itself, then, an orbiting planet, asteroid or spacecraft in a perfectly circular orbit around the Sun maintains a constant speed and constant distance from the Sun; but that perfect circle is an ideal, and in general closed orbits take the form of ellipses. As the body approaches the Sun it speeds up and as it recedes away from the Sun it slows down, thus both its distance from the Sun and its speed are constantly changing as well as its position in space.

If we ask what feature of an orbiting body remains constant, the answer is the specific mechanical energy of that body. That is, if we take its kinetic energy at any point and its gravitational potential energy at that same point, add the two and divide by the body's mass, we arrive at the total energy per unit mass of any body in that orbit (as well as in a family of related orbits). So long as the orbit does not change, that figure remains constant: the energy is constantly being exchanged between kinetic and potential forms, but the total energy is conserved.

Gravitational potential energy is usefully defined to be negative everywhere, while kinetic energy is always positive. This convention leads to the following result: if the body's total energy works out to be negative (gravitational greater, after removing the minus sign, than kinetic), then that body is in a closed orbit (circle or ellipse) around the Sun. If the total energy is zero or positive (kinetic equal to or greater than minus potential), then that body is not bound gravitationally to the Sun and, after a single fly-by of the Sun, it will recede forever on a parabolic or hyperbolic trajectory (the difference being that on a parabola, with zero total energy, the body approaches zero velocity at infinite distance from the Sun, while on a hyperbola it is moving faster, even when theoretically infinitely distant).

To be precise: if it has velocity v and distance r from the Sun, whose mass times the universal gravitational constant is GM, then a body's specific mechanical energy E/m is:

$$E/m = 0.5\ v^2 - (GM/r)$$

This, then, is how a trajectory specialist would think of crossing into interstellar space: not by flying past some boundary line in space, but by accelerating to positive total energy relative to the Sun. On Earth location is all-important, but in space that role is played by energy.

So when can we say that Voyager 1 has left the Solar System? When it was accelerated by its fly-by of Jupiter, which raised its total energy above zero. Whether Voyager is in or out of the Solar System is not a function of location but of energy. A spacecraft could be closer to the Sun than our Earth is, but if it is moving too fast to remain bound to the Sun, then it is already in interstellar energy space, like Arthur C Clarke's fictional starship Rama. An icy asteroid may be drifting up to a light year away from the Sun, but if it is in a closed elliptical orbit around the Sun then it is still in interplanetary energy space.

THE SATURN LINER

One consequence of this reality is that the first human travellers to voyage in interstellar energy space may be making merely interplanetary journeys within the Solar System.

Consider the following case: economic development has allowed the growth of a tourist trade between Earth and Saturn. Using the materials found in Saturn's rings and satellites, space hotels have been constructed offering stunning views of the planet and its famous ring system. These would need to be fully self-contained space colonies, based on a century or two of incremental development of increasingly self-sufficient space habitats in the inner Solar System, with a design heritage going all the way back to the present-day International Space Station, or even further back to Skylab and to the Soviet Mir and Salyut stations of 1971 to 2001.

Saturn, however, is on average 10 AU from Earth, thus 1.5 billion kilometres. The Cassini spacecraft took seven years to make that crossing (1997 to 2004). Wealthy tourists would not want to spend years of their lives in transit. A high-speed Saturn liner would be necessary if such a trade were to exist.

Current designs for magnetoplasma engines of the VASIMR pattern project a maximum exhaust velocity in the region of 300 kilometres per second (some estimates go as high as 500 kilometres per second). Let us imagine that a liner is provided with such an engine and that it departs from high orbit around Earth and arrives at high orbit around Saturn (perhaps with auxiliary high-thrust chemical propulsion at departure and arrival to minimise the gravitational losses associated with its low-thrust main engine). Refuelling facilities are provided at both Earth and Saturn, and the mass ratio for each one-way journey is kept close to 4.9, at which point the rocket dynamics offer optimum energy efficiency.

The cruising speed that results is 240 kilometres per second (affected only slightly by the Sun's gravity over the course of a journey) and the journey time is cut to less than three months, comparable with the lengths of voyages across Earth's oceans in the age of sail. Given that the passengers on such a ship will have good communications with the rest of the Solar System (albeit with a round trip time delay of up to three hours) and will therefore be able to continue to work normally at many computer-based jobs, such a crossing time is not at all unreasonable for a working adventure break that takes less than a year in total.

A speed of 240 kilometres per second near or beyond Earth's position in the Solar System is, however, very much greater than is needed in order to remain in orbit around the Sun. If the liner fails to use its engine to decelerate near Saturn, it will continue on out into near interstellar space on a hyperbolic trajectory, slowing only slightly to 236 kilometres per second at extreme distances from the Sun. It will have become an interstellar spacecraft, although a very poor one, taking over a millennium and a quarter to travel each light year away from the Sun.

The travellers on the Saturn liner will be voyagers in interstellar energy space during their cruise and must drop down back into interplanetary energy space if they are to complete their voyage and arrive at their planetary destination.

CONCLUSIONS

A choice of terminology depends very much upon the purpose one has in mind
for it. The underlying facts are as follows:

(1) There is no clear geographical boundary to the Solar System. We cannot
think of the Solar System as we do of a country on Earth, with borders marked
on a map and a passport office and a customs post set up on the border: the Solar
System is not like that. There is nowhere we can set up a notice saying: "You are
now entering the Solar System (twinned with Tau Ceti). Please fly carefully."

(2) The Solar System possesses a heliosphere, which is the domain where
the Sun's magnetic field is stronger than the interstellar field and controls the flow
of charged particles emanating from the Sun. The distance to the outer boundary
of the heliosphere, the heliopause, is variable in time and direction and, at 128
AU distance in one particular direction, Voyager 1 has now become the first
probe to cross that boundary.

(3) The Solar System possesses a region of interplanetary space between
the orbits of the major planets Mercury and Neptune. This space can be
completely enclosed within a flat cylinder 30 AU in radius and 1.9 AU in depth,
aligned with the ecliptic plane. Thus Pioneers 10 and 11 and Voyagers 1 and
2 left interplanetary space some time ago, travelling on hyperbolic trajectories.
The NASA/ESA Ulysses probe also left interplanetary space in 1992, while still
remaining in an elliptical orbit around the Sun, through a Jupiter gravity assist
that put it into a highly inclined orbit and which took it several AU north and
south of the ecliptic. Ulysses has passed into and out of interplanetary space at
regular intervals ever since.

(4) The Solar System possesses a region beyond interplanetary space, one
which is vastly larger, in which stable heliocentric orbits are possible but no
major planets have been found. It is populated with dwarf planets and cometary
bodies: principally objects belonging to the Kuiper Belt and the Oort Cloud.
These speculatively exist out to a distance of a light year or so, but identifying
the outermost one down to any particular size is not practical because of their
extreme remoteness and faintness.

Thus rather than crossing directly between interplanetary and true interstellar space, a spacecraft must cross a vast and poorly defined intermediate region, the nearer portion of which we may call (following Powers) 'ultraplanetary space', and the further portion beyond the heliopause 'near interstellar space'. The outer boundary of near interstellar space is not demarcated precisely, for it merges imperceptibly into far, true or deep interstellar space at somewhere on the order of a light year's distance. In that region, orbiting bodies are certain to be found that are only weakly bound to the Sun. These are progressively more likely, with increasing distance, to be disturbed out of the Sun's gravity well over the lifetime of the Solar System by occasional encounters with other stars.

(5) There does exist, however, a precise distinction between a body that forms part of the Solar System and one that is not bound to the Solar System or is only passing through. That distinction depends not on location but on energy. Pioneers 10 and 11, Voyagers 1 and 2 and now New Horizons have become our first interstellar spacecraft in this physically meaningful sense, because all five have accelerated above solar escape velocity. All now possess so much kinetic energy that they no longer remain bound to the Sun.

Thus if one is interested in the immediate environment of a spacecraft in terms of particles and fields, the distinction that matters will be whether the vehicle is within the heliosphere, crossing the heliopause, or outside the heliosphere.

If the focus of interest is on the dynamics of a spacecraft's trajectory, then the distinction which matters will be whether it is bound to the Solar System in interplanetary energy space, or unbound in interstellar energy space.

If the focus is on the relationship of a spacecraft to the bodies orbiting the Sun or another star, then attention will be directed towards whether that vehicle is in interplanetary space, in ultraplanetary or near interstellar space, or in deep interstellar space.

Let us celebrate Voyager 1's crossing of the heliopause without too much rhetoric about it 'leaving the Solar System' or 'entering interstellar space'. In the sense of its trajectory, dependent upon its energy, it left the Solar System already after its Jupiter fly-by: it was no longer bound to the Sun. Later it passed

Neptune's orbit and thus departed from known interplanetary space and has now passed through the heliopause and thus departed from solar ultraplanetary space. It will be a long time yet while it traverses solar near-interstellar space, and the date of its final crossing into deep interstellar space must remain unspecified.

NOTE

A shorter version of this chapter first appeared as the online blog article, Astronautical Evolution 91, online at www.astronist.demon.co.uk/astro-ev/ ae091.html and published 1 April 2013, last accessed 20 July 2014.

REFERENCES

[1] M Swisdak, J F Drake, M Opher; 'A Porous, Layered Heliopause', *Astrophysical Journal Letters*, 774, 1 (2013).

[2] *NASA Voyager Status Update on Voyager 1 Location* (online) http://voyager.jpl. nasa.gov/news/voyager_update.html (last accessed 20 July 2014).

[3] *NASA Spacecraft Embarks on Historic Journey Into Interstellar Space* (online) http:// www.jpl.nasa.gov/news/news.php?release=2013-277 (last accessed 20 July 2014).

[4] D A Gurnett, W S Kurth, L F Burlaga, N F Ness; 'In Situ Observations of Interstellar Plasma with Voyager 1', *Science*, 341, pp1489–1492 (2013).

[5] P Gilster; 'Looking Back From Deep Space', Centauri Dreams, 25 February 2013 (online) www.centauri-dreams.org/?p=26629 (last accessed 20 July 2014).

[6] L G Despain, et al.; 'Scientific Goals of Missions Beyond the Solar System', *The Outer Solar System*, *Advances in the Astronautical Sciences*, 29, part II, pp 597–626 (1971).

[7] A Bond; 'Project Daedalus: Target System Encounter Protection', *Project Daedalus: The Final Report on the BIS Starship Study*, pp S123–S125, British Interplanetary Society (1978).

[8] C Sagan; *Pale Blue Dot: A Vision of the Human Future in Space*, Headline (1995), pp151–152.

Section II

Building a Bridge to the Stars

Introduction by Kelvin F Long

I n this section we explore the many things that will have to happen before we can become a true star-faring civilisation. First in this chapter, **Jonathan Brooks**, a teacher and industrialist, looks at what lessons we can learn from history in how mega-scale projects of the past were accomplished. His view, that the past has something to teach us about the future, are well considered and worth reading.

Rachel Armstrong, a Professor of Architecture, argues from the point of sustainability and development, and how it is important that whilst we are pursuing the technologies for space, we also maintain our developments here on Earth. Her initiative for 'Starship Cities' is visionary and inspiring in its own right. After all, how can we build sustainable ecologies on world-ships, if we can't even practice ecology properly on Earth?

The key technology for getting us into space today is chemical rocket propulsion, but if we are to become an interstellar capable society we cannot rely on the current technological paradigms. **Bill Cress**, a business entrepreneur and engineer, actually visited the factory of United Launch Alliance (ULA) and showed how they continue to accomplish launches every year into Low Earth Orbit. ULA were kind enough to support his research and it really shows how difficult it is to achieve even current space exploration today. Then **Richard Osborne**, a real rocket scientist, looks to the future and asks what comes next in the evolution of rocket technology, from the next generation of launchers to single-stage-to-orbit (SSTO) space-planes and space elevators. Both of these chapters provide unique insights and contrasts to allow a glimpse into how rocket technology is likely to progress.

In my own chapter, I look at the possibility for interstellar precursor missions that will explore the Kuiper Belt and the Oort Cloud, and I argue this is how exploration has always been conducted: reconnaissance first, followed by exploration and colonisation. After all, this is how it was done before Christopher Columbus explored the New World.

CHAPTER 4

BUILDING A STARSHIP FROM THE LESSONS OF HISTORY

JONATHAN BROOKS

As a teacher, I educate people from all backgrounds and of all ages, from all over the world. Occasionally, I come across young people who have the vision and the potential to change the world. Now, it is proposed to build a starship, for which the technology has not yet been invented and which will require funding of sums greater than are even imaginable, and to achieve this within 100 years. This is a project that in terms of its scale and imagination is breath-taking. Is this an impossible dream? I think not. Similarly ambitious projects have been undertaken in the past! As a result, we are free to dare to reach for the stars.

One of the handicaps that humankind has begun to overcome is rigid and inflexible thinking. The main subject I teach is Electrical Installation and this is constantly being updated to accommodate new and innovative technologies that are appearing at astonishing speed. To put some perspective on this, a good place to see state of the art technology is in modern ships and comparing them to, say, HMS Victory, which fought at the Battle of Trafalgar and where there was not a single item of electrical technology. By 1920, one in four households had an electrical installation, and 60 years later we had space shuttles. That is how fast that technology is developing. The development seems exponential.

The word 'electricity' was first coined by Thales in 686 BC. This ancient Greek gentleman was a sort of 'Indiana Jones' character who travelled around having adventures. He went to Egypt and measured the height of the Pyramid of Cheops using trigonometry. He was an astronomer and at one point he predicted a solar eclipse. On his travels, he came across a farm where a spinning wheel was being used to create yarn from dry fibres. The spinning wheel had been fashioned from wood and the bearing was made from a hard substance called amber. This is fossilised tree sap, and it can be whittled down to make a smooth surface just like a white metal shell bearing. As the wheel was spinning, Thales saw the bits of flux leaping up and sticking to the bearing. He was very excited and recognised that a great and mysterious force was involved. He decided to experiment and procured various materials and types of cloth to rub them with. He was able to demonstrate the phenomenon of static electricity with various degrees of success. Metals produced completely negative results. Unfortunately, he failed to correctly model his observations. He did not realise that electricity was a flow of particles that would disappear into the Earth at the first available opportunity! He incorrectly concluded that the phenomenon was something to do with the amber and he thought that different amounts of amber were present in the tested materials in differing amounts. He thought that there was no amber at all in metals and hence no electrostatic properties. Had he correctly modelled the problem, presumably, electrical technology would have begun to emerge in ancient times.

It took until 1707 to overcome this consequence. At that time a businessman from Kent, Charles Grey, who was also a keen amateur astronomer, was experimenting with his telescope, and he had two corks on either end of a metal wire. One was being chafed and he observed particles of dust sticking to it. However, although the other was not being chafed, it too was attracting dry particles. To explain the phenomena he coined the terms 'insulator' – from Latin, meaning islands of charge – and 'conductor', meaning to channel the passage of charge. This appreciation of the exact nature of the phenomenon was supplemented and added to by leading and highly imaginative young people of the day, who were termed the 'electricians'! The way in which we model or review problems can lead to astonishing new conclusions. There may be things we have missed.

World history is illustrated with epic construction and design and development programmes. In ancient Egypt, the building of pyramids fit for a Pharaoh to spend eternity led to some fairly remarkable projects. However, that culture became subverted and progressively superseded by others. The pyramids did not produce a benefit to the contemporaneous society, although tourism is an enduring legacy. On Easter Island, the determination of the local population to build stone statues involved the felling of the indigenous forest and has left an enduring legacy, a lesson that societies that embrace huge projects which do not achieve any purpose, which destroy the environment and which consume an impossible portion of the output of the society, may lead to failure of the society itself.

Other societies have embraced huge engineering and construction projects, which have served to transform and propagate society according to a perceivable game plan. Lost, in the mists of time are the full details of the truly epic construction of stone and other circles and megalithic monuments. These astronomical calendars involved unimaginable effort in co-ordination, construction and planning. They mark, apparently, the transformation of society from one of nomadic hunter gatherers to a season orientated agricultural society, holding and nurturing the land.

Of far greater and more discernible effect is the implementation of all that follows from what we call 'civilisation'. I do not mean the colloquial meaning, but rather I mean that everything that has followed from the adoption of the Roman Law. These sage and incredible juristic statements embrace every problem in modern society and offer an insight and wisdom that, if accepted and applied, would solve many of today's contemporary miseries. They are best preserved in the form of the 12 tables to the Pandect's and the Romans inherited source materials, which they modified and, to their cost, subverted. The Roman Law is an incredible source of wisdom and its institutions may well be capable of building a starship.

Any society must have the freedom to dare to dream of such endeavours like those we now contemplate and it is within its mechanisms that the economics to enable such ambition might be found. Many societies would not permit private individuals to even contemplate such a daring project. Therefore, a free society must be a prerequisite to success for such a project.

In 1625, so as to find a means of developing an international law that would put an end to the constant and recurring wars that marred and set back the whole world, Hugh De Grote proposed to use the Roman Law as the foundation of a system of international laws. He collated the common elements of cultures worldwide and showed that these were in fact ideologically identical to the Roman Law. His epic *De Jure Belli Ac Pacis* [1], meaning 'The Rights of War and Peace', became the foundation of international law.

With the arrival of the Normans in the British Isles came the cathedral-building programme and the foundation of the ideology and culture that enabled the United Kingdom to stand alone and then mobilise incredible industrial achievements. It is from these parallels that the construction of a starship might be emulated.

The cathedrals are engineering miracles and, although the designs existed with the aspiration to build a miracle in stone the like of which the world had never seen, the technology to do so had yet to be invented.

The cathedrals were built over a 200 year period and equally inspiring local churches often took 40 years. Despite being damaged during war, most of these astonishing miracles have survived for the best part of a thousand years, yet when they were built there were, on occasions, catastrophes. Towers collapsed and the geometry and engineering had to be re-evaluated. Just like in the history of building rockets, there were crashes and setbacks.

When NASA introduced their bold plan to send a man to the Moon in the 1960s, they harnessed civil management techniques from centres of academic and industrial excellence worldwide, to draw together many skills and bring multiple minds to bear on engineering solutions. This produced huge spin-offs and benefits in social and technological development, so that the endeavour enriched and benefited civil society and was not a burden and a drain upon it.

It is clear from the lessons of failed endeavours that mighty projects that span generations must achieve interim paybacks and cannot be a simple non-productive burden upon society. Such a mighty project would certainly fail unless it could produce rewards in the short term. It is clear from the Apollo programme and its successors that the spur of challenge to the industrial and academic western world encourages and propagates technological innovation, and these endeavours enhance and enrich the whole of western society. The spin-offs from such a programme have the potential to provide solutions to other pressing world problems. Many people believe that wealth should only be used to solve immediate social problems. In fact, dynamic change is affected by multiple vectors and an aspiration that develops new wealth and new technology will produce a beneficial vector, in the same way that a sailing ship of old must moderate its course to take advantage of prevailing trade winds.

The cathedral building projects gave rise to a support infrastructure. These were the great cities; York, Preston, Coventry, Liverpool, Winchester, Bath, London and Salisbury, to name a few. Not only were hundreds of individual crafts, skills and trades required, but the people to undertake this employment had to be recruited and trained. There was no magic pot of wealth to pay for these endeavours, the wealth had to be created as an aspect of the staged project. Analogously, to make buildings you must first make bricks. This was a considerable handicap in relation to the starting position of NASA, who was funded by the USA's existing wealth.

Exploitation of Roman Law and the building of a society with institutions to implement its jurisprudence, enabled all of this to be accomplished, both in the short and middle term. As a result of the project, not only were the great trading and wealth-generating cities created, which went on to become the engines of commerce and industry, but the cathedrals themselves served as training centres by which illiterate slaves would be endowed with skills and turned into free men, saved from slavery by redemption. In short, they functioned as technical colleges. Over seven year periods, highly skilled craftsmen were trained and so motivated as to produce sustainable industrial output for the rest of their lives, creating wealth, which was regulated by an innovative and original economic system.

The methodology of the cathedral builders was straightforward. They were a tiny band, supported by a military force that might be compared to a military regiment, concealed within a military order. The landowners were allowed to keep the land, in exchange for a pledge of allegiance. From about 1087 onwards, the cathedral building programme was conceived and, around these construction sites, a walled city in support was set up.

An accurate view of the western world is afforded by Adam Smith's *Wealth of Nations* [2]. In this nineteenth century book, the word used again and again is 'apprenticeship'. This reflects the true values of the cathedral builders. In common with many cultures, they believe that the wealth of a nation is in its children. A form of economics that encourages and facilitates essentially the farming of moral, law abiding, productive, happy, well-adjusted and able children, trained and enabled to exploit their full potential. Children free from the horrors of war, and insurrection. These are the dreams, aspirations and methodology of the western world. Through understanding these economics, I believe a starship could be built and the research and staged development would have the benefit of solving many of the world's contemporaneous problems as spin offs.

More or less, a great State such as the United Kingdom is a trading entity and the de facto sovereign nations within it are stake holders and essentially shareholders. What the State needs is workers who pay taxes and through, their benign and industrious endeavours, create value that may be exchanged for those of others, either using tokens, called money, or by exchange. This enables trade. If a State is beset with dissent and conflict, its output will deteriorate and it will sink into an abyss of misery.

In the absence of liquid cash, various forms of point schemes where the parties do not intend to enter into legal relations and are obligated to each other as members of a family, may be developed to notionally apportion reward for values contributed. There is nothing that would prevent in theory the formation of a de facto sovereign nation or a collection of rocket scientists and others whose objective is to build a starship.

There has been much talk about whether the efforts of initiating endeavours, such as the NASA/DARPA 100 Year Starship, should be charities. I believe that this is the wrong approach. In principle a western world business should do nothing but good. It is essentially a charity that is sustainable. The guiding light of the western world is value for money. Such a project as seriously envisioned by organisations with the credibility of NASA and DARPA is so vast that whether or not a starship is built, it will change the perception and the values of people worldwide. Therefore, their sensibilities and concerns need to be recognised. People believe that rocket scientists are clever. If people are clever enough to actually fulfil the role of NASA and DARPA, reasonably the public should expect that they can also generate the wealth needed for the project, by earning an honest living. In short; giving value for money. The free young people that believe their aspiration can be achieved using a charity are removed from commercial and economic reality. What will drive such a project forward is a sound, well-conceived commercial plan. The first requirement would be to exploit the asset achieved by the formation of the association. The first product is a think-tank of highly able and talented scientists. Such a group capacity is a marketable resource. This force can also be used to train and educate youngsters who have the potential to achieve any of the skills required to build or conceptualise the starship. The objective would be to set up centre-of-excellence training schemes that enable youngsters from all over the world to make a contribution, whether as bricklayers, cleaners, astronauts or rocket scientists. Everybody's contribution is valuable. Training centres are marketable commodities and contributors can be paid on a course to course basis.

Secondly, launch systems will be required. These can be applied to solve other existing problems. One such problem for which huge funds are being deployed non-productively is that of the burgeoning nuclear waste industry. Such a centre of excellence could take nuclear waste and arrange its safe and monitored storage at a site where ultimately, it would be launched to orbit and sent off for disposal by burning in the Sun. If this contribution could be made to humankind, every penny spent will ultimately be a saving and the project will do massive good. The skills to achieve this latter objective are encompassed within the same endeavour.

Just as a person with the skills to make bricks is essential to every building project, so people with the skills to create wealth are essential to such a project, because the wealth required is so vast that it must be created. The project must encompass wealth creation under its umbrella of operations and this wealth must be created by giving value for money, through fair, honest tax-paying trade. Such transactions would be sustainable and beneficial to every party, directly or indirectly connected. Because of the scale of the operation required, it may be ultimately necessary to invent new forms of commercial vehicle, enabling multiple subscription, or to adopt the mechanisms within international law.

I believe that to implement this project, an interim objective would be to acquire a training resource, such as an estate, and a stable means of income that is able to generate wealth to pay decent salaries as a precursor to the conceptualisation and development of technological systems. The economy of the western world has a seasonal nature and in a recession, like a winter, there is no money available for projects that are hundreds of years away from fruition. Right now, we need to set up something that pays the bills. The estate would be used to operate training schemes, which would be of unparalleled excellence. Young people would be rendered fit using methods gleaned from the military and would be taught skills that enable them to generate wealth in the short term. From such a starting position, a huge array of money-making propositions could be operated, which collectively would operate as a commonwealth. They would contribute to a wholly owned centre of excellence research facility solving the problems of building a starship. It is by this approach that a starship-based society can be realistically achieved.

REFERENCES

[1] Hugo Grotius; *De Jure Belli Ac Pacis (on the Laws of War and Peace)*, 1625.
[2] Adam Smith; *An Inquiry into the Nature and Causes of the Wealth of Nations*, 1776.

Chapter 5

In the process of human development we are reverse terraforming the Earth. In other words, our modes of thinking, our methods, technologies and the kind of design and engineering projects that we use to advance our civilisation are reducing our capacity for long-term survival, not enhancing it.

Our propensity for environmental destruction is an extremely poor indicator of our potential to establish viable space habitats. The importance and urgency of our practices are best highlighted by our lack of success in designing biosphere experiments, although we have inevitably learned much from the endeavours. Biospheres are technologically constructed, closed and controlled experimental ecosystems. They are designed to investigate fundamental ecological principles to help us better manage Earth's resources and potentially enable us to establish ecosystems beyond Earth's surface. For example, the Soviet BIOS-3 series of experiments that ran from 1972 to 1984 examined how functional closed biological systems might support a terrestrial 'ark' that could be sustained by careful management alone. BIOS-3 supported a community of three people using an algae cultivator and a 'phytron', which grew wheat and vegetables using sunlight. Yet, BIOS-3 was not a 'closed' biosphere as dried meat and energy were provided from external sources and human waste was stored instead of being recycled back into the system [1].

In the 1990s the mission was attempted again with Biosphere II in Arizona. This project aimed to understand how people in close confines, in a closed ecological system, could work together over a sustained period. Yet it was quickly clear that despite being equipped with a desert, rainforest and ocean, it was going to be very difficult to create a sustainable environment. Oxygen levels steadily fell, the ocean acidified, internal temperatures rose, carbon dioxide levels fluctuated, and vertebrates and pollinating insects died, while the crew became depressed, dysfunctional and lost weight. Only the cockroaches and ants thrived. The first Biosphere II mission supported a crew of eight and lasted for two years from 1991 to 1993 before its exquisitely designed and monitored ecosystem collapsed [2]. Indeed, this important study clearly demonstrated our inability to design self-sustaining environments that promote and enhance life's persistence, even when we are taking great care of them under closely monitored conditions on the surface of a life-bearing planet that we are already evolved to inhabit. Sadly, these results bode very poorly for our long-term survivability in extreme environments beyond Earth's atmosphere.

However, our survival is not just a question of our environmental suitability. It also depends on how we adapt and respond to environmental pressures. The extreme challenges of non-terrestrial habitats invoke a range of strategies that enable the human body to exist in configurations better suited for survival than our natural form. For example, we use space-suit designs to form barriers against hostile environments, while the extreme performance artist Stelarc imagines the human body becoming harder and drier, more like a machine, to survive the lifeless terrain of space [3]. Others, such as transhumanists[1], look to a range of technologically mediated body modifications including genetic engineering and even mind uploads to transcend the environmental challenges of space habitation [4]. Yet, while technology can offer significant survival advantages, there is still no escaping the problem of the long-term viability of existence posed by

---

[1] Transhumanism is based on secular humanism and atheism. Transhumanists very often use spiritual, mythological or parapsychological concepts such as the immortality of the mind/soul/spirit and transcendence of the body, and attribute their material realisation not to ephemeral forces but to specific technologies such as, nano- and bio-technologies, robotics and information technologies [5].

environmental poverty and barren terrain. However, this chapter does not intend to deal with the rich palette of anatomical modifications that have been proposed to increase our survivability in extreme habitats. Rather, over the following pages we will examine environmental design principles that may help us construct habitats that are innately life-promoting and self-replenishing.

THE BUILT ENVIRONMENT AS A TERRESTRIAL WORLD-SHIP LABORATORY

Architectural designer Richard Buckminster Fuller's notion of 'Spaceship Earth' creates an image of our world with a finite amount of resources that cannot be replenished from the impoverished void of the cosmos but must be regenerated, using our Sun's energy to work with Earth's living systems [6]. The design and engineering principles of human habitation may therefore be explored through the discipline of architecture that examines spatial and temporal relationships through technology and the programmatic design of our living spaces [7]. The built environment offers a critical and experimental forum where it is possible to theoretically and practically critique the fundamental principles of human development through biosphere construction. By examining the principles of producing terrestrial habitats in ways that deal with resource shortages (as in our 'mega-cities', a term that describes vast urban conurbations of more than ten million people) or extreme environments (such as deserts akin to Magnus Larsson's 'Dune' project, which imagines a bacterial printer turning sand into stone [8]), we can identify a set of propositional and testable criteria for the construction of space habitats and world-ships, which are artificially constructed ecosystems like biospheres, but rather than being located on Earth, they propose to support human colonies beyond the terrestrial environment, in space. Indeed, the living interiors of world-ships and our resource-constrained mega-cities share many common challenges with respect to the kinds of infrastructures that enable their populations to thrive. Visionary designer Buckminster Fuller literally considered cities as biospheres whose parameters could be controlled [9] and constructed geodesic domes as closed system habitats around various sites, such as the Montreal Biosphere (1967) – a museum dedicated to the environment, in his ambition to improve the housing of humanity and its relationship with the environment. Fuller proposed to construct a geodesic dome two miles across to span midtown Manhattan and regulate weather and reduce air pollution. This was designed to reduce the demand for energy by creating a more stable atmospheric system. Such biomes not only speak of the environmental constraints imposed by resource shortages for twenty-first century mega-cities where water, sanitation, food and even the availability of living space dictate the liveability of our habitats, but they also describe the context in which world-ship interiors are imagined. An accessible and transferrable platform may be constructed that equally advances our understanding of both mega-cities and the design of world-ships by drawing on Fuller's legacy and engaging with principles drawn from 30 years of advanced biotechnologies to construct 'synthetic' ecologies[2]. Our cities may therefore be regarded as accessible laboratories to explore new, transferrable principles in the design and engineering of ecological networks, which may ultimately enable us to thrive both on Earth as well as amongst the stars.

SUSTAINABILITY

Sustainable development is itself a relatively new concept [10] that has been intensified by an awareness of the impact of global industrialisation on our environment and invites questions about how the current generations can meet their own needs as well as securing the welfare of future generations. While Earth is a resilient system that has taken 4.5 billion years to evolve its incredible robustness and capacity to evolve, our own attempts at deciphering and reconstructing what is necessary to make an ecosystem are exposed by our inability to resurrect ecosystems, such as after forest clearing [11] or, indeed, even when these are designed from scratch as in Biosphere II. At this point in time,

---

[2] Synthetic ecology refers to the rational design and engineering of environmental networks and can be thought of as the 'systems' version of synthetic biology.

despite many advances in biotechnology, we are still very far from being able to replicate the efficiency that natural systems can provide and our understanding of how ecological systems work is at a very early stage of development [12].

Current notions of sustainability are centred on a varied set of frameworks that possess unique and often mutually exclusive forms of internal logic. These work in many ways that are less damaging to our ecological systems such as the Passive House [13], which addresses issues of material conservation and improving the mechanical efficiency of systems [14] and environments, such as Paolo Soleri's notion of architectural ecologies (arcologies) where cities are formally entwined with their landscape [15]. Other forms of sustainable design propose environmental support systems, such as green walls and roofs that nominally speak of transforming our desert-like built environments made from inert materials into richer ecological landscapes. These artificial landscapes serve functional and aesthetic roles within our cities by adorning our living spaces with elements that provide rural vistas within concrete landscapes [16] and which often pander to our cultural preconceptions of nature rather than dealing with the actual materiality of our urban living spaces that is based on inert matter such as bricks, glass and concrete [17]. Yet, nature-inspired forms and functions are used to infer our connections with the relentlessly material natural world. For example, biomimicry abstracts specific details of natural systems that are engineered through industrial processes to metaphorically represent practices that are more 'like' nature [18]. Of course, each of these approaches to produce 'sustainable' forms of practice is valid when framed by their internal logic. Yet, when considering a long-term pathway for establishing models of sustainable development, it is worth remembering that the concepts and practices that form sustainable narratives are not mature solutions and are still very much evolving.

From an engineering perspective, these current models of sustainability do not take us any closer to constructing mature ecologies than we are to making life from scratch in the laboratory. Our sustained future in a synthetic environment actually depends on our ability to identify the principles of 'ecopoiesis'[3], which refers to transforming a lifeless planet that is at effective equilibrium, such as Mars, into a dynamic elemental system that might evolve complex ecologies through further design and engineering interventions. The fundamental principles of this process equally apply to the construction of worldship habitats and 'sustainable' cities, which require their own elemental cycles, weather and even seasons for prolonged, healthy living.

The technology of ecological systems does not, however, arise from the same platform as machines. Instead, ecological systems are founded on the principles of process, organising hubs and material flows within an environment. The science of working with systems is also much more recent and does not have a set of formal tools and practices like mechanical technologies do. Therefore, our challenges in designing environments are subject to conceptual as well as practical challenges. While the field of cybernetics has been examining how objects may relate to each other through different orders of feedback [20], their design and engineering conventions limit our engagement with complex systems. Specifically, the relationships, networks of exchange and flows of matter that shape complex systems do not simply behave as bounded objects with emergent properties – as in the 'Game of Life' where each cell exists in one of two states – but can be transformed in their very nature from one set of agents into another. This is particularly marked when 'tipping points' are reached within the system and, to date, can only be demonstrated using chemical models, for example periodic oscillations that cause colour changes in chemical systems, such as the Belousov Zhabotinsky reaction [21], or interactions between individual dynamic droplets that produce novel, synchronous forms and behaviours that cannot be predicted from a knowledge of the systems' components alone [22]. These transformation differ, for example from Skylar Tibbits' 4-D self-assembly system that proposes to grow complex structures from simple objects using the principles of self-organisation

---

[3] The term 'ecopoiesis' was first coined by Robert Haynes in 1984 with regard to space exploration and the process of 'terraforming': "Ecopoiesis is my neologism. The term refers to the fabrication of a sustainable ecosystem on a currently lifeless, sterile planet, thereby establishing a new arena in which biological evolution ultimately might proceed independent of further human husbandry," [19].

while not undergoing a constitutional change in their fundamental nature [23]. Likewise, it cannot be assumed that providing basic ingredients for world-ship ecosystems, typified by O'Neill cylinders [24], which require vital infrastructures, such as light, water and an atmosphere to produce a habitable environment, implies their inevitable, spontaneous, self-assembly into an ecosystem that is able, for example, to produce water and nutrient cycles. Indeed, the relationships between elements may well need to be managed like a form of farming for artificial environments to become self-regulating ecological systems [25]. Even on Earth, ecological systems require management if they are to actively support human survival. This in itself is not entirely impossible on Earth's surface through land management, environmental technology and soil fertility techniques that are based on intensive practices that also draw from the open nature of Earth's Sun-soaked biosphere [26]. We do not as yet know whether it is possible to boot up ecologies in a non-terrestrial environment using these approaches.

Despite our experience of farming over many millennia, the way that we might continue to thrive long-term on our native planet is unclear. Indeed, all human societies use a range of technologies to develop unique relationships with their environments that come with an even greater set of associated concepts about how they might be used. It would be woefully premature to formalise what a sustainable approach to the construction of a world-ship may mean to a space faring community. In designing an ecological system for a world-ship, it is therefore more important to establish a range of different frameworks that are open, capable of evolving and that can withstand rapid cultural, technological and even environmental change across generations. Such approaches will require the carefully orchestrated collaboration of multiple disciplines where unity in the vision for mutual survival is adopted through a diverse set of approaches.

Developments in the field of biotechnology over the last 30 years have highlighted the importance of ecological design as being not only relevant to our external spaces but also to our interior ones. New techniques including DNA sequencing have enabled us to identify a range of processes that enable us to work with such precision and such small scales that we can think of living processes as being a kind of technology. They have also revealed that we are actually composed of many nested ecologies of collaborating human and non-human bodies. Indeed, only 10 percent of the cells in our body have human DNA, the rest construe a bacterial organ that we have come to recognise as the 'microbiome', which shapes our physiology [27]. Our bacterial microbiome provides many essential functions such as providing an effective immune system, producing mood altering molecules and digesting our food. Even inside our cells we are not exclusively human but are collections of co-operating bodies such as mitochondria in our cell matrices. Even our genetic code is packed with bacterial, viral and non-coding regions of DNA, which have also been called 'junk' or biological 'dark matter' [28]. Indeed, the human body is not a machine – a trope that has captured our cultural imagination since the Enlightenment – but an ecosystem with very different design and engineering principles. So, if we are to develop a survival system to support humans, it is even more essential that principles for building and nourishing ecosystems are established. Unlike machines, ecosystems are not built from the top-down but are evolved from the bottom-up as interacting bodies forge networks of interactions and couple together to produce tissues and organs, or within an environmental context form rich, fertile fabrics such as soils and coral reefs. These structures offer the infrastructure that increases the probability of more sophisticated, complex events taking place, underpinning the evolution of multi cellular life and biodiversity.

There is much to learn about the practical aspects of building environments lively enough to enable ecological systems to thrive and proliferate. Our current practices assume that our ecologies are stable and constant so there is little consideration given to their advancement. Indeed, we are more likely to destroy, cull or suppress the 'unruly' activities of nature in our cities than promote them, as they erode inert surfaces and thrive uninvited in hidden niches. A new era of construction is necessary where we learn to build systems and ecologies by working with non-equilibrium systems, ones that are restless, promote material exchanges and facilitate flows to resist the inevitable increase in entropy of a starship. Such resistance, defined by Erwin Schrödinger in his essay *What is*

Life?, characterises the nature of living systems [29]. Indeed, these are the very conditions that architects are currently addressing when they propose sustainable designs to increase the availability of fundamental resources while increasing the resilience and liveability of our habitats. Despite many engineering successes, architecture does not propose to have solved these issues and we remain vulnerable to the whims of central power grids, supply chains, economic pressures and natural disasters, which may reduce the quality of our lives.

Indeed, our recent biotechnological advances offer more than the potential to engineer lifelike processes, but to also combine these with other technological platforms along the lines of dynamic chemistries and robotics and computer algorithms. Such interweaving of technologies is referred to as the construction of 'post natural' fabrics, which also includes lawns, green roofs, greenhouses, green walls, agricultural crops and gardens. Indeed, if we are not to wait several billion years for starship ecology to evolve to sufficient maturity to support its living communities, then it is vital to begin to learn how to effectively fabricate post natural environments as a synthesis of spontaneous processes with technologically mediated ones.

PRINCIPLES OF WORLD-SHIP DESIGN

Since the current approach to designing and engineering our terrestrial habitats assumes that the environment is a constant, our efforts become centred on the production of the vessel, the geometries that forge its surfaces and the infra-structural systems that form its 'standing reserve' [30]. This approach imagines worldships as 'arks' that are loaded up with provisions (the standing reserve) and then sailed into the unknown. If left unaddressed by the inability to top up supplies en route, then resource shortages in an ark lead to tipping points not only in physical survival but also to the psychological and social stability of the system [31].

A more challenging way of thinking about the provisions on a world-ship is to design recycling into the system so that ecosystems may spontaneously operate and thrive under their local conditions. In this design scenario the indigenous culture is not simply harvesting but is also contributing to the resilience and fertility of the system. Within world-ships, ecologies are assumed as a kind of constant and it is unclear how material regeneration that persists on Earth can actually be accomplished in non-terrestrial environments. For example, how do Gerard O'Neill's communities deal with rubbish or human waste? Indeed, how exactly does the 'biot' ecosystem on Rama in Arthur C Clarke's 1973 novel *Rendezvous With Rama* metabolise through the mineral laden sea [32]?

Ecosystem technologies

As a means of exploring the kinds of materials, technologies and approaches that may underpin the design and engineering of ecological systems, it is necessary to consider what kind of technologies exist within nature. The impacts and effectiveness of these 'living' technologies, which possess some of the properties of living things but are not given the full status of being 'alive' [33], produce qualitatively different outcomes to machines. For example, they are robust, adaptable, and resilient, they can transform at tipping points and are capable of generating unpredictable events. Their performance may also be assessed through their success in promoting ecological relationships by increasing fertility and orchestrating the flow and transformation of matter through dynamic, elemental systems, rather than by quantifying the resources and energy they have consumed.

To develop such habitats to the stage where they are meaningful to starship design collaborations between architects, scientists and construction engineers are necessary. Lessons learned from the construction, implementation and use of these systems might inform space-engineering requirements and help decide whether the kinds of propulsion system for a particular starship is slow or fast, wet or dry. On Earth these experiments will help advance further architectural developments in 'sustainability'. It is also essential to develop appropriate metrics that engage with ecological concepts to monitor the progress of these projects so they are not simply collapsed back into a system of mechanical values. In other words, we not only need to measure the efficiency of ecosystems within design and engineering practices but we must also balance their performance against other livability factors including biodiversity, fertility and recyclability.

ECOLOGICAL TECHNOLOGIES FOR WORLD-SHIP AND CITY CONSTRUCTION

Currently this author is working on a series of projects that serve as examples of very practical projects that may help us take the first near-term steps towards our long-term goals in starship design and engineering. They are at their earliest stages of development and each of them demonstrates some of the principles outlined in this particular chapter. While the technologies themselves are not new, the way they are designed, engineered and applied within an urban context embodies a different kind of approach to the production of architecture.

The Hylozoic Ground Series are installations by architect Philip Beesley, for which this author designed a range of 'chemical organs' [34]. The dynamic chemistries are entangled in the cybernetic system as a series of notional 'glands' that slowly change colour in the presence of dissolved carbon dioxide, like artificial smell or taste organs. The work embodies the idea of 'living' spaces and buildings that are not inert but instead are responding to many cues from their inhabitants, ultimately becoming part of an intimate ecology of exchanges. Such interactions and networks propose a paradigm shift in the way that architecture is practiced where the architecture is responsive to cues in its surroundings and is also able to produce a tiny amount of matter that records its environmental relationships. This award winning work exists as a permanent installation in the atrium of the Leonardo Building in Salt Lake City [35].

The Algaeponics Project at the University of Greenwich's School of Architecture, Design and Construction is the first permanent algae installation in the UK and was built by California-based engineers Sustainable Now Technologies. A research station that will be installed in 2014 provides data on the performance of a local strain of algae to quantify the amount of biofuel and biomass that such a structure can produce [36]. The longer-term aim of this installation is to develop a biofuel station for local river traffic on the Thames.

At the heart of the Algaeponics Project is a flexible bioprocessing unit known as the Greenstone Device, which is being developed by Sustainable Now Technologies and Astudio architects with an eye towards its use in urban applications. For example, the device is being situated within a sixth form college in Twickenham, England, as a pedagogical system that is centred on an environmental technology for capturing carbon, producing energy and matter and promoting 'green' business activity to the next generation of entrepreneurs.

Each Greenstone Device is a modular system made up of a hundred 10-gallon 'alpha' bioreactor units containing water and algae and which are circulated by a small solar-power driven motor. The rate of water flow can be sensed and controlled using a digital computer system that also runs a set of software programs as a basic 'brain' for monitoring and controlling the bioreactor through any digital device with a modern web browser. Food in the form of carbon dioxide is bubbled into the system and is drawn from the surrounding air. The unit may also be fed by connecting it to a flue so that it scrubs waste gases from industrial burners. Effectively, each unit is an aquarium that just houses tiny green organisms that capture light and carbon dioxide – like little green cows – as they are pumped through the aquarium to produce biomass. The water does not have to be fresh and microalgae will thrive even in grey water. The products of the system can be decanted off and processed in a variety of different ways, from 'cracking' the algae by using an electrical current to release oils, to burning the biomass.

In the short term, bioreactors can reduce the running costs of corporate buildings. For example, the HSBC building at Canary Wharf in London could house around 65 Greenstone units, which would produce about 19,000 tonnes of biomass using carbon dioxide each year and potentially reduce total fuel consumption by up to 8.5 million litres of diesel per year. Greenstone devices can also reduce fuel consumption by providing natural shading and cooling as a result of the algae capturing solar thermal heat, decreasing the need for air conditioning.

In domestic settings Greenstone devices fit in the back of a garage and can produce around a gallon of biodiesel a week. This means a biodiesel-converted car could take local trips around town, at an average fuel efficiency of 30 miles per gallon, without ever having to top up at a petrol station.

STARSHIP CITIES

The Initiative for Interstellar Studies has begun holding a series of workshops such as Starship London [37], which began to establish contexts in which it may be possible to collectively design our cities differently so that they are not imagined within the context of a constant environment. By imagining our favourite location within our home city as part of a world-ship environment, Starship Cities provokes its participants to consider how it would be necessary to design our living spaces differently so that we can promote the ecological relationships that have supported us for thousands of years. The speculative nature of the Starship Cities project aims to develop a model for constructing ecologies that can deal with top-down (starship to city) and bottom-up (city to starship) approaches in parallel to enable more effective preparation for our journey to the stars, as well as catalysing necessary ecology-promoting changes within today's cities. Drawings, models and prototypes of these proposals offer a very immediate set of actions that can be taken towards starship development and they could be tested by interweaving them with the fabric of our living spaces, for example by building prototypes and installations that apply living technologies such as algae bioreactors, anaerobic digesters, bioluminescent light sources, heat absorbing and emitting substances (e.g. sodium thiosulphate) and even hygroscopic materials (e.g. calcium chloride) in sites where ecological relationships may be augmented. These are examples of the kinds of architectural strategies and systems that may begin to change the operating systems that underpin the way we live.

These mutually reinforcing approaches to starship design may help us better understand how to design and engineer a world by developing life-promoting supra- and infrastructures. Systems that examine the principles through which we may build ecological networks that promote the liveability of our habitats may be prototyped through this approach. Creating the conditions for liveability as we travel to the stars may be achieved in a very immediate and practical sense by weaving starship fabrics into our cities, which have been considered and designed during these open workshops that invite experts and non-experts to collaborate and construct our shared futures both on Earth and amongst the stars. The outcomes of these experiments may be measured differently to the way that machine-based technologies are evaluated as successful, in terms of energy efficiency and resource conservation. Rather, living technologies and the post-natural fabrics that are promoted in the outputs of Starship Cities may be assessed through their ecological relationships, for example by evaluating biodiversity or soil fertility. It may even be necessary to define the terms of evaluation locally so they are meaningful to each site, rather than using a globally-abstracted standard that is not relevant to any particular site, ecology or space. Such interventions may not only help us increase the liveliness of our surroundings but also establish how we may appreciate them and understand more about the fundamental building blocks that are necessary for constructing worlds.

Indeed, unless we are unable to do this sustainably on Earth our chances of supporting a multi-generational colony within a synthetic environment in a world-ship are extremely slim. Like the construction of a world-ship, the experiments associated with Starship Cities are likely to take more than a generation to develop into a formal practice. However, great architectural projects may span more than a century before their completion, such as Antonio Gaudi's La Sagrada Familia, which is a cathedral that is only just being finished by modern architectural methods.

Just imagine starting another La Sagrada-style project today, but one that is distributed throughout every city in the world. Not only would our grandchildren see beautiful installations and prototypes within our cities that increase the fertility of their environment but they would also experience what it is like to be in a city that is being positively terraformed. Their home-place may be as blooming as the countryside in that our homes may directly provide renewable sources of vital substances for the continued survival of our future generations – whether they are at home on Earth or journeying amongst the stars.

REFERENCES

[1] Russian CELSS. 1983-2013. Russian CELSS Studies [online] Available at: www.permanent.com/russian-celss.html. [Accessed 5 September 2013].

[2] M Nelson and W Dempster; 'Living in Space: Results from Biosphere 2's Initial Closure, an Early Testbed for Closed Ecological Systems on Mars', *American Astronautical Society: Science & Technology Series*, 86, pp363–390, AAS 95–488, 1996.

[3] Stelarc; 'The Hum of the Hybrid,' Earlier Statements [online]. Available at: http://stelarc.org/?catID=20317 [Accessed 23 November 2013].

[4] A Andreadis; 'Dreamers of a Better Future, Unite!' [online]. Available at: www.starshipnivan.com/blog/?p=60 [Accessed 15 February 2015].

[5] J Guga, 'In the machine we trust', *Beyond Artificial Intelligence: Artificial Golem Intelligence – Proceedings of the International Conference Beyond AI, in Pilsen, Czech Republic, 12–14 November* 2013, pp 76, 2013.

[6] R B Fuller; *Operating Manual for Spaceship Earth*, Lars Muller Publishers, 2008.

[7] R Armstrong; 'Lawless Sustainability', [online]. Available at: www.architecturenorway.no/questions/cities-sustainability/armstrong/ [Accessed 16 April 2013], 5 December 2012.

[8] M Larssen; *Dune*, in W Myers and P Antonelli's. *Bio Design: Nature, Science, Creativity*, Thames and Hudson, 2013, pp42-45.

[9] J Carlson, 'The 1960 Plan to Put a Dome Over Midtown Manhattan,' Gothamist [online]. Available at: http://gothamist.com/2012/03/08/the_1960_plan_to_put_a_dome_over_mi.php [Accessed 23 November 2013], 8 March 2012.

[10] Our Common Future, Report of the World Commission on Environment and Development, World Commission on Environment and Development, 2 August 1987. Published as Annex to General Assembly document A/42/427, Development and International Co-operation: Environment [online]. Available at: www.un-documents.net/a42-427.htm. [Accessed 5 September 2013].

[11] K P Timoney and K Peterson; 'Failure of Natural Regeneration After Clearcut Logging in Wood Buffalo National Park, Canada,' *Forest Ecology and*

Management, (87), 1–3, pp89–105, 1996.

[12] S Richer de Forges; 'The Ecological Development Revolution,' Eco-Business.com. [online] available at: www.eco-business.com/opinion/ecological-development-revolution/ [Accessed 22 November 2013], 19 November 2013.

[13] International Passive House Association: 'The Passive House Difference,' [online]. Available at: www.passivehouse-international.org/index.php?page_id=238 [Accessed 23 November 2013].

[14] D Turrent; *Sustainable Architecture*, London: RIBA Publications, 2008.

[15] P Soleri; *The Bridge between Matter and Spirit Is Matter Becoming Spirit: The Arcology of Paolo Soleri*, Anchor Books, 1973.

[16] E Ambasz; 'The Universitas Project: Solutions for a Post-Technological Society,' Museum of Modern Art, 2006.

[17] T Morton; *The Ecological Thought*, Harvard University Press, 2012.

[18] J Beynus; *Biomimicry: Innovation Inspired by Nature*, William Morrow, 1997.

[19] R H Haynes; 'How Mars Might Become a Home for Humans?' [online] Available at: www.users.globalnet.co.uk/~mfogg/haynes.htm [Accessed 22 November 2013], 1993.

[20] N Wiener; *Cybernetics – Or Control and Communication in the Animal and the Machine*, John Wiley & Sons, 1948.

[21] B P Belousov; 'A Periodic Reaction and Its Mechanism,' *Compilation of Abstracts on Radiation Medicine*, 147, p145, 1959.

[22] R Armstrong and M M Hanczyc; 'Bütschli Dynamic Droplet System,' *Artificial Life Journal*, 19 (3–4), pp331–346, 2013.

[23] L Tischler; 'MIT's New Self-Assembly Lab is Building a Paradigm Shift to 4-D manufacturing,' Technology, Fast Company. Available at: www.fastcompany.com/3006145/mits-new-self-assembly-lab-building-paradigm-shift-4-d-manufacturing. [Accessed 26 December 2013], February 26 2013

[24] G K O'Neill; *The High Frontier: Human Colonies in Space*, William Morrow & Company, 1977.

[25] C Schurig et al; 'Microbial Cell-Envelope Fragments and the Formation of Soil Organic Matter: A Case Study From a Glacier Forefield', *Biogeochemistry*,

vol 113, issue 1–3, pp595–612, May 2013. Also: A Miltner et al, 'SOM Genesis: Microbial Biomass As a Significant Source' *Biogeochemistry*, vol 111, Issue 1–3, pp41–55, November 2012. Helmholtz Centre for Environmental Research – UFZZ, 'Fertile Soil Doesn't Fall from the Sky. Research for the Environment. The contribution of bacterial remnants to soil fertility has been underestimated until now, Research for the Environment http://www.ufz.de/index.php?en=31184 [Accessed 23 November 2013]. 14 December 2012

[26] J Zaleski; 'Five Innovations that are Boosting Soil Fertility. Indian Agrarian Crisis,' [online]. Available at: http://agrariancrisis.in/2011/11/29/five-innovations-that-are-boosting-soil-fertility/ [Accessed 22 November 2013], 29 November 2011.

[27] J Glausiusz; 'Your Body is a Planet. Discover' [online]. Available at: http://discovermagazine.com/2007/jun/your-body-is-a-planet#.Uij4bRbAY04. [Accessed 5 September 2013], 19 June 2007.

[28] G D Zalakb; 'The Challenge of Microbial Biodiversity: Out on a Limb', *Nature*, 476, pp20–21, 2011.

[29] E Schrödinger; 'What is Life? The Physical Aspect of the Living Cell'. Based on lectures delivered under the auspices of the Dublin Institute for Advanced Studies at Trinity College, Dublin, in February 1943. [online] Available at: http://whatislife.stanford.edu/LoCo_files/What-is-Life.pdf. [Accessed 16 April 2013], 1944.

[30] M Heidegger; 'The Question Concerning Technology'. In William Lovitt and David Farrell Krell eds. Martin Heidegger, Basic Writings. Revised and expanded edition, pp287–311, 1993.

[31] L R Rosenberger; 'The Strategic Importance of the World Food Supply', *Parameters*, pp84-105, 1997.

[32] A C Clarke; *Rendez-vous with Rama*. London: Gollancz, 1973.

[33] M Bedau; 'Living Technology Today and Tomorrow', *Technoetic Arts A Journal of Speculative Research*, 7 (2), pp199–206, 2009.

[34] R Armstrong and P Beesley; Soil and Protoplasm: The Hylozoic Ground Project, Protocell Architecture', *Architectural Design*, volume 81, issue 2, March/April 201, pp 78–89, 2011.

[35] P Beesley; Architect Inc, Hylozoic Veil. [online] Available at: http://philipbeesleyarchitect.com/sculptures/1016_The_Leonardo/index.php [Accessed 23 November 2013], 2011.

[36] R Armstrong, Living Buildings for Tomorrow's Cities, BBC.com [online]. Available at: www.bbc.com/future/story/20130520-greening-the-cities-of-tomorrow. [Accessed 5 October 2013], 21 May 2013.

[37] M Brown; 'Starship London: From Docklands to the Stars,' The Londonist [online]. Available at: http://londonist.com/2013/10/starship-london-from-docklands-to-the-stars.php. [Accessed 23 November 2013], 7 October 2013.

CHAPTER 6

BUILDING HEAVY LIFT ROCKETS:
HOW UNITED LAUNCH ALLIANCE DOES IT

BILL CRESS

If one were to try to launch a fully assembled starship from Earth, the fuel requirements to get the vehicle off the ground and into low-Earth orbit (LEO) make the process economically and practically impossible. Therefore, the construction and assembly of any future starship will need to be accomplished in space, specifically in LEO and getting the components there will be accomplished using heavy-lift rockets with multiple launches over an extended period of time.

This chapter will provide a glimpse into what is required to construct rockets and launch them successfully, showcasing how the process is being done today by one company with an amazing track record.

UNITED LAUNCH ALLIANCE

Building rockets is not for the faint of heart but one company consistently gets it right. In December 2006, United Launch Alliance (ULA) was formed through a joint venture between two giants of the industry, Lockheed Martin and the Boeing Company. They use Atlas and Delta rockets to launch payloads into space for the United States Government as well as NASA, the National Reconnaissance Office and others. The Atlas and Delta families of rockets have been enormously successful for more than half a century, taking into orbit everything from weather and telecommunications satellites to scientific missions designed to study the Universe at large. The specific rockets used

The ULA facility in Decatur, Alabama. Image reprinted with permission of United Launch Alliance © 2013.

by ULA, namely the Atlas V, Delta II and Delta IV rockets, sport an unparalleled 100 percent mission success and are launched from two sites on opposite coasts, Cape Canaveral Air Force Station in Florida and Vandenberg Air Force Base in California [1].

The Atlas V is a single-stage rocket system and the successor to previous Atlas rockets developed in partnership with the US Air Force Evolved Expendable Launch Vehicle (EELV) programme. Meanwhile the Delta rockets have been in operation since the 1960s, with the first Delta rockets evolving from modified Thor intermediate range ballistic missiles that were developed in the mid-1950s for the US Air Force. Like the Atlas rockets, the Delta rockets are single-stage vehicles. Today the Delta IV is the most advanced of the Delta family, having been designed and constructed in association with EELV and is capable of launching virtually any medium- to heavy-size payload into orbit [2].

The experience, personnel, facilities, and capabilities belonging to ULA position them at the forefront of successful space access, both now and in the future. The next generation of heavy-lift rockets that will allow humankind access to the stars will undoubtedly be assembled in ULA's facilities and launched under the control and guidance of their teams.

Why is this company so good at what they do? "ULA has a continual focus on mission success, making sure our rockets work right the first time," says Dan Caughran, ULA's Vice President of Product Operations. "This mindset is embedded in every process that we do."

To achieve this, says Caughran, ULA's 'Perfect Product Delivery Ethic' is to understand every step of the customer–supplier relationship and to strive for perfection in every step [3]. "That's how we define Perfect Product Delivery – building that reliability and the ability to deliver mission success by having perfection, or as close as you can get to it, with the resultant excellence you achieve along the way."

Of course, there will be no heavy-life rockets to launch components for a starship in the future if there are no future engineers. ULA recognise this and have begun outreach programmes to excite the next generation of engineers, aiming specifically at K-12 (i.e., the entirety of primary and secondary school education) science and mathematics education. The importance of these outreach and educational activities is clear to see when one considers that the US currently lags behind other more-developed countries when it comes to the number of undergraduates earning degrees in the STEM subjects. They acknowledge the deficiencies and they are promoting progress. This mindset carries throughout the company and is a primary driving force behind the rockets that ULA are manufacturing, assembling and successfully launching on a consistent basis [4].

The development and component building process for an interstellar vehicle with a proposed budget in the billions, if not trillions, of dollars, will require a company like ULA and their support sub-contractors and suppliers to implement the designs and create the finished components that will eventually be launched into space for assembly into a completed starship that will venture to the stars.

ULA'S FACILITIES

As research for this chapter, the author spent time at ULA's Decatur facility in Alabama, where approximately 1,700 people are employed. The facility is enormous. Approximately 35 acres of floor space are found under one roof, which equates approximately to 457,000 square metres (1,500,000 square feet) of which nearly 381,000 square metres (1,250,000 square feet) is the footprint on the main floor and the balance is on multiple floors within the building. It appears that the tallest point of the structure is approximately 40 metres (~130 feet) from ground level to the roof line with 33.5 metres (110 feet) clear hook height inside the building at the tallest point where friction stir welding of vertical components takes place. Gantry cranes are prevalent everywhere ranging from 10- to 50-ton capacity and are spread throughout the building. They are used to elevate and move components along the various process lines safely and efficiently.

Visiting the building and checking in through stringent security is the preliminary step for accessing ULA. Initially, visitors are greeted by model

High-pressure testing at ULA's Decatur facility. Image reprinted with permission of United Launch Alliance © 2013.

displays of the various rockets for which ULA manufactures and assembles components. The tour starts with an introduction to some of the components and processes that are used for both the manufacturing and assembly of the rockets created at Decatur. Looking at individual displayed samples, one is impressed with the quality of the products and components and the assembly techniques used to create that quality.

The Decatur facility is composed of the following departments:

1) Skin fabrication, rings and domes;

2) Chemical processing;

3) Atlas integrated assembly and checkout;

4) Centaur weld area;

5) Weld centre;

6) Delta IV upper stages;

7) High pressured testing;

8) Spray-on foam insulation;

9) Delta IV integrated assembly and checkout;

10) Composites;

11) Metallic.

These are the process areas that result in finished final stages that are then delivered to the launch sites. We will take a close look at these manufacturing and assembly operations later in this chapter.

The calm atmosphere and demeanor of the people working on various projects within the facility is striking. There is no hustle or bustle. Everything appears very well defined and controlled and the technical expertise of the personnel is apparent.

Decatur is not ULA's only facility. Other locations include:

• Washington DC, where ULA is able to keep in close contact with customers and key legislators;

• Harlingen, Texas, where the Atlas V payload fairings and adaptor fabrication and assembly is performed at the Harlingen production facility;

• Denver Colorado, where ULA's programme management, engineering, test emission and support functions are headquartered;

• Pueblo Colorado, where the Delta II storage is located;

• California Vandenberg Air Force Base, where launch operations on the US west coast include launch support personnel at three launch pads; the Atlas V space launch complex 3W, the Delta IV space launch complex-6 and Delta II space launch complex-2.

• Cape Canaveral Air Force Station in Florida is the site of ULA's east coast launch operations where they have the use of four launch pads including the Atlas V space launch complex-41, Delta IV space launch complex-37 and Delta II space launch complex-17A and 17B.

DELTA IV ROCKETS: THE MOST POWERFUL SPACE LAUNCH VEHICLES TODAY

The Delta IV Family of rockets is assembled in three different configurations, referred to as medium capacity, medium-plus capacity and heavy-lift capacity.

To date, there have been a total of 22 Delta IV rockets manufactured and successfully launched by ULA. These rockets have put the heaviest payloads into space and continue to be the workhorse of choice for dependable heavy-lift requirements.

The Delta IV Heavy has an overall height of 72 metres (236.2 feet); it is taller and carries heavier payloads than any other rocket in the world. It is capable of placing 12,757 kg (28,125 lbs) into geosynchronous transfer orbit. This is substantially more capable than any other competing rocket today.

They are built so that the bottom two-thirds of the rocket comprises the fuel tanks and main engines, known as the first stage. The top third of the rocket is called the second stage and this is where cargo, other electronics, secondary engines, and fuel tanks are integrated. The Delta IV has been frequently upgraded over the past 50 years and incorporates many of the technologies that were previously used, such as the main fuel tanks and payloads fairings. One of its advanced capabilities is the use of liquid hydrogen fuel in its first stage, making it relatively inexpensive but reliable to operate.

The first stage of the Delta IV heavy is comprised of three Common Booster Cores (CBC) that are powered by a rocketdyne RS-68 engine that burns liquid hydrogen and liquid oxygen as the fuel source. The RS-68 became the first large liquid fuelled rocket engine designed in the US since the space shuttle main engine in the 1970s. The goal for the RS-68 was to reduce the cost versus the Space Shuttle Main Engines (SSMEs). On the Heavy, the main CBCs' engines throttle back to 58 percent rated thrust at approximately 50 seconds after lift-off, while the strap-on CBCs remains at 102 percent. This allows the main CBC to conserve propellant and burn longer. After the strap on CBCs separate, the main CBCs engine throttles back up to 102 percent before it throttles back down to 58 percent, prior to the main engine cut-off [5].

The RS-68 engine is mounted to the lower thrust structure of the vehicle by a four-legged (quadrapod) thrust frame and enclosed in a protective composite conical thermal shield. Above the thrust structure is an aluminium isogrid (a grid pattern machined out of the inside of the tank to reduce weight) liquid hydrogen tank, followed by a composite cylinder called the centre body, an aluminum isogrid liquid oxygen tank and a forward skirt. Along the back of the CBC is a cable tunnel to hold the electrical and signal lines, and a feed line

An exploded cross-section of the Delta IV. Image reprinted with permission of United Launch Alliance © 2013.

111

to carry the liquid oxygen from the tank to the RS-68. The CBC is of a constant five-metre diameter [6].

The precision machining and mating of the components in this stage were observed to be of the highest quality obtainable in a commercial manufacturing operation. The structure assembly is virtually a work of art; there are no flaws, gouges, scratches, splatters or anything of a nature that could be observed. This is a significant testament to the quality and training of the workforce and the attention to detail that ULA delivers.

The cryogenic second stage of the Delta IV in the heavy rocket has a five-metre diameter liquid hydrogen tank and a further lengthened liquid oxygen tank. The second stage engine is a Pratt and Whitney RL-10B-2 that features an extendable carbon–carbon nozzle to improve specific impulse. A cylindrical interstage is used on the five-metre version that is built from composites to mate the second stage to the first stage [7].

The L-3 communication Redundant Inertial Flight Control Assembly (RIFCA) features a six ring laser gyroscope and accelerometers, each to provide a higher degree of reliability.

To encapsulate the payload a five-metre composite fairing is used for the five-metre variant of the heavy rocket. There is also a Boeing-built Titan IV five-metre aluminium isogrid payload fairing available, if payloads require it.

The Delta V Heavy uses two additional CBCs as strap-on boosters that are separated earlier in flight than the centre CBC; the Delta V heavy also features a stretched five-metre composite payload fairing.

Orbital capacities of the Delta V Heavy include:
• Geosynchronous transfer orbit (GTO) 13,130 kilograms (28,950 lb.);
• Geosynchronous orbit (GEO) 6,275 kilograms;
• Escape orbit 9,306 kilograms;
• C3 performance of $30\text{km}^2\ \text{s}^{-2}$, 5,228 kilograms;
• C3 performance of $60\text{km}^2\ \text{s}^{-2}$, 2,521 kilograms.

The total mass of the Delta IV heavy at launch is approximately 733,000 kilograms [8].

ATLAS V: AMERICA'S FIRST CHOICE FOR DEEP SPACE MISSIONS

Classified as a medium-heavy launch vehicle with heavy lift capabilities, the Atlas V stands 58 metres (191 feet) tall on the launch pad and weighs 334,500 kilograms (737,400 lb) fully fuelled. This rocket has been in use since 2002 and has a 100 percent success record.

The rocket was originally built by solely by Lockheed Martin but is now produced by ULA. It has lifted various notable payloads into space including the New Horizons mission to Pluto and the Curiosity rover on Mars, and it continues to be used for high-profile launches.

The rocket can support between one and five solid rocket boosters with a higher number enabling a heavier payload to be taken into LEO or beyond. Future iterations may also allow an additional Centaur stage to be attached to the rocket that would mean even bigger payloads could be taken into space, such as crewed capsules.

The largest payload taken to space with this rocket was the military communication satellite MUSO-1 with a mass of 29,402 kilograms (64,820 lbs). The Atlas V is also part of the Air Force's EELV programme that means none of the rocket is reusable. The solid rocket boosters and core engines are discarded after take-off.

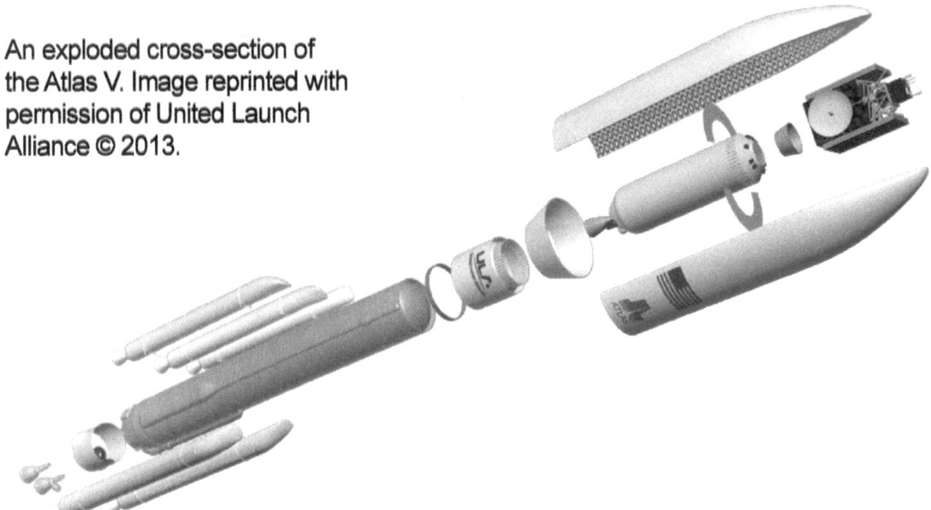

An exploded cross-section of the Atlas V. Image reprinted with permission of United Launch Alliance © 2013.

The Atlas V has set a record for the fastest ever spacecraft to leave Earth's atmosphere, reaching 57,615 kilometres per hour (35,800 miles per hour). This rocket will be in service many years into the future based on its past performance [9].

The unassembled Atlas V at ULA's Decatur facility. It is eventually assembled at their Harlington facility. Image reprinted with permission of United Launch Alliance © 2013.

THE PROCESS OF BUILDING HEAVY-LIFT ROCKETS

The complex process of building rockets requires a very large facility, a substantial work force, easy access for subcomponents to be delivered and staged from sub-contractors, substantial material handling equipment, manufacturing and the associated machinery, a large assembly area and complex testing and quality control programmes that ensure the outcome of the final product.

ULA's Decatur facility constantly has many types of rocket components in various stages of manufacture, assembly, staging, testing and finishing. The process begins by acquiring plate stocks of aluminium, sheets of stainless steel, tubing stock and other materials. The first manufacturing step begins in the Fabrications and Components Department where the aluminium plates are milled into isogrid patterns and then formed into a curved condition that will ultimately become the fuel tank walls or the barrels of the fuel tanks. Machines in this area will also form and machine rings of aluminum that will allow the joining of the pieces of the fuel tanks. Other components of the rocket are also manufactured and then joined together in this department. Propulsion tubing and ducts are made that will carry the fuel down to the engines, while pneumatics, gasses and helium bottles that maintain constant pressure in the tanks are fabricated as well. Fabrications and Components also manufacture electrical and avionic harnesses that are the nervous systems of the rocket itself. These carry the commands, data and information to different components and navigation units. Structures, panels, tubes and harnesses are all built during the fabrication and components process.

There are many proprietary processes that are well beyond the scope of this chapter but some that are in the public domain and stand out in the author's mind are the various welding processes involved for numerous components as well as the outer skin of the rocket and the fabrication of fuel tanks.

The outer shell of the rocket starts out as flat aluminum sheets three metres by fifteen metres (10ft × 50ft) and 1.9 centimetres (three-quarters of an inch) thick. Those sheets are then processed through a flatbed CNC milling machine that precisely removes the majority of the aluminum material from the sheet. In the case of the Delta IV rocket, they start out with 4,535 kilograms (10,000 lbs) of aluminum and when milled to create the isogrid pattern the final results weigh only 500 kilograms (1,100 lbs). Approximately 90 percent of the original stock is removed in the milling process. Once completed the resulting sheet is basically an isogrid of rigid aluminum with the majority of space between the isogrid patterns reduced to a quarter-inch thick outer shell of the rocket. The isogrid pattern provides the structural strength for the materials to handle the load of launching the rocket and payloads into space. By doing this, the weight of the outer skin is substantially reduced while being able to maintain the strength and integrity of the sheet because of the reinforcing nature of the isogrid pattern.

Once quality control inspection is completed after the milling process is finished, the sheets are then moved to a large rolling machine that curves them to a predetermined tolerance and arc. Once those tolerances are checked and the mating surfaces of the edge of the sheets verified they are sent through a welding process to weld the vertical and horizontal seams flawlessly. After the work is completed in Components and Fabrications the pieces move forward in two parallel operations.

As one would expect from the name, at the Weld Centre components such as barrels and panels are taken and welded together, while domes are placed and welded to the ends of the barrels, which basically tops-off and makes the fuel tanks. Four different types of welding processes are used in the Weld Centre, the first relating to stainless steel structures and the last three used on aluminium structures:

1. Resistance spot welding used for the Centaur Upper Stages, which are stainless steel;
2. GPAW constant arc-welding, a standard industry-wide welding process;
3) Variable Polarity Plasma Arc (VPPA) welding, another standard process;
4) Friction stir welding, a state-of-the-art process that uses friction and pressure to fuse the pieces of aluminium together. This process has a lot less variables that have to be controlled so it is a lot more reliable and repeatable process, providing much stronger joints and mechanical properties in the joined areas [10].

The friction stir welding produces a superior strength weld and a flawless seam on the materials that are welded together. The rotatable upper shoulder, the

Aluminium sheets fed through a CNC milling machine to create an isogrid of rigid metal. Image reprinted with permission of United Launch Alliance © 2013.

pin device and the lower shoulder of the material will be configured to friction stir weld the work piece. Friction stir welding is a solid state joining process that offers several advantages over fusion welding processes including higher joint strength and lower distortion. Friction stir welding can join alloys that may not be welded by fusion welding processes. These advantages make friction stir welding a valuable joining process in many industries including the aerospace industry [11].

The second process takes place at the Tank Centre where large fuel tanks are fabricated in various ways using both aluminium and stainless steel.

On the second stage of the Atlas V vehicle very thin sheets of stainless steel foil are used to create the fuel tanks because of its corrosion resistance properties. The material thickness ranges from about 0.068 thousandths of an inch at the thickest part down to 0.015 thousandths of an inch at the thinnest part. The tank is a pressure stabilised design that requires pressure on the tank to maintain the structural integrity. As a result, a constant five pounds of pressure is maintained at all times during fabrication, shipping and assembly until fuel is injected into the tank.

The Atlas V is the highest performing vehicle in the world as a result of its weight to payload capability. The advantages include no wasted energy produced by the engines to unnecessarily lift heavier structures. Using stainless steel foil creates an envelope to contain the fuel in a very lightweight tank.

All finished tanks are inspected with X-rays and undergo hydrostatic or pressure tests to make sure the required structural integrity is achieved. The tanks are then sent to a department that applies some type of additional corrosion protection to the tanks. On cryogenic tanks a spray-on foam insulation (SOFI) is used. This provides the thermal protection that keeps the cryogenic liquid hydrogen and liquid oxygen fuel from boiling off too quickly.

Another process takes place in the Structural Assembly Area. This is where aluminum and steel parts are taken and mechanically assembled together. Components like centre bodies that are placed between the liquid oxygen and hydrogen tanks on the Delta IV are assembled together to make that structure. Also, engine sections that go on the aft end of the booster stage are built here. This allows the attachment of the engines like the RD-180 for the Atlas or the RS-68 for the Delta IV. Structures that allow for the integration and attachment of the different fuel tanks, engines and rocket components are also assembled in this department.

The structural assembly of the Delta IV rocket is accomplished in Decatur. For the Atlas V series the structural assembly is accomplished in Harlingen, Texas. The ULA facility in Harlingen is much smaller in size than Decatur; it is approximately only 91,400 square metres (300,000 square feet) but the processes are similar.

To continue the assembly process, tanks from the Tank Centre and structural assemblies from the Structural Assembly Area are brought together in the Integration Assembly and Check-Out (IACO) department, which is the final assembly point. On the floor there are separate final assembly areas for the Atlas and Delta first stages and also second stage assembly. Here tanks

Assembling the Atlas V rocket in a large hangar at Harlington. Image reprinted with permission of United Launch Alliance © 2013.

and structures are mated and bolted together to form the main stages of the structures. Pneumatic and propulsion systems and avionics systems are integrated here. All systems are tested to make sure they function properly. Once completed everything is prepared to pack up and ship out.

Three main pieces of the rocket are shipped completely assembled. The first stage or booster, the second or upper stage and the payload fairing, which is the very top of the rocket that circumscribes the spacecraft itself (for both the Atlas and Delta series a four-metre and a five-metre fairing are produced depending on payload requirements of the client). Those three stages are then shipped fully assembled to the launch site.

At the launch site, the stages are stacked to form the vehicle. The first stage is stood up, the second stage is then placed on top of that, the payload and payload fairing is placed on top of the rocket and then everything is prepared to launch. For Atlas rockets this is all done in a vertical stack. For Delta IV rockets it is a combination of horizontal integration and some vertical stacking.

PLACING AN ORDER FOR A ROCKET

On average, ULA works through a 24-month contracting window starting from the placement of an order. The actual fabrication inside the Decatur facility is approximately 22 months for the overall process, from the start of machining to the delivery of the vehicle. ULA estimate a schedule of three-to-six months for the completion of work at the launch site.

For the Delta IV Heavy, stacking all three of the liquid-fuelled stages together at the launch site requires a six-month process. This then comprises a 28-month schedule from initial machining to launch readiness. In order to move the process along as smoothly as possible, some components from sub-contractors or suppliers are pre-ordered in advance of the overall ULA time frames.

Today, the cost of building rocketsis in the range of hundreds of millions of dollars depending on the configurations required by the client. Vehicles that do not perform properly would have catastrophic results for those clients and a payload loss would dramatically impact programmes, schedules and future budgets. Realising this, ULA's Perfect Product Delivery presents a rocket launch with the best-in-the-business chance for mission success.

Presently there are over one thousand suppliers and sub-contractors involved in the process of building rockets and the components required. Some of the major sub-contractors/suppliers include:

• Pratt and Whitney Rocketdyne (PWR) who provide the following:

  RS-68 – Delta IV booster engine;

  RL10B-2 – Delta IV upper stage engine;

  RL10A – Centaur upper stage engine;

• RD AMROSS, who provide the RD-180/Atlas V booster engine;

• ATK, who provide the following:

  Graphite Epoxy Motor (GEM) 60 – Delta IV solid rocket motors;

  Heat shield, Centaur interstage adapter and boat tail assemblies – Atlas V;

• Aerojet, who provide solid rocket motors to the Atlas V.

ULA uses the Delta Mariner to transport assembled rocket components to launch sites. The Delta Mariner is a cargo ship operated by Foss Maritime specifically for ULA. Its primary role is transporting components for the Boeing Atlas V and Delta rockets to launch facilities at Cape Canaveral Air Force Station in Florida and Vandenberg Air Force Base in California. The ship is designed for shallow inland waterways as well as the open ocean and is capable of carrying up to three 49-metre (160-feet) long Delta IV Common Booster Cores. Some cargos carried by the Delta Mariner were formerly transported by Russian transport aircraft from the manufacturer to the launch site [12].

Loading Atlas V assemblies onto the Delta Mariner for transport to the launch site. Image reprinted with permission of United Launch Alliance © 2013.

GOING INTERSTELLAR

The final design and construction of an interstellar starship and placing the vessel's components into LEO for assembly will require government support and a large group of scientists and technicians. It will be a prodigious task. Components must be built that will operate reliably for 50–100 years, unless disruptive technologies are proven that can provide revolutionary propulsion systems and accelerate the travel times to a star.

ULA business focuses on space and providing transportation to space. They are positioned perfectly to help meet the challenges involved by providing the capabilities of what they do best on the launch vehicle side of those tasks. They welcome the opportunity to offer what they excel at, providing the greatest value to the launch vehicle part of the equation. In this author's opinion they are the obvious choice for the role of effectively getting to space as we know it today.

The ULA has an incredible history of building and launching rockets into space, including launching human payloads. They were a part of the Gemini and Mercury missions. That history has been founded and strengthened through a constant pursuit of 100 percent mission success, which they achieve by focusing on one launch at a time. Every launch has its own unique critical aspects and every launch is treated with the utmost importance it deserves.

The foregoing showcases some of the complexities of building rockets and what is required to successfully produce reliable, dependable vehicles to support space programme initiatives. There are thousands of components that must be joined together to function flawlessly and place payloads into space and ULA manage and control the processes impeccably.

REFERENCES

[1] 'ULA Company Overview', United Launch Alliance, LLC, [online], www.ulalaunch.com/site/pages/About_Overview.shtml (last accessed May 2014).

[2] 'ULA Quick Facts', United Launch Alliance, LLC, [online], http://ulalaunch.com/site/pages/About_QuickFacts.shtml (last accessed May 2014).

[3] Interview with Dan Caughran, Vice President of Product Operations, United Launch Alliance June 2013.

[4] 'ULA in the Community', United Launch Alliance, LLC, [online], http://ulalaunch.com/site/pages/About_Citizen.shtml (last accessed May 2014).

[5] 'Delta IV Heavy Demo Launch Timeline', *Spaceflight Now*, [online], www.spaceflightnow.com/delta/d310/041201launchtimeline.html, 1 December 2004, (last accessed May 2014).

[6] Kyle, Ed; 'Space Launch Report Delta IV Data Sheet', Space Launch Report, [online], www.spacelaunchreport.com/delta4.html 5 September 2010 (last accessed May 2014).

[7] 'ATK Composite and Propulsion Technologies Help Launch National Reconnaissance Office Satellite', ATK [online] www.atk.com/news-releases/ 15 December 2006 (last accessed May 2014).

[8] Delta IV Heavy, Wikipedia [online], http://en.wikipedia.org/wiki/Delta_IV May 2013 (last accessed May 2014).

[9] O'Callaghan, Jonathan; '10 Amazing Space Rockets', *All About Space*, 12 (2013) p65.

[10] Interview with Christa Bell, Communications and Public Relations, United Launch Alliance, June 2013

[11] Eller, M, Z Li and D Roussel; 'Friction Stir Welding Apparatus and Method', US patent 20100140321 A1, to Lockheed Martin, June 2010.

[12] Wikipedia MV Delta Mariner, http://en.wikipedia.org/wiki/MV_Delta_ Mariner (last accessed May 2014).

CHAPTER 7

FUTURE LAUNCH VEHICLES

RICHARD OSBORNE

Currently, the world is well serviced by expendable launchers, with US launchers such as ULA's Atlas and Delta, Orbital Sciences' Antares and SpaceX's Falcon 9, Russia's Soyuz and Proton launchers, Ukraine's Zenit launchers, ESA's Ariane 5 and Vega, China's Long March launcher family, Japan's H-II and India's GSLV. However, all are based on an architecture that has fundamentally remained the same since the dawn of the Space Age, namely the artillery shell/missile approach of ballistically launched vehicles that are thrown away after a single use.

In the short term, launch systems will still be predicated on the artillery shell/missile approach of expendable launch vehicles, although innovative thinking is resulting in actual flight hardware that moves beyond this and towards reusable, vertically launched vehicles. The progress made by companies such as SpaceX on rapid iteration and practical steps towards reusability and two-stage-to-orbit reusable launch systems has been phenomenally fast in comparison with the glacial speed of the space industry and this alone will significantly accelerate the move towards opening up near Earth space. With the advent of reusable multi-stage launch vehicles, the construction of small interstellar probes, comparable to the size of today's interplanetary probes but with long transit times, becomes feasible.

The medium-term sees launch systems based on single-stage-to-orbit (SSTO) reusable launch vehicles enter the fray and it is this development that will really open up near-Earth space in a significant way, since once a vehicle is reduced to a single stage, maintenance costs and refurbishment for reuse should drop compared to multi-stage reusable vehicles. Although SSTO has long been considered by many in the launcher community to be impossible, a small band of diehards has kept the dream of SSTOs alive over the decades and paths to the realisation of single-stage-to-orbit are now becoming clearer.

The path towards SSTOs is littered with the bones of the valiant attempts that have tried: the X-30 NASP, the X-33, the Rotary Rocket Roton and British Aerospace/Rolls Royce HoTOL, as well as concepts such as Douglas's Rombus, Ithacus and Hyperion, SSX, Kawasaki's Kankoh Maru and Pacific Launch Systems Phoenix to name but a few. However, cost reductions and advances in technology and manufacturing processes are making the task more achievable than in the past. The reductions in launch costs offered by single-stage-to-orbit launchers such as Reaction Engines' Skylon promises the ability to build large, in-space construction facilities. With such facilities established, the construction of small interstellar probes with reduced transit times becomes feasible.

By the time large interstellar starships can be built, it is likely that second generation SSTOs will be in service and the launch costs will have reduced sufficiently that construction of astro-engineered megastructures such as space elevators can be considered. By the point at which a space elevator becomes available, transportation to space of assemblies and sub-assemblies of a large interstellar starship will become a significantly lower-cost undertaking and, with a Solar System-sized economy [1], the construction and launch of a large interstellar mission the size of Project Daedalus [2] or Project Icarus [3] seems a much more achievable project.

NASA SLS

The NASA Space Launch System (SLS) [4], whilst far from the concept of a low cost launch vehicle, offers the capability for launching very large payloads with diameters twice that of most launchers. This makes it attractive for launching payloads such as outsized space telescopes, Mars sample-return

SLS Vehicle Configurations

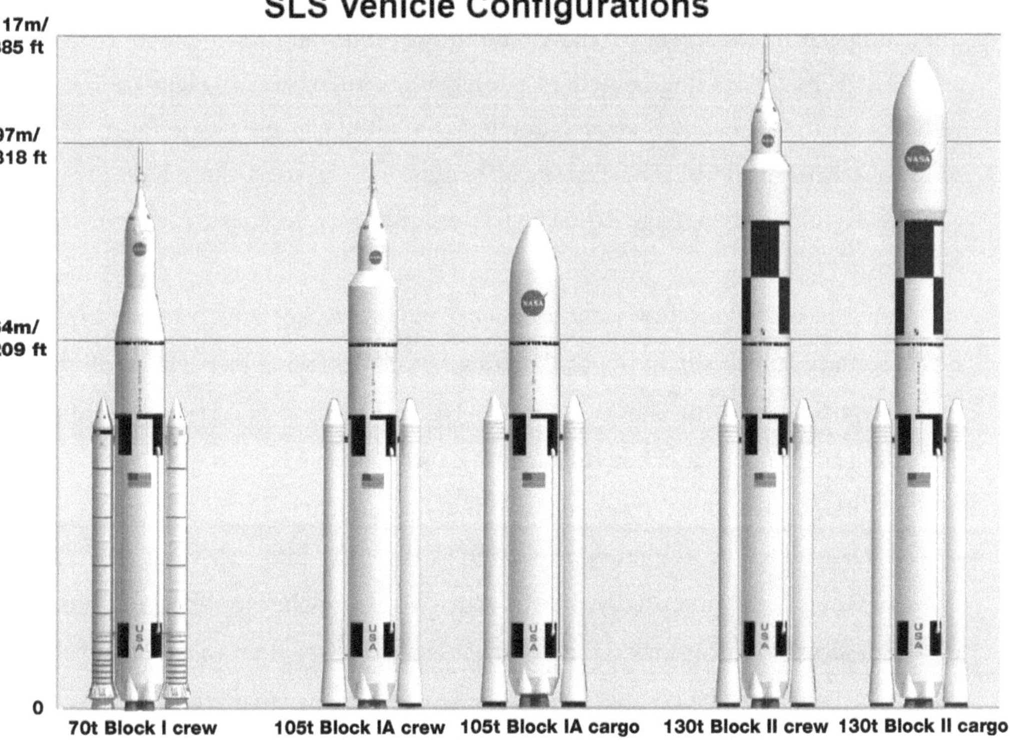

SLS launch configurations. Image: NASA.

vehicles, the proposed Bigelow Aerospace Olympus inflatable space station [5], large upper stages for trans-lunar missions and clusters of large stages for use in interplanetary spacecraft.

The SLS will be the largest launch vehicle to date, with a size and power even greater than the Saturn V and N-1 Moon rockets. The SLS is derived from the space shuttle system, with two large solid rocket boosters, a cluster of high performance, RS-25 liquid hydrogen/liquid oxygen rocket engines developed from the space shuttle main engines and an upper stage also powered by liquid hydrogen and liquid oxygen. Two versions are envisaged: one lifting a payload of 70 tonnes to low Earth orbit (LEO), which will be used for carrying the Orion multi-purpose crewed vehicle; and a second version following on from this, lifting 130 tons to LEO.

ARIANE 6

The European Space Agency Ariane 6 design [6] follows the same conservative approach as many existing launchers, namely to be an expendable launch vehicle – in this case, a two-stage launcher. The initial concepts were for a solid rocket motor propulsion system, although more recently there has been consideration of a two-stage liquid rocket propulsion system, with several small solid rocket boosters.

Ariane 6 would follow a traditional expendable rocket launch ascent from ESA's Kourou launch site in South America, before deploying its payload in orbit. The Ariane 6 design moves away from the Ariane 5 method of carrying multiple large payloads on one launcher, to launching a smaller single main payload.

With the advent of lower cost commercial launchers such as SpaceX's Falcon 9, and the Falcon 9 reusable, as well as the entry into service of the Russian Angara and potentially the Russian Baikal launcher too (both of which will both look at the commercial market), it is hard to see how an expendable such as Ariane 6 can be competitive in the world launcher market.

H-III

The Japanese H-III rocket [7, 8] is an expendable two-stage rocket being developed by the Japanese Space Agency (JAXA) and Mitsubishi Heavy Industries. First given the go-ahead in 2013, the H-III is planned to be launched from 2020 onwards. The H-III will have a smaller payload capability (3,000 kilograms to LEO) than its predecessor, the H-IIA (6,000 kilograms), which is a consequence of striving for cost reductions and efficiencies in the launcher.

The H-III rocket will use liquid hydrogen and liquid oxygen propellants in both its stages. The first stage will be powered by two LE7 liquid rocket engines, while a single LE5 liquid rocket engine will drive the second stage. The H-III will be able to have its thrust augmented by the attachment of up to six solid propellant strap on rocket boosters.

An artist's impression of one of the possible configurations of the Ariane 6 rocket.
Image: ESA/D Ducros.

LONG MARCH 5

The Long March 5 rocket [9] is the next generation of Chinese expendable orbital launcher, designed to be comparable to the US Atlas V, Delta IV and Falcon 9 launch vehicles. The Long March 5 will use liquid oxygen and kerosene propellants and will be capable of orbiting payloads of up to 25,000 kilograms to LEO, with strap-on solid boosters for thrust augmentation. Currently, it is planned for the Long March 5 rocket to be introduced into service sometime before 2020, although this will be subject to successful testing.

FALCON 9 HEAVY

The Falcon 9 Heavy [10] is a development of the Falcon 9 [11], with three Falcon 9 first stages clustered together to provide sufficient thrust to enable the Falcon 9 Heavy to deliver the largest payload capability to orbit since the Saturn V launcher. The Falcon 9 Heavy will launch from either Cape Canaveral in Florida, or Vandenberg Air Force Base in California.

ANGARA

The Angara launcher [12] is planned to be the next generation of Russian launcher. An expendable launcher, using liquid oxygen and RP-1 for its propellants, the Angara is designed to be modular so it can be used in various configurations from light payload through to heavy payload launch, with variants planned in one core module, three core and five core configurations, as well as a possible seven core version.

The first stage of the Angara was successfully used as part of the first South Korean orbital launcher, the K-1. The Angara first stage was mated to a South Korean designed and built second stage and, although the Angara worked flawlessly, the second stage suffered a launch failure. The subsequent success of a following launch of the K-1 rocket has put the Angara back on track in terms of demonstrating its capability.

The first orbital launches of the Angara with both of its stages are scheduled to take place in Russia in the 2014–2015 timeframe, with a prior test of a two-stage Angara launcher on a sub-orbital trajectory having been successfully completed in July 2014.

BAIKAL

Baikal is a semi reusable version of the Angara rocket, with either a single or multiple fly-back, first stage boosters based on the Angara first stage that detach after their liquid oxygen and RP-1 propellants are expended, and deploy wings for a return to the launch site, augmented by a small embedded jet engine. The second stage then carries on to deliver the payload on an orbital trajectory before burning up on re-entry.

A full size mockup of the Baikal fly-back booster was displayed at the Paris air show several years ago, sporting a deployable scissor wing. After a period of inactivity there are indications that Baikal is again progressing. It is possible that the progress of SpaceX in the US with the Falcon 9 reusable has sparked a resurgence of interest in developing Baikal.

STRATOLAUNCH

Stratolaunch [13] takes a different approach to the other launch systems whereby, instead of launching a rocket from a launch pad, it carries an expendable launch vehicle on what will be the world's largest plane – the dual hulled Stratolauncher, currently being constructed by Scaled Composites with a mating system constructed by Dynetics.

The Stratolauncher is based on two Boeing 747 airframes, with a central wing joining them and with the expendable rocket launch vehicle slung between them. Stratolauncher takes off from a runway with the fully fuelled rocket, climbs to an altitude of around 10–15 kilometres and releases the rocket, which then ignites its propulsion system and climbs to orbit, dropping its stages during the ascent and then deploying a payload of up to 6,000 kilograms.

The original plan was to deploy a SpaceX Falcon launch vehicle from the Stratolauncher aircraft, but early in the development process Stratolaunch moved to using another rocket vehicle, going with the Orbital Sciences Antares rocket.

Stratolaunch Systems was founded in 2011 by Burt Rutan, who is the founder of Scaled Composites and Paul Allen, the co-founder of Microsoft, both of whom have collaborated previously on SpaceShipOne, the world's first private human-carrying vehicle to go into space. Stratolaunch has its corporate headquarters located in Huntsville, Alabama.

BLUE ORIGIN

Unique amongst launch developments, Blue Origin [14] has maintained a very low profile – so low in fact that many are unaware that they even exist! Overlooking them would be unwise, however. In what little information is known about them, they have demonstrated a capability to design, build and test fire a large liquid hydrogen/liquid oxygen rocket engine, the 100,000lbf thrust BE-3, as well as test launch one of their sub-orbital launchers, work on liquid oxygen/methane rocket engines and repeatedly launch large rocket-powered platforms to demonstrate hovering and translation.

Blue Origin is privately funded by Jeff Bezos, the founder of Amazon, although Blue Origin has also worked with NASA several times on a range of development projects and was awarded \$3.7 million by NASA in 2009 as part of the first Commercial Crew Development (CCDev) programme, followed by \$22 million in funding from NASA in 2011 for the CCDec Phase Two programme.

FALCON 9 REUSABLE

When Elon Musk, CEO of Space Exploration Technologies (SpaceX), suggested he would be developing a reusable version of the Falcon 9 launcher, many in the space industry scoffed. However, step-by-step, the critics have been proven wrong with actual flight achievements many said could not be achieved. Falcon 9 Reusable is, as the name suggests, the reusable version of the currently operational Falcon 9 expendable launch vehicle, using the same Merlin ID gas generator cycle rocket engines, powered by RP-1 and liquid oxygen propellants. The Falcon 9 reusable [15] is designed so that both its first stage and second stage are not only recovered, but fly back to land at the launch point.

The SpaceX Grasshopper reusable first stage test vehicle during a test launch. Image: SpaceX.

The development programme to date has seen several launches of full-size first stages (named Grasshopper 1 and Grasshopper 2), to altitudes of up to one kilometre from its test site in Texas, demonstrating hovering and translation of large first stage-sized vehicles, before returning to pinpoint landings with the accuracy of a helicopter onto their launch pads.

Additionally, on two recent orbital launches, SpaceX have successfully achieved powered re-entry manoeuvres of the first stage, along with stable descent and powered landings of their first stage on the ocean to retire risk (with the second ocean landing improving on the hard landing of the first landing). These are considerable achievements as well as significant steps in terms of moving launch vehicle development forward, since although the capability has existed to achieve these for a long while, only SpaceX have been bold enough to pursue this approach.

It is not unreasonable to predict that SpaceX will succeed in reusability of their first stage and thus a lowering of their launch costs, although returning to the launch site is a major task still to achieve. Achieving reusability of a second stage is a much harder proposition however, since the stage is moving far faster and requires both a heat shield and the ability for what has to be a light stage to also be robust enough to maintain sufficient structural integrity during re-entry to enable re-use. This would be likely to take considerably longer to achieve, if it is indeed viable in its current form.

Whilst the second stage will be far harder to convert to reusable operations, this is not the case for the SpaceX crewed capsules, the Dragon and Dragon V2. As part of the overall operations of the Falcon 9 reusable launch vehicle, the Dragon capsule with its parachute recovery and the Dragon V2 with its hot helicopter, propulsive landing return, will demonstrate another area where costs can be driven down by virtue of them being reusable at all.

Mars Colonial Transporter Booster

The splendidly named Mars Colonial Transporter is a SpaceX concept for a future spacecraft designed to transport over 100 tonnes of payload to the surface of the planet Mars. However, to get it into space, a very large launch vehicle is required to deliver part, or all, of the Mars Colonial Transporter.

Any proposal is currently mere speculation, but the type of configuration that has been extrapolated for the proposed launch vehicle would be a reusable launcher, ten metres in diameter and using nine of the next generation SpaceX Raptor full-flow staged combustion rocket engines, powered by liquid oxygen and methane. It is conjectured that the launcher may even consist of three of these boosters clustered together to make a very large launch vehicle.

Skylon

Reaction Engines' Skylon [16] is currently the only SSTO vehicle under development. Providing a capability that leapfrogs the expendable launch vehicles and reusable two-stage-to-orbit launchers in terms of payload launch costs and launch availability, this spaceplane offers aircraft style runway operations and very flexible launch opportunities.

A depiction of Skylon departing the International Space Station. Image: Adrian Mann/ Reaction Engines Ltd.

A standard Skylon mission would see a payload of up to 15 tonnes loaded into the Skylon payload bay in a spaceport loading area, much like a conventional cargo aircraft. Skylon would then be towed out to the end of a runway, filled with its propellants and then take off like an aircraft, climbing on an air-breathing ascent before transitioning to rocket mode to achieve LEO. It would then deliver its payload,

Above, re-entering Earth's atmosphere. Below, Skylon emerging from its hangar. Images: Adrian Mann/Reaction Engines Ltd.

such as a satellite or cargo for a space structure such as a space station, propellant depot or orbital dockyard. Skylon would then prepare to de-orbit, re-entering Earth's atmosphere and, after re-entry, follow a gliding descent back to its spaceport of origin.

Skylon's advantage is that it uses an air-breathing rocket engine with a pre-cooler to deeply cool, but not liquify, incoming air before injecting that air into the combustion chamber with hydrogen fuel. It uses this air-breathing mode of operation from take-off until high in the atmosphere. At an altitude of 26 kilometres (almost three times the altitude of a commercial passenger jet) and at a hypersonic speed of Mach 5.4, Skylon transitions to pure rocket mode, where it uses its onboard supply of oxygen for the remainder of the ascent to orbit.

Currently, development work has focused on the innovative engine that powers Skylon, the SABRE (Synergetic Air Breathing Rocket Engine). Skylon has two engine nacelles, one on the tip of each wing, each containing a SABRE. Development work has seen successful ground testing of a hydrogen heat exchanger – the most efficient heat exchanger ever built – ground testing of the pre-cooler modules down to cryogenic temperatures and for full duration operating times, as well as testing of a range of altitude compensating expansion/deflection rocket nozzles, testing of a liquid oxygen cooled combustion chamber, as well as wind tunnel testing of the design and comprehensive trajectory and re-entry modelling.

The Skylon spaceplane also underwent a thorough analysis by the European Space Agency, together with launcher experts from around the world, successfully addressing any technical concerns.

Second generation SSTOs

With the advance of technology, most notably with the practical development of carbon nanotubes and graphene for structural elements, aerogel-type thermal shielding, 3D printing for advanced fabrication, lightweight and ultra small sensors for more thorough distributed diagnostics, as well as lighter weight engine technology, improved heat exchangers and advanced battery technology, we can see paths towards a second generation of SSTOs slowly revealing themselves, even now.

By the time the first generation of SSTOs become well established later in the first half of the twenty-first century, design will have started on the next generation and, undoubtedly, there will be breakthroughs of which we cannot even imagine that will provide significantly enhanced capabilities for second generation SSTOs, delivering larger payloads with enhanced reliability and for even lower costs. This is one place the notorious conservatism of the space industry will work in its favour, since many of these technologies will be well established and proven by the time they would be considered for use in second generation SSTOs.

As well as the anticipated air-breathing SSTO spaceplane designs and vertical take-off and vertical landing rocket powered SSTO designs, we may be able to add such configurations as laser driven vehicles [17], where either Earth-based or space-based lasers will focus energy on a combustion chamber on an Earth-to-orbit vehicle to enable another form of SSTO transportation.

SPACE ELEVATOR

The ultimate step in launching is not to launch at all! The way this counter-intuitive concept makes sense is through the use of a space elevator; an artificial structure that enables high-tech, space-capable versions of elevator cars to ascend and descend cables extending all the way into space. Whilst pushing the limits of materials technology to a level that makes such a venture currently impractical, by the time second generation SSTOs are operational in the second half of the twenty-first century, there should have been sufficient development in advanced materials so that construction of a space elevator can be seriously contemplated.

First conceived by Konstantin Tsiolkovsky as long ago as 1895, the concept was developed further by Yuri Artsutanov in the late 1950s, advanced by various American engineers in the 1960s and 1970s, and popularised by Sir Arthur C Clarke in his 1979 novel *The Fountains of Paradise*. The space elevator [18] is a technology that could be achieved with future developments in material strength and mass decrease.

A breakthrough on the road to development of a space elevator was the discovery of carbon nanotubes, which offer an excellent combination of light weight but excellent tensile strength. Subsequent to the discovery and use of carbon

An artist's impression of a station situated at the top of a space elevator. Image: Adrian Mann.

nanotubes, the development of another carbon-based material, graphene, has caused excitement as a result of it also offering a similarly good combination of light weight and great strength. More recently, the development of Carbyne also looks promising.

Progress in using such materials is still in its infancy and there is no doubt that future developments of these materials and others yet to be fabricated will bring the construction of a space elevator closer to conception. Whether these will be the actual materials necessary, or whether further breakthroughs will be required, is yet to be seen and will only become clear as the technologies are further proven. It is sometimes postulated that because these materials offer tensile strength that may lead to a space elevator, that this means that a space elevator is possible in the near-term future. This is not the case however, because the various materials need to be proven to be up to the challenge and possible to fabricate on a large scale first, much less reach commercialisation and production.

In addition to the development of strong, lightweight materials, work has already been carried out on tethers for proving other technologies related to realising space elevators, with a number of tethered challenges demonstrating climber and beamed power technologies, such as the NASA Centennial Challenge. The LiftPort Group is working towards a lunar space elevator since, as a proof of concept, this is a technically more achievable goal than an Earth space elevator, albeit with the need to transport materials out to lunar space first.

DISRUPTIVE DEVELOPMENTS

Whilst developments such as a technological singularity are largely speculation at this point, one of the developments often associated with such a transition is advanced molecular nanotechnology [19] that would, if introduced at any point in the future evolution of space launch technology, be a major disruptive technology if applied to the development and fabrication of new, lighter weight, stronger materials for space access systems and, indeed, interstellar spacecraft.

Molecular nanotechnology is a technology that offers advantages both for fabrication of space-based vehicles and structures on Earth, and for in-space construction. It could enable breakthroughs even using existing launch vehicle technology, since once a molecular nanotechnology fabricator payload is delivered to orbit the overheads should be significantly less than for conventional construction and would offset the launch costs.

Nuclear fusion is another technology that would be decidedly disruptive, since at the point fusion is achieved sustainably and relatively cost effectively, it will have applications for interstellar spacecraft propulsion systems. There are currently several diverse approaches to nuclear fusion that are under development, ranging from the large Tokamak-style International Thermonuclear Experimental Reactor (ITER) and the Lawrence Livermore National Labs' Inertial Confinement Fusion (ICF) approach, to the smaller, private ventures such as the Lawrenceville Plasma Fusion Project, the Inertial Electrostatic Confinement (IEC) EMC2 Project, the Tri-Alpha Project, Helion Energy Project and General Fusion Project. Whilst a nuclear fusion propulsion system is slightly different to a nuclear fusion reactor, a breakthrough with any of these fusion projects could be a very (positively) disruptive development for deep space/interstellar propulsion.

Another technology often seen as disruptive is artificial intelligence (AI), but having AI is not necessary to mount an interstellar mission; rather it is more of a 'nice to have'. Existing expert systems such as IBM's Watson, or Google's autonomous cars, represent a level of computational capability that would certainly be more than adequate for controlling an interstellar mission. In fact, our experience with launching planetary probes shows that existing computational capabilities are adequate for many interstellar spacecraft IT requirements. With the rapid advances in reducing the size of electronics, the ability to fit the computational hardware in Google's autonomous cars, or even IBM's Watson, into a small enough volume to launch as an interstellar probe control system will arrive within the next decade, which will provide a significant capability for interstellar probes.

As a purely space-related disruptive development, a breakthrough with propellantless propulsion would undoubtedly be the single, greatest advance that could be achieved for interstellar travel, but the theory behind the science and a path to technical development is still in its infancy.

CONCLUSIONS

The trend in launchers leads inexorably towards reusable launchers. Although there are a number of expendable launchers that will stubbornly resist the trend for another decade or two, it is clear that with SpaceX making breakthroughs in orbital flights with a recoverable first stage, as well as the testing of elements of the Skylon SSTO spaceplane and a resurgence of interest in Baikal-type flyback launchers in Russia, the dam has broken and there is no way back for detractors of reusable launch vehicles. This is a very positive development, because it will finally usher in lower cost access to space, followed by cheap access to space and, at that point, humanity can start to consider construction of the in-space infrastructure needed for affordable, relatively large, interstellar missions.

The point at which access to space is sufficiently mature to enable interstellar missions to be launched is almost here for the slower, sail-based or nuclear electric-based interstellar missions. The launch capability is there; it is just a case of the interstellar spacecraft technologies needing to be further developed to enable early interstellar missions to be mounted. Certainly the technologies to mount precursor interplanetary missions exist now and, within a decade and given concerted effort, the first slow interstellar missions could be launched. The drawback, however, is that the launch costs are still sufficiently high that the volume of launches has to go up to enable the launch costs to come down sufficiently to enable affordable interstellar missions.

With fully reusable launch systems the launch costs will decrease significantly and will enable larger, more capable interstellar vehicles to be constructed in Earth orbit, perhaps with much more powerful propulsion systems such as pulsed nuclear fusion.

With the advent of a space elevator, the game will change significantly yet again and, with the cost reductions of space access that could be achieved by such a system, humanity will be able to realise more affordable interstellar missions as part of a Solar System-wide infrastructure. With such space infrastructure, humanity can more easily contemplate interstellar missions of the scale of Project Daedalus and Project Icarus and, finally, within human life-spans, we will be able to realise missions to the stars.

REFERENCES

[1] Parkinson, R C; *Citizens of the Sky*, 2100, (1987).

[2] Long, K F and P Galea (editors); *Project Daedalus: Demonstrating the Engineering Feasibility of Interstellar Travel*, BIS, (2011).

[3] Long, K F, Richard Obousy, M Fogg, A Tziolas, R Osborne, A Presby, A Mann; 'Project Icarus: Son of Daedalus – Flying Closer to Another Star', *Journal of the British Interplanetary Society*, 62, 11/12, pp403–414 (2009).

[4] NASA Space Launch System [online]. Available at www.nasa.gov/exploration/systems/sls/ (last accessed 13 July 2014).

[5] Bigelow Aerospace [online]. Available at www.bigelowaerospace.com (last accessed 13 July 2014).

[6] Launch vehicles: Ariane 6 [online] Available at www.esa.int/Our_Activities/Launchers/Launch_vehicles/Ariane_6 (last accessed 13 July 2014).

[7] 'JAXA Plans to Test New Large Rocket From 2020', *The Japan Times* [online]. Available at www.japantimes.co.jp/news/2013/12/25/national/science-health/jaxa-plans-to-test-new-large-rocket-from-2020/6 (last accessed 13 July 2014).

[8] Clark, S; 'Japan Moves Forward with Replacement for H-2A Rocket', *Spaceflight Now* [online]. Available at www.spaceflightnow.com/news/n1403/04h3rocket/#. U3K174GSzcx (last accessed 13 July 2014).

[9] 'Long March 5 Will Have World's Second Largest Carrying Capacity', *Space Daily* [online]. Available at www.spacedaily.com/reports/Long_March_5_Will_Have_World_Second_Largest_Carrying_Capacity_999.html (last accessed 13 July 2014).

[10] SpaceX Falcon Heavy [online]. Available at www.spacex.com/falcon-heavy (last accessed 13 July 2014).

[11] SpaceX Falcon 9 [online]. Available at www.spacex.com/falcon9 (last accessed 13 July 2014).

[12] Sokolov, O; 'The Russian Way to a Creation of Reusable Aerospace Systems: The Concept of Using Fly-back Boosters', Commercial Space Technologies Ltd Report (2013).

[13] Stratolaunch Systems [online]. Available at www.stratolaunch.com (last accessed 13 July 2014)

[14] Blue Origin [online]. Available at www.blueorigin.com/ (accessed 13 July 2014).

[15] 'Reusability: The Key to Making Human Life Multi-Planetary', SpaceX [online]. Available at www.spacex.com/news/2013/03/31/reusability-key-making-human-life-multi-planetary (last accessed 13 July 2014).

[16] Hempsell, M; 'An Analysis of the Skylon Infrastructure', *Journal of the British Interplanetary Society*, 63, pp 129–135 (2010).

[17] Myrabo, L N and D G Messitt; 'Ground and Flight Tests of a Laser Propelled Vehicle', American Institute of Aeronautics and Astronautics, AIAA 98-1001 (1997).

[18] Edwards, Bradley C and Eric A Westling; *The Space Elevator: A Revolutionary Earth-to-Space Transportation System*, Spageo Inc (2002).

[19] Drexler, K Eric; *Engines of Creation: The Coming Era of Nanotechnology*, Doubleday (1986).

CHAPTER 8

THE ROAD BEFORE THE STARS:
INTERSTELLAR PRECURSOR MISSIONS

KELVIN F LONG

Before we send starships to the worlds around other stars, we must validate the technology, its performance and our capabilities in space, whilst properly mapping out the path between here and there. History has shown that the discovery of lands across the Western Atlantic was preceded by similar trips that achieved just this. Enter the domain of interstellar precursor missions; robotic spacecraft scouts designed not to reach the stars, but to go to intermediate locations outside of our Solar System and deep within interstellar space. In this chapter, we discuss these important missions and some of the historical concepts proposed in the literature, which are necessary precursors as we attempt to expand beyond the frontier of our origin.

THE EXPLORATION OF INTERSTELLAR ISLANDS

Visionary space enthusiasts would have us believe that we can jump straight to the construction of giant starships and embark on our conquest of the stars, paying little attention to the vital discoveries and critical exploration that must first form a part of any enduring strategy. These visions manifest themselves in concepts like the Enzmann starship [1], which was adopted as part of an extensive roadmap for interstellar exploration by G Harry Stine in his 1973 *Analog Science Fiction and Fact* article [2]. Stine argued that a mission to the stars would be completed as part of a full programme rather than as a one-off mission. This plan would include the identification of an astronomical target, the launch of unmanned probes to the destination and finally the launch of a full expedition fleet of ten starships to the target system.

In his 1976 paper *A Programme for Interstellar Exploration* [8], the physicist Robert Forward laid out a more realistic roadmap. His plan set forth an optimistic 50-year programme for (manned) interstellar exploration consisting of five phases. This involved sets of mission definition studies, launching automated payloads and developing the critical technologies. This would be followed by ultra-high propulsion system feasibility studies. Then would come the development of human-rated vehicles leading up to the launch of the first crewed exploration starship.

These sorts of programmes are exciting ideas for how interstellar colonisation may take place, but the reality is that you firstly have to prove the vital technologies as well as to map out the space inbetween the origin and the destination. This would be enabled by firstly exploring the outer reaches of our Solar System and then proceeding on to any other objects that lay in our path, carefully refining the technology and proving our capabilities until we have the confidence to take that next giant leap across the 'Sea of Suns'. This suggests the launch of many interstellar precursor missions, beyond the frontier of our Solar System and to the many 'islands' in their various forms dotted across space.

In order to plan for the eventual colonisation of interstellar space, it is wise to first learn the lessons of history. There is no greater example of the triumph of exploration than those adventures that led to the colonisation of America. The period between 1400–1600 is known by scholars as the 'Age of Discovery'. Much of the coast of Africa had been explored by the Greeks, the Portuguese and even the Phoenicians, who may have circumnavigated the entire coast of Africa from 609 to 593 BC. The Atlantic Ocean lay ahead of them to the west of Europe, thought by the Hellinistic Greeks and the Romans to be the vast sea that separated them from Asia to the west. Seen as the stormiest and most turbulent of seas, it presented a formidable and dangerous challenge to any sailor that dared to cross it and it was known by many different names including that given to it by the Medieval Arabs: the 'Green Sea of Darkness'. Medieval Europeans did not just see the Atlantic Ocean as a vast watery void; they filled it with mysterious islands that had apparently been reached by sailors of the past. Some had ventured out as far as Iceland and Greenland, particularly the Norse between 860–930. But prior to the fifteenth century medieval Europeans gave the Atlantic Ocean little attention. Instead, they confined their trade routes to the coastal lines of Europe and the British Isles. Then in the fourteenth century fishermen from England and Germany began to visit the fishing territories of Iceland while the French began to fish near the south-west of Ireland. These voyages allowed sailors to build up their experience, knowledge and capabilities for sailing in the Atlantic Ocean. The idea of crossing the Atlantic to find new lands or new routes slowly began to emerge.

At the time much trade was happening between India and the eastern Mediterranean. This included items such as ivory, spices, gems, pearls, food and tortoise shells. The main trade route to India from the mouth of the Red Sea involved going along the coast of southern Arabia. The route tended to be perilous with hostile Arab traders not keen on these new arrivals. Later the ships sailed across the open waters of the Indian Ocean. The Greek ships were built heavier and sturdier than those of the Arab vessels so that they could withstand the turbulence of the buffering seas. They even discovered the secret of the famous monsoon winds, strong winds caused by climatic conditions that blow steadily from the south-west or the opposite direction, depending on the time of year, which enabled ships to sail from the mouth of the Red Sea across the Indian Ocean, to the western coast of India. This improvement in the Greek ships enabled them to make regular trading voyages to India. Some Greek ships even sailed out past the Straits of Gibraltar to explore the western coast of Africa, although not always with success.

It was much later that the Genoese interest in Atlantic exploration began to open up the passage to the west, with many citizens pooling their resources to outfit an expedition to India. The two galley fleet was equipped with supplies sufficient to last ten years. Departing in 1291, they sailed passed the Straits of Gibraltar, soon sighting Cape Nun, rediscovering the Canaries in the post-classical era. Although they did not succeed in establishing a successful trade route to India, the Italian presence in the Atlantic grew from here, with merchants of Venice and Genoa journeying across to southern England and Portugal. Then, sometime prior to 1340, the Genoese began to visit the Canary Islands, which were also known as the 'Fortunate Isles'. Later, expeditions were mounted to explore these islands by Genoese, Florentines and the Portuguese. Because of the lack of wealth discovered, some of the interest in these islands was soon lost and was not rekindled again until 1420 with Prince Henry, the Navigator of Portugal whose activities eventually led to the opening of a sea route to India. However, the Majorcans and the Catalans quickly continued the exploration of the Canary Islands and became interested in expanding their activities into the Atlantic Ocean. By the start of the 1490s the Europeans, in particular the Portuguese, had been exploring the western coast of Africa and the Atlantic Ocean for many decades. Several expeditions out into the Atlantic were soon thwarted by adverse winds and currents in the latitudes of the Azores.

It was in 1492 that Christopher Columbus set out to see the western passing to India, sailing south of the Azores and experiencing a smooth voyage that took him to unknown lands. As eloquently described by Ronald Fritze [28], "It was the beginning of a series of geographical revelations that would profoundly transform the entire world politically, economically, socially and culturally." Columbus set out on four bold voyages across the 'Ocean Sea' (the Atlantic) between 1492 and 1503, attempting to establish permanent settlements and later initiating the Spanish colonisation of the New World. His trip was funded by King Ferdinand II of Aragon and Queen Isabella I of Castile, but only after much petitioning for the funds. On firstly assessing the proposals of Columbus, the savants of Spain had reported that he had grossly underestimated the distance to Asia and they pronounced the idea impractical. On later consideration, his proposals were supported, with half of the financing coming from private Italian investors. On completion of the trip Columbus was to be given the rank of Admiral of the Ocean Sea; he would also be appointed Viceroy and Governor of these lands and he would receive ten percent of all the revenues from the new lands in perpetuity. It did not quite work out quite so glamorously for the bold Columbus, but his place in history was at least secured.

The vision of interstellar flight is also a bold and noble undertaking. It will require the maturation of critical technologies, the capabilities of large-scale project management and the organisation of world governments, industry and academia like no other project in history. It will also likely be an expensive endeavour and should not be entered into lightly. In studying the Age of Discovery, we see many important elements behind Columbus' expeditions that may also have relevance for interstellar flight. Firstly, it was backed by investors and a national state (monarchy). Second, there was a clear business reason for the trip, including Columbus being promised rewards on his success. Thirdly, the voyages were preceded by earlier trips out into the Atlantic and the Canary Islands, which established knowledge, capability, confidence and experience before the full Atlantic crossing was attempted. What will be the Fortunate Isles of space that lead to the exploration of interstellar space and to venture to worlds around other stars?

If we learn anything from history, then we realise that rather than our space-aspiring civilisation constructing large, Columbus-style missions to the stars, it is more likely that before we do venture out to other stars such missions will be proceeded by adventurous explorers – both robotic and human – discovering the lands of new islands in space that go beyond our Solar System and away from the warmth of our own Sun. This is the domain of nearby interstellar space, the wide emptiness of void, bathed in cosmic rays and photons of light traversing the ages of the cosmos.

As we ponder the construction of these pre-starships, where might we choose to send them? The first destination is the myriad dwarf planets that ring our Solar System, some yet to be discovered. The dwarf planet Pluto's furthest point from the Sun is a distance of around 48 AU (7.2 billion kilometres – one AU, or astronomical unit, is the distance between Earth and the Sun, 149 million kilometres) and is one of these destinations. A NASA spacecraft named New Horizons is already on its way there, to arrive in 2015. So, our future missions should go beyond this distance. Other dwarf planets and smaller 'Plutinos' have been discovered. Many of the dwarf planets discovered so far reside within the Kuiper Belt of icy objects beyond Neptune, named after Gerard Kuiper who popularised the possibility of the belt in 1951, although the concept had been first raised by several astronomers, most notably the Irish astronomer Kenneth Edgeworth. The Kuiper Belt is believed to be the source of most short-period comets, defined by taking less than 200 years to orbit the Sun. Short-period comets travel along the ecliptic plane where most of the planets are located. It extends from the orbit of Neptune outwards past Pluto to the 'Kuiper Cliff', where there is a significant drop-off in objects at 50 AU (7.5 billion kilometres).

The orbits of some of the objects out here actually take them beyond the Kuiper Belt, such as the dwarf planet Eris, which was only discovered in 2005 and was the final nail in the coffin for Pluto's planetary status. Fittingly, Eris is named after the Greek goddess of discord. Eris vies with Pluto to be the largest object beyond Neptune discovered so far – it has a diameter between 2,326–2,400 kilometres, while Pluto is believed to be around 2,360 kilometres. Its perihelion distance is 38 AU (5.6 billion kilometres), within the Kuiper Belt, but interestingly its highly elliptical orbit take Eris to an aphelion distance of 97 AU (14.5 billion kilometres) from the Sun. It has an orbital period of around 560 Earth years. It is an interesting world, thought to be made of mainly rock (it contains a greater mass than Pluto) as well as reflective water-ice and methane-ice that makes it shine brighter than Pluto. Eris even has its own moon, named Dysnomia.

Another distant object with an even more elliptical orbit is Sedna, named after the Inuit goddess of the sea. It is around 1,000 kilometres in diameter, or a third the size of Pluto, and its orbit, which is tilted to the plane of the ecliptic by 12 degrees, is entirely located beyond the Kuiper Belt and in a region known as the 'scattered disc'. At perihelion it is at 76 AU, or 11 billion kilometres from the Sun, while at its most distant it reaches 937 AU, which is an enormous 140 billion kilometres from the Sun. It is also three times further from the Sun than Pluto and, until 2014, it had the distinction of being the most distant object in the Solar System. Then another Sedna-sized object was announced in 2014, designated 2012 VP113, which has an orbit even more eccentric than Sedna's. Although its aphelion is 'only' at 449 AU (67 billion kilometres) its perihelion is 80.5 AU (12 billion kilometres), which is the most distant perihelion of any object known. The discoverers of VP113, Chad Trujillo and Scott Sheppard, suspect that a much larger object lurks even farther from the Sun and its gravity is causing the orbits of Sedna and VP113 to become elongated, as well as creating the Kuiper Cliff. They calculate that its mass could be even larger than Earth's, putting it in the super-earth realm. However, its existence is based solely on conjecture at this point.

The outermost reaches of our Solar System form an interesting place to explore. It is around here that the heliopause is located, which is the outer boundary of the heliosphere, a region of space dominated by the magnetic field of the Sun dragged out into space by the solar wind. The Voyager 1 spacecraft crossed the heliopause and entered the interstellar medium that lies beyond in August 2012. The interstellar medium is composed of the particles, gas, dust and magnetic fields that exist between the stars in the Galaxy. It is bathed in cosmic rays, which are high-energy particles originating most likely from supernovae remnants.

The exact abundance of particles in the interstellar medium is not well known, but it is primarily formed of hydrogen. A spacecraft moving through this region of space will help to correctly characterise these properties and improve our understanding of the interstellar medium's intermediate role between stellar and galactic scales. Measurements taken along any flightpath through the interstellar medium could include the density of the interstellar gas, the ionisation state of the gas, the density and size distribution of solid interstellar dust grains and the strength of the interstellar magnetic field. A modern review of interstellar medium properties is given by the astronomer Ian Crawford [6].

The next destination for our precursor probes is the gravitational lensing point, located between 550 AU and 1,000 AU from the Sun. This is where a spacecraft can be positioned to use the Sun as a gravitational lens according to the

General Theory of Relativity to magnify the light of nearby stars and planetary systems on the line-of-sight precisely opposite the Sun from the spacecraft and make them visible, including exoplanets, halo discs and Oort clouds far beyond the capabilities of any other ground or space-based observatories. The reason a range of distances is given for where the focal point of the Sun's gravitational lens is located is because unlike an optical focus, a gravitational focus does not have a specific distance to which all of the light rays are brought together. Instead all the points in the line away from the Sun are foci, with 550 AU (82.3 billion kilometres) being the minimum distance. For any spacecraft passing through this line of sight, the effects created by the Sun's corona – which works to deflect light away from the Sun – diminish and the image enhancement is improved. This effect becomes less pronounced the further from the Sun you go, up to around 1,000 AU (149.6 billion kilometres). The lensing point could also be used for Search for Extraterrestrial Intelligence (SETI) experiments, in particular using the Sun's ability to magnify radio waves from the hydrogen line at a frequency of 1420 MHz to the emission from hydroxyl molecules at 1,666MHz (the wavelength range between which is colloquially known as the 'water-hole' for interstellar communications). One of the most active researchers studying a gravitational focus mission is the Italian scientist Claudio Maccone, who advocates the FOCAL space mission [31].

Heading out further still, our probes will venture into the Oort Cloud. Named after the Dutch astronomer Jan Hendrik Oort, the Oort Cloud is an immense spherical cloud surrounding the Solar System. It is located outside the Kuiper Belt and extends from around 2,000 AU (300 billion kilometres) to 50,000 AU (7.5 trillion kilometres, or 0.8 light years) from the Sun. The outer boundary of the Oort Cloud therefore marks the true edge of the Solar System, where the gravitational influence of the Sun becomes so weak that objects can easily escape out into deep space. The Oort Cloud is largely made up of icy objects that are dispersed at distances of tens of millions of kilometres from each other. When another star comes near, its gravitational influence can affect the orbits of the Oort Cloud objects, sending some of them towards the inner part of our Solar System; this is the source of what we call long-period comets. It is not known for sure how many objects reside in the Oort Cloud, but the total mass is predicted to be around 40 times the mass of Earth. When a spacecraft goes beyond the Oort Cloud, it would truly be an interstellar voyager.

The next possibility that we may encounter in the search for 'islands in space' is free-floating planets. These are not in orbit around any star but perhaps have been ejected out into deep space. Alternatively, the more massive (Jupiter-like) of them may have originated in interstellar space and formed in a similar manner to stars. The number of unbound planets could be so common as to exceed the number of stars in the Galaxy, but this is speculation. Most will consist of low mass rocky or icy planetary embryos ejected from their host star systems, often through the migration of other planets towards their suns. As well as free-floating planets, it is entirely possible that the first star system our robotic probes will visit will in fact be a brown dwarf system. Although none are currently known to be closer than Alpha Centauri, theories predict that brown dwarfs are so common in the Galaxy that there should be one closer than 4.2 light years. Brown dwarfs are non-luminous 'failed' stars that, with only around eight percent the mass of the Sun, are not massive enough to ignite their hydrogen fusion fuel properly. Instead we search for them at infrared wavelengths, looking for their dim thermal emission.

Finally, there is much unresolved physics that these precursor probes may yet be able to help us understand. This includes the nature of dark energy, dark matter or helping in the detection of gravitational waves by becoming one arm of a long range interferometer. These probes can also help to improve our measurements of astrometry, detect supernova explosions or conduct scans for SETI. Some have suggested the possibility of searching the Kuiper Belt for infrared signatures associated with intelligent life [17].

As we reach towards the next great frontier of becoming an interstellar civilisation, it is clear that we must first lay the roadmap by demonstrating our capability in deep space missions. Robotic interstellar precursor missions have the potential to achieve just that, whilst increasing our knowledge of the cosmos and all that it contains. Fortunately, our efforts to begin this programme have already begun.

THE FIRST STARSHIPS: THE VOYAGERS

The Voyager spacecraft are not what would normally be associated with a starship, but they are the closest thing to one that we have so far launched into space. The two spacecraft are now at the outer reaches of our Solar System, having crossed the heliopause, the edge of the Sun's magnetic influence and this at least justifies the category of interstellar precursor spacecraft.

As described in a recent JPL report [4], the Voyager 1 and 2 missions became possible thanks to the coincidence of three unlikely variables. The first was the occurrence of a rare planetary alignment of Jupiter, Saturn, Uranus and Neptune, which would enable the Grand Tour trajectory utilised to its maximum by Voyager 2 to take place. The second was the co-incidental maturation of our technological readiness, to be able to complete such a mission. The third variable was the discovery of the gravity assist manoeuvre in 1961 by Michael Minovitch [5]. Another big factor which was instrumental in bringing the Voyager mission to reality was the Pioneer programme; in particular the Pioneer 10 and 11 spacecraft.

The NASA Pioneer programme (which included Pioneers 6 to 11) is one of the most successful in space history. In particular the Pioneer 10 and 11 probes are the first robotic explorers to visit the outer planets and to travel beyond the orbit of the dwarf planet Pluto.

Pioneer 10 was launched in March 1972 and became the first interstellar spacecraft to reach escape velocity from the Solar System, passing Neptune in June 1983. It was overtaken in terms of distance from the Sun by Voyager 1 in February 1998, at a distance of 69.4 AU (ten billion kilometres). As of today the spacecraft is at a distance of over 106 AU (15.8 billion kilometres) from the Sun. Currently travelling at a speed of around 2.6 AU (389 million kilometres) per year, it will pass within the region of the present position of the star Aldebaran, 68 light years away in Taurus, in two million years time.

The main purpose of the Pioneer 10 mission was to study interplanetary space as well as the solar wind interaction with magnetic fields. Because the

154

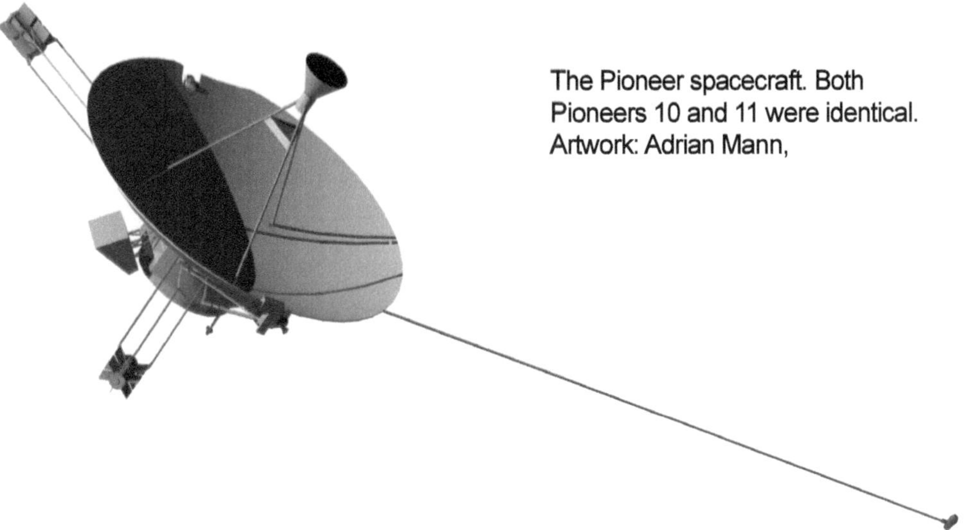

The Pioneer spacecraft. Both
Pioneers 10 and 11 were identical.
Artwork: Adrian Mann,

spacecraft was performing a fly-by of Jupiter it also studied the Jovian atmosphere
and satellite system. Contact with the Pioneer 10 probe was lost in 2003 as a
result of power reduction and long-range communications reliability.

Pioneer 11 was launched in April 1973 and also used a Jupiter gravity
assist to pick up velocity while obtaining dramatic images of the Great Red Spot
of Jupiter as well as accurately determining the mass of the Jovian moon Callisto.
Pioneer 11 had similar mission objectives to the Pioneer 10 probe, although
the mission also included a close fly-by of the planet Saturn and it obtained
spectacular images of the ring system, discovering a new ring as well as a new
Moon. Unfortunately communications were lost in November 1995, twenty-
two years after launch. Pioneer 11 is expected to be around 90 AU (13.5 billion
kilometres) from the Sun at the time of writing.

The Pioneer spacecraft both had a mass of 258 kilograms of which
around 29 kilograms comprised the eleven science instruments. The Pioneer
probes were each powered by a radioisotope thermoelectric generator (RTG),
providing around 155 watts of power during the launch and 140 watts by the
time of the Jupiter fly-by encounter. The spacecraft also had both a medium-
and high-gain antenna for communications. The Pioneers 10 and 11 spacecraft
also carried a message to any would-be space travellers that might encounter

the probes, in the form of a 120 gram, 349 centimetre square pictorial plaque constructed from gold and anodised aluminium and included information depicting our knowledge of the atom, our position in the Galaxy relative to 14 distant pulsars and controversial drawings of a naked male and female human silhouetted by the spacecraft.

The lessons learned from the Pioneer missions were important in the design of the Voyager missions. The Voyager spacecraft were NASA and JPL initiatives, officially approved in May 1972. Voyager 1 was launched in September 1977 and Voyager 2 the month before. It is estimated that a total of $865 million has been spent on the Voyager missions using a total work force of 11,000 people. So far, a total of five trillion bits of information have been sent back to Earth from the Voyager probes, the equivalent of over 7,000 CDs.

The primary mission for the Voyager probes was the exploration of the gas giants, Jupiter and Saturn. Voyager 2 also went on to explore Uranus and Neptune and it remains the only spacecraft to ever have visited those planets. The primary mission was completed in 1989 when Voyager 2 passed the planet Neptune. At the completion of the primary mission Voyager 1 was at a distance of around 40 AU (six billion kilometres) and Voyager 2 at a distance of 31 AU (4.6 billion kilometres). The mission was then extended to the 'Voyager Interstellar Mission' that had the objective of exploring the outermost edge of the Sun's domain and beyond. These missions helped to characterise the outer Solar System environment and look for the heliopause boundary, where the Sun's magnetic fields and outward flow of the solar wind come to an end. This extended mission has three phases: (1) The penetration of the termination shock, which is the point where the solar wind slows from supersonic to subsonic speeds and the spacecraft begin to detect large changes in plasma flow direction and magnetic field orientation; (2) the exploration of the heliosheath, which is the region between the termination shock and the heliopause; and (3) the exploration of interstellar space. Voyager 1 has, as of August 2012, entered phase three.

The key to the success of the Voyager mission was the aforementioned rare geometric arrangement of the outer planets that allowed for a four-planet

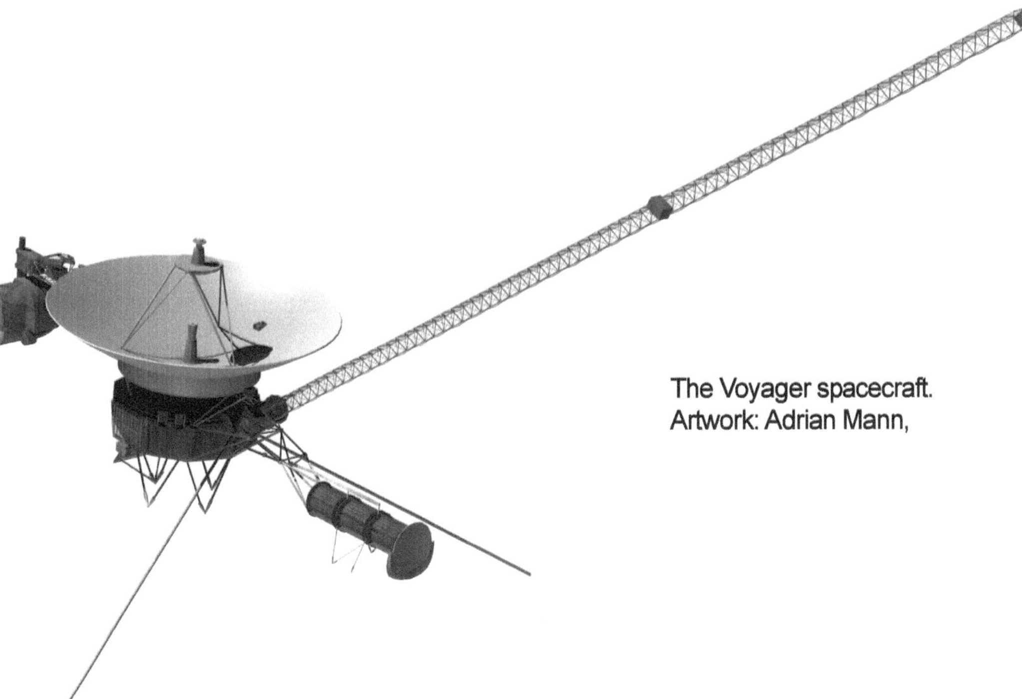

The Voyager spacecraft.
Artwork: Adrian Mann,

tour using a minimum of propellant and trip time. The planetary arrangement of Jupiter, Saturn, Uranus and Neptune only occurs every 175 years. It allows for a spacecraft to fly by one planet and onwards to the next without the need for onboard propulsion. The gravitational field of each planet flown past bends the spacecraft's flight path and increases its velocity enough to deliver it to the next destination. This 'gravity assist' technique was first demonstrated with the NASA Mariner 10 Venus/Mercury mission in 1973–1974. As a result of the fly-by technique, the total flight time to Neptune was reduced from 30 to 12 years. More than 10,000 different trajectories were studied to choose the right path for the Voyager missions.

When the Voyager spacecraft passed near the Jupiter system they were exposed to hard doses of radiation. This required very special shielding requirements to protect the sensitive electronic components.

Data from the Voyager probes is transmitted through the Deep Space Network (DSN). This is a global tracking system operated by JPL and there are antenna complexes in the California Mojave desert, near Madrid in Spain and in Tidbinbilla, which is near Canberra in Australia. The data is sent from the Voyager probes at a transmission rate of 160 bits per second and are received with a 34-metre diameter radio dish, although high transmission updates are also sent regularly at 152 kilobytes per second, requiring a 70-metre diameter dish. All data received and transmitted by the two spacecraft are via their 3.7-metre high-gain antenna. The signal from the Voyager spacecraft at their extreme distance is very weak, so that the power hitting the antenna is only 10^{-16} watts, approximately 20 billion times less than a standard digital watch.

The Voyager spacecraft are identical and have a mass of 773 kilograms each, with 105 kilograms being allocated for scientific instruments. The spacecraft were powered by three 315 watt RTGs.

There is a variety of science under study with the interstellar phase of the mission. This includes the strength and orientation of the Sun's magnetic field, the composition, direction and energy spectra of the solar wind particles and interstellar cosmic rays, the strength of the radio emissions originating from the heliopause, the distribution of hydrogen within the outer heliosphere. There are five science teams participating in the Voyager Interstellar mission, examining the data from six operating instruments on each Voyager probe.

Both of the Voyager probes carry a special greeting card for any intelligent life that may find it. This is called the 'Golden Record' and is a continuation from the earlier Pioneer 10 and 11 probes that carried small metal plaques identifying their time and place of origin for the benefit of any other space farers. This message is carried by a phonograph record, a 12-inch gold-plated copper disc containing sounds and 115 images selected to portray the diversity of life and culture on Earth. The sounds include music selections from different cultures and eras, spoken greetings from people of Earth in over 55 languages (ranging from the 6,000 year old Akkadian to the modern Chinese dialect Wu). It includes sounds such as sea surf, wind, thunder, birds, whales and other animals. The contents were chosen for NASA by a committee chaired by the Cornell University astronomer Carl Sagan. Of the Golden Record, Sagan has said that the record will be found and played only if there are advanced space-faring civilisations in interstellar space, but nevertheless that "the launching of this bottle into the cosmic ocean says something very hopeful about life on this planet" [40]. The voyage of the Voyager probes is likely to be one of the most remembered missions by the generations ahead.

By early summer 2014, Voyager 1 had reached a distance of nearly 128 AU (19 billion kilometres), escaping the Solar System at a speed of 3.6 AU per year, 35 degrees out of the ecliptic plane to the north. Voyager 2 was at a distance of 104 AU (15.6 billion kilometres) and is escaping from the Solar System at a speed of 3.3 AU per year, 48 degrees out of the ecliptic plane to the South. Voyager 1 crossed the termination shock at 94 AU (14 billion kilometres) in December 2004 and Voyager 2 crossed it at 84 AU (12.5 billion kilometres) in August 2007. The reason for the difference in distance is that the heliosphere is not symmetrical and the termination shock also moves inwards and outwards depending on solar activity and the strength of the solar wind. The spacecraft have since gone on to explore the heliosheath with Voyager 1 passing through the heliopause, the outer boundary of the Sun's magnetic field and solar wind, at a distance of 121 AU (18 billion kilometres). Eventually, the probes will pass other stars. Voyager 1 will drift within 1.6 light years of the star AC+79 3888 in the constellation of Camelopardalis in around 40,000 years' time. In the same time frame Voyager 2 will make a 1.7-light year pass of the star Ross 248 and then in 296,000 years it will drift within 4.3 light years of the star Sirius.

By the year 2017 the Voyager probes will have been in space for 40 years. The probes have enough electrical power and thruster fuel to operate to at least until the year 2020 and may survive until 2025, nearly half a century after their launch. The reason their mission will end is because of their diminishing power and hydrazine levels that prevent further operation and for the Voyager probes to be able to get a target lock on Earth for signal transmission. Their longevity is a remarkable achievement for Earth's first robotic ambassadors to go beyond the Solar System. They are moving out into parts of space where no person or technology from Earth has ever flown previously. Whilst we on Earth go about our daily businesses of living, the Voyager achievements go largely unnoticed by the general public. Yet, every day they break a new record and push the human frontier that bit further outwards.

One could argue that the Voyager probes are not really starships, but merely 'star-probes'. Whatever we call them, it is worth keeping in mind that they are the furthest we have sent our robotic ambassadors and they are embarking on a journey towards the distant stars. For spacecraft that were launched in the 1970s and are still operating, perhaps we can give the Voyagers special dispensation and deservedly honour them as humanity's first starships.

DESIGNING THE INTERSTELLAR PRECURSOR MISSIONS

If we wanted to send probes to the islands of space, what would they look like? How fast would they go? How would they be propelled? Although national space agencies have made no serious considerations beyond the drawing board for sending probes to go beyond the Voyager distance, this has not stopped creative engineers and physicists from considering how those ambitious spacecraft missions might be one day achieved by coming up with innovative concepts. Let us consider some of these. It is helpful to set a time limit on what we might think is appropriate for these interstellar precursor missions, from launch to mission completion. Assuming that the working career of a scientist might be one of the criteria, then a journey time of around 50 years seems a reasonable requirement. If we assume that the acceleration time is only months and can be neglected, then the trip times to various targets are shown in Table One, assuming this outer limit. For such long duration space missions the biggest problem has always been the propulsion necessary for the required high performance. To send a space probe to a distance of 200 AU in a time scale of 15, 20, 30 or 50 years requires an average velocity of approximately 13, 10, 7 and 4 AU per year respectively. This is equivalent to 19, 33, 47 and 63 kilometres per second, respectively, and compares to the moderate 17 kilometres per second or around 3.3–3.6 AU per year experienced by the Voyager probes. To put this into perspective, the nearest stars are over four light years away, which is approximately 270,000 AU (40 trillion kilometres). To make that trip in one century would require a cruise velocity of around 2,700 AU (404 billion kilometres) per year, which is three orders of magnitude higher than the Voyagers. Because this velocity are so large, conventional propulsion systems based upon chemical engines cannot meet the performance requirements. Hence alternative technology is required.

Table One: Cruise speeds for 50-year interstellar precursor missions

Destination	Distance (AU)	Cruise speed (AU/yr)
Pluto	48	1.0
Eris	97	1.9
Heliopause	120	2.4
Sedna	76–937	1.5–18.7
Gravitational lens point	550–1,000	11–20
Oort Cloud	2,000–50,000	40–1,000

THE INTERSTELLAR PROBE MISSION

In the late 1990s NASA and JPL organised a project to design a space probe to exceed a distance of 200 AU distance within 15 years as a minimum goal. The Interstellar Probe Science and Technology Definition Team led the study and the project was called the Interstellar Probe (ISP) [11]. It would embark on a journey outside the solar heliosphere and study the connection between the Sun, Earth and interstellar space. The 150-kilogram probe would have been propelled via a 200-metre radius solar sail with a required area density of one gram per metre squared, designed to achieve a velocity of around 70 kilometres per second or 14 AU per year. This is around five times the speed of the Voyager 1 and 2 spacecraft. The key reason for selecting a sail for the mission was the mission requirement of getting to 200 AU within only 15 years. This meant that a propulsion system was required that could rise to cruise velocities that are several multiples of those of the Voyager and Pioneer probes. Hence after some discussion the team decided that sails were the only technology available that offered both the performance and near-term technological maturity, with some initial investment. The mass of the probe and sail combined would be around 246 kilograms. The initial trajectory would see the spacecraft head into the inner solar system and fly within a quarter of an astronomical unit of the Sun in order to take advantage of the enormous radiation pressure at this distance. This would give the craft sufficient acceleration to boost out of the Solar System and the sail would be ejected at around 5 AU on the way out when the acceleration becomes inconsequential and to ensure there is no interference with the onboard instruments. The spacecraft antenna dish would serve as the main structure and three struts around the 11-metre central hole in the sail would support this. Control of the spacecraft would be achieved by moving the central mass with respect to the sail's centre of mass by changing the length of the three struts and it would be spin stabilised when in sail mode.

The main objective of the mission would have been to study the nature of the interstellar medium, its influence on the Solar System and its chemical evolution. Another objective was to explore the outer Solar System to look for clues to its origin and understand the nature of other planetary systems should they exist. A significant penetration into the interstellar medium was seen as a key mission goal. The probe would carry a 25 kilogram payload onboard, with a large suite of instruments designed to measure the properties of energetic particles, cosmic rays, magnetic fields and the dust at the boundary between the heliosphere and the nearby interstellar medium. Additional instruments considered for the mission included the use of a telescope to search for Kuiper Belt objects or an instrument to identify organic modules in space. More exotic possibilities for the mission included the measurement of low energy antiprotons emitted from primordial black holes or a search for so called Weakly Interacting Massive Particles (WIMPs) that could make up the missing dark matter in cosmological models. Although the probe was designed to go to 200 AU or more, it was capable of going out as far as 400 AU (60 billion kilometres) in 30 years and indeed this was one of the goals of the designers. This really would have been an ambitious yet achievable project if it had been supported past the planning stage.

The Innovative Interstellar Explorer Mission

The Innovative Interstellar Explorer (IIE) was a NASA proposal to launch a 34 kilogram payload to around 200 AU (30 billion kilometres). It largely grew out of an earlier study for a Realistic Interstellar Explorer (RIE) proposed by space engineer Ralph NcNutt, who has been evolving the concept of an interstellar precursor mission in collaboration with JPL for some years [22].

McNutt has argued the case for an interstellar precursor mission as a high priority for science. The initial plan for the mission was for it to be launched in 2014 and arrive at its destination in 2044, around 30 years after launch. By the year 2025 the Voyager 1 and 2 spacecraft would have ceased transmitting and be at around 150 and 125 AU from the Sun, respectively. So although they may have passed the heliopause it is unlikely that they will have reached the undisturbed interstellar medium, which is thought to be beyond 200 AU. Hence the proposed IIE spacecraft would have been able to reach it and transmit data using its high gain antenna. By the year 2057 the spacecraft would have been expected to reach 300 AU (45 billion kilometres), some 43 years after launch. By the year 2147 the spacecraft would reach 1,000 AU (150 billion kilometres) from the Sun and be the furthest and fastest human-made object ever launched into space. This would be a century after the planned launch date. One century for 1,000 AU sounds like a long time and if we multiply this by the distance to alpha Centauri at 272,000 AU, such a probe would take many tens of thousands of years to reach the nearest star to the Sun.

One of the primary science goals of the mission was to reach and measure the properties of the interstellar medium as well as the exact location of the termination shock. It would also measure the properties of the solar magnetic field and incoming cosmic rays, which are important measurements for future spacecraft that leave our Solar System. These measurements would

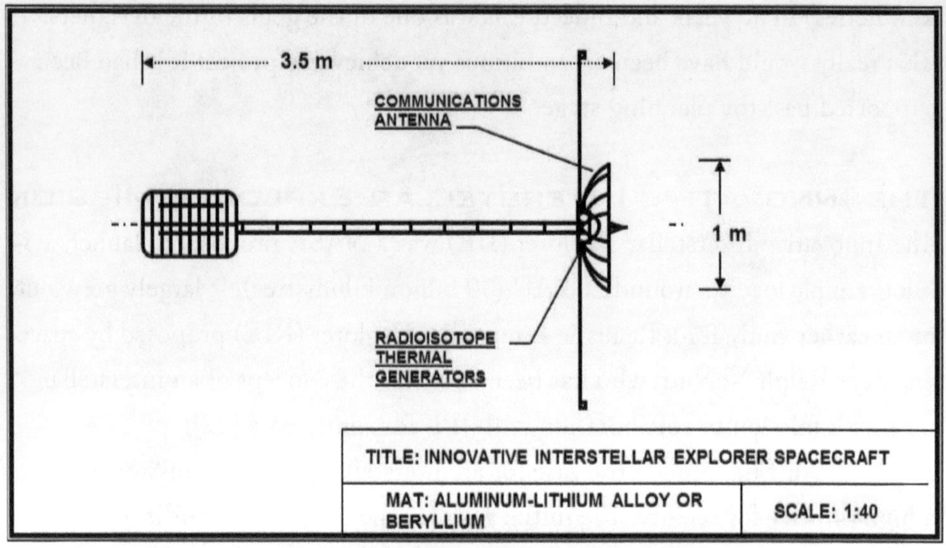

A schematic of the Inniovative Interstellar Explorer spacecraft. Image: Kelvin F Long,

have tremendous implications for understanding the path of our supersonic Solar System through the intense cosmic ray environment of our Milky Way. Our Solar System is protected from these rays by the solar heliopause, the boundary between the solar wind generated by our own Sun and interstellar space. Innovative Interstellar Explorer measurements may also have shed light upon the role of cosmic rays on biological mutation rates of life on Earth or even whether cosmic rays affect global climate change. Another science goal could be the measurements of big bang nucleosynthesis in terms of the abundance of helium-3, deuterium and lithium-7. Exotic physics issues such as the detection of gravitational waves or further tests of our theories of gravity would also be desirable objectives.

The vehicle would use a kilowatt-power ion engine that burns xenon propellant and is powered by an RTG, similar to the Voyager probes. This is necessary for long duration missions and to achieve the required vehicle speed. Any power source that reduces the overall vehicle mass is also a bonus. In earlier versions of the mission plan, the trajectory would have obtained velocity by dropping deep into the Sun's gravitational potential and then executing a velocity burn at perihelion. However, this was abandoned in favour of a Jupiter fly-by where the gravity assist would boost the spacecraft velocity by around 25 kilometres per second or 5.2 AU (778 million kilometres) per year. It would obtain a maximum cruise velocity of 37.5 kilometres per second or 7.8 AU (1.17 billion kilometres) per year.

THOUSAND ASTRONOMICAL UNIT MISSION

During the 1970s and 1980s scientists from JPL had discussed the concept of an interstellar precursor mission that would last between 20 and 50 years and would explore the heliopause. In its original incarnation it was known simply as the Interstellar Precursor Mission (ISP). After further consideration the concept eventually became known as the Thousand Astronomical Unit (TAU) mission. The idea was to design an interstellar precursor mission that was restricted to using only technology that had actually been tested and that a good distance to aim for that would stretch that technology to its limit was deemed to be 1,000 AU, hence the name of the mission. The mission was for a planned launch date around the year 2000. Although solar sails were considered an option for propulsion if unfurled sufficiently close to the Sun, the final concept used nuclear electric propulsion, where a nuclear reactor electrically ionises the gas of a low thrust/high specific impulse ion engine. This would allow the vehicle to reach a cruise velocity of around 100 kilometres per second after a ten-year burn, sufficient to escape the Solar System. After jettisoning the main engine, the vehicle would then cruise to its destination in under 50 years from initial launch.

The 60-tonne TAU spacecraft would carry a 1.2-tonne science payload including spectrometers and employ 12 xenon ion electric thrusters, each with a specific impulse of 12,500 seconds and burning for around two years with a specific mass of four kilograms per kilowatt. The vehicle systems would be powered by a one-megawatt nuclear reactor with a specific mass of 12.5 kilograms per kilowatt. The vehicle was about 25-metres long and contained a 15-metre diameter main antenna for communications, located near the front of the vehicle. A nice addition was a Pluto Orbiter, which could be dropped

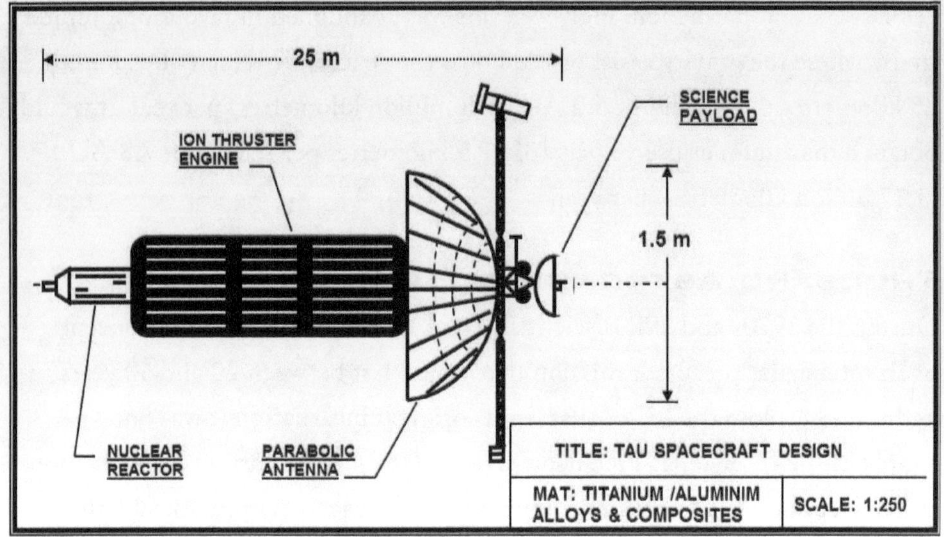

A schematic of the Thousand Astronomical Unit mission concept.
Image: Kelvn F Long.

into Pluto orbit as the spacecraft went past, as well as making observations
of Pluto's moons including Charon. Meanwhile TAU itself would study any
Kuiper Belt objects it came across, as well as the gravitational lensing point and
aspects of Einstein's General Theory of Relativity by providing a long baseline
from Earth for searching for gravitational waves. It would also perform long
baseline parallax measurements to improve the accuracy of stellar distances. It is
a shame that space agency managers did not realise the enormous potential in
such a credible spacecraft mission, especially since it could have been built and
launched by the year 2000 as proposed. The spacecraft would have been well
into the Kuiper Belt by now.

Beam Lightsail mission

In 1998 a group of physicists assembled to consider the ambitious potential of a
beamed light-sail power system as part of a roadmap for interstellar exploration
[27]. This included physicists such as Geoffrey Landis and Robert Forward, two
of the experts in this field. The team came up with what they called three 'straw-

man' missions that progressively built towards a full interstellar mission. The team identified interstellar dust as a possible mechanism for performance degradation and measurements of the interstellar dust size distribution would be an important science driver in any precursor missions. In order to bring about such an interstellar precursor roadmap a simple sail demonstration was seen as the first crucial step. Future tests would include the deployment of lasers in space and the use of laser launched rockets. Laser driven sails clearly do provide good potential for future interplanetary, precursor and interstellar missions, provided such a programme was implemented and the key technological steps demonstrated.

The interstellar roadmap examined three mission scenarios involving a Kuiper Belt mission, an Oort Cloud mission and an interstellar fly-by. The Kuiper Belt mission would be launched to a distance of 100 AU (15 billion kilometres) taking just over five years to complete the journey and travelling at a cruise velocity of around 100 kilometres per second. It would be propelled by a laser beam, focused through a one-kilometre diameter lens in the inner Solar System and fired at the craft's one-kilometre diameter sail. The total spacecraft mass would be around 200 kilograms and it would carry a 66 kilogram payload.

The Oort Cloud mission, on the other hand, would be to a distance of 10,000 AU (1.5 trillion kilometres) taking nearly 18 years to complete the journey while travelling at a cruise speed of around 3,000 kilometres per second. It would use the same sail diameter but the lens required for this mission would be 100 kilometres in diameter. It would have the same spacecraft mass and payload mass as the 100 AU mission. The full interstellar mission, launched to the nearest stars 4.2/ 4.3 light years away, would build on from this capability. It would take around 42 years to complete the journey travelling at a cruise velocity of 30,000 kilometres per second, or 10 percent of the speed of light. The lens diameter would be 200 kilometres in diameter. The spacecraft mass would also be 200 kilograms but the payload mass would be reduced to 33 kilograms. These mission architectures were credible, especially since the concept does not require the spacecraft to carry on board its own reaction mass, a disadvantage of other propulsion schemes.

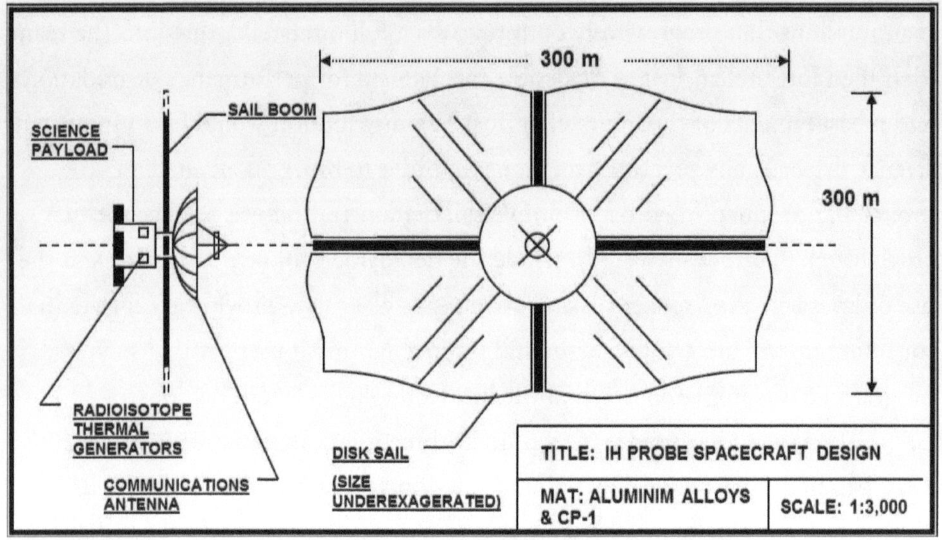

A schematic of the Interstellar Heliosphere Probe mission concept.
Image: Keliv n F Long.

Icarus Pathfinder and Starfinder mission

In 2011 members of an interstellar design team, called Project Icarus, designed some concepts that would pave the way towards energetic propulsion systems to enable interstellar flight. These were based on the British Interplanetary Society's Project Daedalus architecture [7] from the 1970s, but scaled down. The Icarus Pathfinder spacecraft was to reach a distance of 1,000 AU in a time frame of around 12.5 years, moving at a cruise velocity of 474 kilometres per second. It would be a technology demonstrator, testing out some of the vital engineering technology that would be required for a fusion powered vehicle. A plasma drive was chosen as the propulsion system for this concept. If successful, this would be followed up by a later mission called the Icarus Starfinder probe, which would be sent out into the Oort Cloud to a distance of 10,000 AU. The Starfinder probe would be a full fusion propelled vehicle, taking just over 49 years to complete the trip and travelling at a cruise speed of around 1,500 kilometres per second. The exhaust velocity associated with these energetic propulsion systems ranges from an upper theoretical limit of 500 kilometres per second (Pathfinder) to around

2,000 kilometres per second (Starfinder). A more ambitious version of Starfinder was also designed (MkII) that would go to the edge of the Oort Cloud at 50,000 AU (7.5 trillion kilometres) after a trip of one century and travelling at 3,000 kilometres per second.

Ion Drive mission

The British propulsion engineer David Fearn has considered the application of ion engines to an interstellar precursor mission that could be completed in 25–30 years [12]. He discussed an advanced form of a four-gridded ion thruster that can provide a specific impulse as high as 150,000 seconds with thrust levels in the Newton range and may have a velocity increment exceeding 37 kilometres per second. Fearn identified the requirement for a nuclear power source with a mass to power ratio of 15–35 kilograms per kilowatt and an output power of several tens of kilowatts. Such a cruise velocity would allow a mission out to 200 AU (30 billion kilometres) with a performance competitive against alternative propulsion schemes such as solar sails.

Project Dragonfly

Finally, it is worth mentioning the Initiative for Interstellar Studies' (I4IS) Project Dragonfly, which is an initiative to send a small probe outside of the Solar System propelled by laser beams, perhaps to 200 AU, overtaking the Voyager probes. The probe itself may be very small utilising ChipSat and NanoSat, The initial phase of the project will be to define the likely trade space for where such a mission would be plausible, although it is thought that a version of Dragonfly as an interstellar precursor mission would have a payload mass in the range 1-1,000 grams and would require 10-1,000 megawatt power levels. The limiting factor on the mission is likely to be the power and communication systems and the team already has some novel solutions to address this, involving using swarms of laser-propelled probes, a flight of Dragonfly's, interconnected by their own equivalent of a wireless network. Although the project is in its early phases, it does make for a credible interstellar precursor mission to begin the process of the slow human expansion towards those distant stars.

CONCLUSIONS

The concepts discussed in this chapter are listed in Table Two to illustrate the range of innovative ideas proposed to date. These concepts encompass a suit of propulsion methods. For context, they are also compared against the Pioneer and Voyager missions. Many others continue to advocate for near-term interstellar precursor missions. NASA even has an In-Space Propulsion Technology Programme to invest in technologies that have the potential to revolutionise the robotic exploration of deep space. This includes advanced chemical propulsion, solar electric propulsion and solar sails [39]. Electrodynamic tethers also have applications for interstellar missions and could be used in near-term precursor missions [18].

Table two: Cruise Speeds for Interstellar Precursor Missions

Probe (propulsion)	Distance (AU)	Cruise speed (AU/yr)
Pioneer 10 (gravity assists)	100 in 31 years	2.6
Pioneer 11 (gravity assists)	100 in 22 years	2.6
Voyager 1 (gravity assists)	150 in 45 years	3.6
Voyager 2 (gravity assists)	130 in 45 years	3.3
Interstellar (Sail) Probe	200 in 15 years	13
Project Dragonfly (beamed)	200 in 20 years	10
Innovative (Ion) Explorer	200 in 15–30 years	7–13
Fearn (Ion) Engine	200 in 24–30 years	8
TAU (nuclear electric)	1,000 in 50 years	21
Beamed (Laser Sail) Kuiper Belt	100 in 5 years	21
Icarus Pathfinder (plasma)	1,000 in 12.5 years	100
Icarus Starfinder (fusion)	10,000 in 49 years	320
Beamed (Laser Sail) Oort Cloud	10,000 in 18 years	632

Some argue for nuclear powered, laser driven, plasma propulsion systems that could see human missions to the Oort Cloud in a human lifetime [33]. Others push for the use of microwave beam technology that has enormous

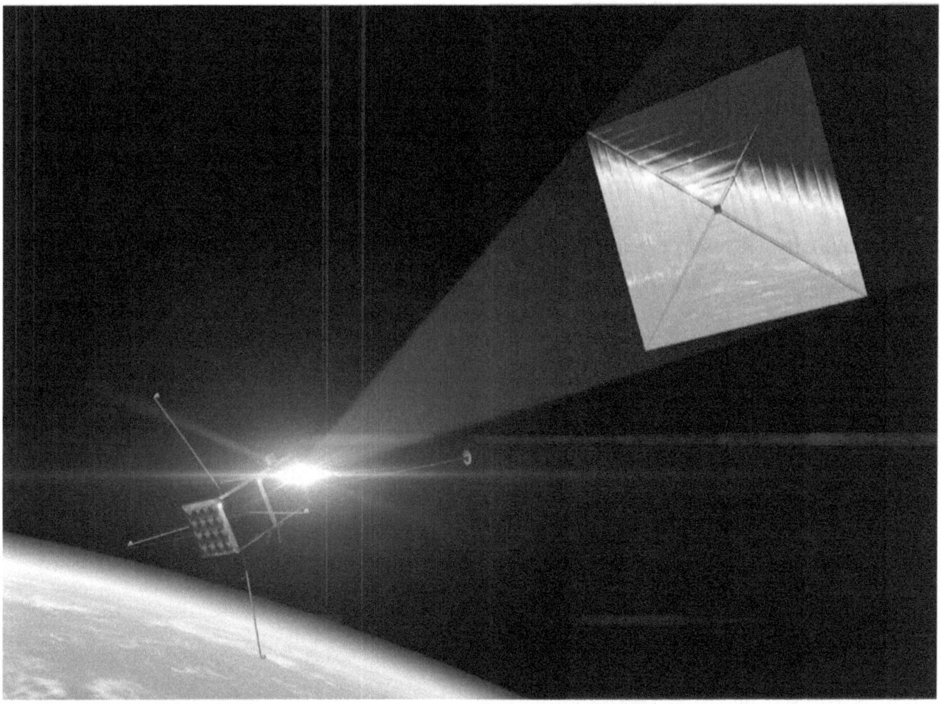

Above, Project Dragonfly under laser-sail propulsion. Below, a Dragonfly 'swarm' could help ease communication difficulties. Artwork: Adrian Mann.

potential to propel spacecraft beyond our Solar System [8]. The physicist
Robert Forward had pioneered the use of microwave beaming technology
in his now famous Starwisp concept [9]. Gregory Matloff took this one step
further and proposed the launch of a one-kilometre mesh, 160-gram payload
called 'Robosloth' that could in theory reach up to 20 percent of the speed of
light, reaching the nearest stars in around two decades [16]. When one reviews
the literature, it is quite clear that there is no shortage of ideas for how to
send spacecraft out into deep space. We know how to do it, all we lack is the
motivation and the finance for mounting the mission.

The International Academy of Astronautics holds a regular meeting
called Symposiums on Realistic Near Term Advanced Space Missions, with
the sub-title 'Missions to the Outer Solar System and Beyond'. These meetings
are held in Italy and demonstrate the importance that the interstellar precursor
missions are given by the scientific space community. One of its organisers, the
scientist Giancarlo Genta, says: "The frontier can surely be expanded at least to
include the whole Solar System and very likely can go much further... our present
civilisation has all the possibilities to accomplish this task in a few generations
but nothing can be given for granted."[35]

Reaching out past the bounds of our Solar System to the icy worlds of the
Oort Cloud seems entirely feasible, with either sail-based technology or high
energy density propulsion systems. All we are lacking is the will to go. As we
strive to become a space-faring civilisation and aspire to boldly send starships
to other stars, it is clear that we must send our robotic ambassadors to targets
of closer scientific interest first. This will prove the technological capability,
whilst laying the foundations for the interstellar railroad ahead. These missions
are entirely viable and if we begin the work today, we could have spacecraft
approaching the Oort Cloud by the middle of the current century. Once this is
achieved, we can finally assemble our Columbus-style missions to the stars and
begin a new age of discovery.

REFERENCES

[1] Crowl, A, K F Long, R Obousy; 'The Enzmann Starship: History and Engineering Appraisal', *JBIS*, 2012.

[2] Stine, G H; *Analog Science Fiction and Fact*, 1973.

[3] Hale, A; 'Nearby Solar-Type Stars as Candidates for Interstellar Robotic Missions', *JBIS*, 49, 4, pp150–154, April 1996.

[4] Nosanov, J, A Shapiro and H Garrett; 'The 34 Year Starship', 100 Year Starship Presentation, Orlando, Florida, October 2011.

[5] Minovitch, M A; 'Method for Determining Interplanetary Free-Fall Reconnaissance Trajectories', Jet Propulsion Laboratory Technical Memo TM-312–130, 1961.

[6] Crawford, I A; 'Project Icarus: A Review of Local Interstellar Medium Properties of Relevance for Space Missions to the Nearest Stars', *Acta Astronautica*, 68, 7–8, pp691–699, April–May 2011.

[7] Bond, A and A Martin; 'Project Daedalus', *JBIS*, 1978.

[8] Forward, R L; 'A Programme for Interstellar Exploration', *JBIS*, 29, 10, pp611–632, October 1976.

[9] Forward, R L; 'Starwisp: An Ultra-Light Interstellar Probe', 22, 3, pp345–350, May/June, 1985.

[10] Forward, R L; 'Roundtrip Interstellar Travel Using Laser-Pushed Lightsails', *J Spacecraft and Rockets*, 21, 2, pp187–195, March–April 1984.

[11] Forward, R L; 'Feasibility of Interstellar Travel: A Review', *Acta Astronautica*, 14, pp243–252, 1986.

[12] Fearn, D G; 'Technologies to Enable Near-Term Interstellar Precursor Missions; Is 400 AU Accessible?', *JBIS*, 61, pp279–283, 2008.

[13] Walker, R, D Izzo and C de Negueruela et al; 'Concepts for Near-Earth Asteroid Detection Using Spacecraft with Advanced Nuclear and Solar Electric Propulsion Systems', *JBIS*, 58, 7/8, pp268–279, July/August 2005.

[14] Matloff, G L; 'Earth Interstellar Precursor Solar Sail Probes', *JBIS*, 44, 8, August 1991.

[15] Matloff, G L; 'The State of the Art Solar Sail and the Interstellar Precursor

Mission', *JBIS*, 37, 11, pp491–494, November 1984.

[16] Matloff, G L; 'Robosloth: A Slow Interstellar Thin-Film Robot', *JBIS*, 49, 1, pp33–36, January 1996.

[17] Matloff, G L and A R Martin; 'Suggested Targets for an Infrared Search for Artificial Kuiper Belt Objects', *JBIS*, 58, 1/2, pp51–61, January/February 2005.

[18] Matloff, G L and L Johnson; 'Applications of the Electrodynamic Tether to Interstellar Travel', *JBIS*, 58, 11/12, pp398–402, November/December 2005.

[19] Jaffe, L D, C Ivie, J C Lewis, R Lipes, H N Norton, J W Steans, L D Stimpson and P Weissman; 'An Interstellar Precursor Mission', *JBIS*, 33, pp3–26, 1980.

[20] Jaffe, L D et al; 'An Interstellar Precursor Mission', *JBIS*, 33, 1, pp3–26, March 1980.

[21] Jaffe, L D et al; 'An Interstellar Precursor Mission', JPL Publication, 77–70, 1977.

[22] McNutt, R L, Jr et al; 'A Realistic Interstellar Explorer', *Advances in Space Research*, 34, 2004.

[23] Swinney, R and K F Long and A Hein; 'Project Icarus: Exploring the Interstellar Roadmap Using the Icarus Pathfinder and Starfinder Probe Concepts', *JBIS*, 65, 7/8, July/August 2012.

[24] Long, K F and R Obousy; 'Starships of the Future', *Spaceflight*, 53, 4, pp140–144, April 2011.

[25] Long, K F; *Deep Space Propulsion: A Roadmap to Interstellar Flight*, Springer, 2011.

[26] Benford, G and J Benford et al; 'Power-Beaming Concepts for Future Deep Space Exploration', *JBIS*, 59, 2006.

[27] Landis, G A; 'Beamed Energy Propulsion for Practical Interstellar Flight', *JBIS*, 52, 1999.

[28] Fritze, R H; *New World's, The Great Voyages of Discovery 1400-1600*, Sutton Publishing, 2002.

[29] Nock, K T; 'TAU: A Mission to a Thousand Astronomical Units', AIAA-87-1049, 1987.

[30] Noble, R J; 'Radioisotope Electric Propulsion for Robotic Science Missions to Near-Interstellar Space', *JBIS*, 49, 9, pp322–328, September 1996.

[31] Maccone, C; *The Sun as a Gravitational Lens: Proposed Space Missions*,

Colorado Springs IPI Press, 1997.

[32] Cassenti, B N; 'Robotic Interstellar Missions and Advanced Nuclear Propulsion', *JBIS,* 49, 9, pp357–360, September 1996.

[33] Kammash, T; 'Nuclear Powered Laser Driven Plasma Propulsion System', *JBIS*, 58, 11/12, November/December 2005.

[34] Kammash, T and M J Lee; 'A Fusion Propulsion System for Near-Term Space Exploration', *JBIS*, 49, 9, pp351–356, September 1996.

[35] Genta, G; 'The Challenge of Very Deep Space Exploration: How Far will the Frontier Be?', *JBIS*, 59, 2, pp43–47, February 2006.

[36] Gruntman, M; 'Solar System Frontier: Exploring the Heliospheric Interface from 1AU', *JBIS*, 59, 2, pp54–58, February 2006.

[37] Gruntman, M, R L McNutt Jr and R E Gold et al; 'Innovative Explorer Mission to Interstellar Space', *JBIS*, 59, 2, pp.71-75, February 2006.

[38] Gruntman, M and R L McNutt, Jr et al; 'Innovative Explorer Mission to Interstellar Space', *JBIS*, 59, 2, 2006.

[39] Johnson, L; 'NASA's In-Space Propulsion Technology Program: A Step Toward Interstellar Exploration', *JBIS*, 59, 3/4, pp99–103, March/April 2006.

[40] NASA; What is the Golden Record? The Voyager Interstellar Mission website (online), http://voyager.jpl.nasa.gov/spacecraft/goldenrec.html (last accessed 13 October 2014).

Section III

Interstellar Technologies and Solutions

Introduction by Kelvin F Long

In this section we jump into the various technological ideas for how we can start to address the interstellar challenge. We start with an excellent review of developments in advanced electric propulsion from **Angelo Genovese**, a spacecraft engineer who works on this technology for his living.

Next, we get a tour de force review by **Adam Crowl**, an interstellar advocate, on the current thinking about the many ways by which we can send spacecraft to other stars. Although there are many methods not discussed here, such as fission-fragment rockets, fusion propulsion, antimatter annihilation systems, Adam examines laser-sails and pellet propulsion specifically.

Tiffany Frierson, a physics student, takes us into the world of faster than light travel with proposals for warp drive-based systems and the use of exotic matter and negative energy technologies. She is followed by **Remo Garattini**, himself a Professor of Physics, who explores the physics of wormholes. He takes us on a tour of this mathematical construct from the perspective of Einstein's General Theory of Relativity.

Interstellar travel is not just all about propulsion of course, so next we have two excellent chapters about interstellar communications, the first by **Divya Shankar**, a spacecraft post-graduate student, followed by physicist and astronomer **David Fields** who examines the possibility of communication through quantum entanglement. To round off this chapter, we have an excellent introduction to artificial intelligence by **Jeremy Clark**. Perhaps the Heuristically programmed ALgorithmic (HAL) computer is on its way after all.

CHAPTER 9

ADVANCED ELECTRIC PROPULSION
FOR INTERSTELLAR PRECURSOR MISSIONS

ANGELO GENOVESE

The main difficulty in achieving interstellar travel is the vast void that has to be covered between the Solar System and the nearest stars. This means that a very great speed and/or a very long travel time are needed. The fastest space probe built by humankind, Voyager 1, has covered (as of 30 June 2014) a distance of 128 AU from the Sun in 37 years since its launch. Voyager 1's current relative velocity to the Sun is 17 kilometres per second (61,200 kilometres per hour). At this velocity, 74,000 years would pass before reaching Proxima Centauri, were the spacecraft travelling in the direction of that star. These incredible numbers clearly explain why interstellar travel is practically impossible with conventional means (chemical propulsion and gravity assist manoeuvres).

The key propulsion parameter to enable interstellar exploration is the specific impulse, defined as the propellant exhaust velocity divided by the gravitational acceleration constant g; basically it indicates how efficiently a thruster consumes propellant. In order to reduce the propellant mass and consequently the starship mass to reasonable values, the specific impulse must be as high as several million seconds (see Table Three on page 203), which is more than three orders of magnitude higher than the specific impulse of the best chemical thruster. These ultra-high values of specific impulse can be obtained only with a nuclear fusion propulsion system, an antimatter thruster or even more exotic propulsion concepts, like propellantless propulsion devices where specific impulse is infinite, such as *Star Trek's* iconic propulsion system, the warp drive.

However, as David Fearn wrote in 2000 [1], "many breakthrough propulsion concepts have been proposed, but no conclusive results offering a near-term solution to the problem have so far emerged. It is therefore realistic to assume that such solutions, if any, are probably decades away from implementation. It is thus prudent to consider extra-Solar System missions employing extensions of existing technologies". Fourteen years later this statement is still up-to-date; hence, it is prudent to look for the few existing technologies that can be extended in order to enable interstellar precursor missions. Electric propulsion (EP) is one of them, probably the most promising one together with the solar/light sail concepts.

ELECTRIC PROPULSION BACKGROUND

Electric propulsion is a technology which allows for much higher exhaust velocities than conventional chemical propulsion, resulting in a major reduction of the propellant mass for a certain space mission. This leads either to a significant decrease of the launch mass of a spacecraft or to much larger payloads. In general, electric propulsion comprises all types of propulsion in which a certain amount of propellant is ionised and then accelerated by electric or magnetic fields, or both.

Electric propulsion was first conceived more than 100 years ago by the American physicist Robert H Goddard (1882–1945), who as early as 1906 addressed the problem of producing "reaction with electrons moving with the velocity of light," and wrote down his thoughts on this problem in his notebook [2]. He was considering electrons and not ions because the concept of the ion, as an atomic-sized particle possessing a net positive charge, had not yet been fully established at that time. His visionary ideas culminated in a US Patent ('Method and means for producing electrified jets of gas', number 1,163,037, filed in 1917) that represents the world's first documented electrostatic ion accelerator intended for propulsion.

Almost at the same time, on the other side of the world, Konstantin Eduardovitch Tsiolkovsky (1857–1935), a self-taught school teacher from Kaluga in Russia, derived the most fundamental mathematical expression in the field of space propulsion. The Tsiolkovsky Rocket Equation gives central importance to the rocket exhaust velocity, thus laying the basis for electric propulsion (*Exploration of Universal Space by Means of Reactive Devices*, 1903). In 1911 he published his first statement on the possibilities of electric propulsion: "It is possible that in time we may use electricity to produce a large velocity for the particles ejected from a rocket device." However, he did not attempt any analytical study of the application of electricity to rocket propulsion, acknowledging that EP was at that time just a dream. His attention turned towards more prosaic problems as liquid propellant chemical rockets [3].

The third space travel visionary who independently developed the idea of electric propulsion was Hermann Oberth. Born in Romania, he studied physics in Germany and, in 1929, he published the all-time astronautics classic, *Wege zur Raumschiffahrt* (*'Ways to Spaceflight'*). The entirety of the last chapter, 'Das elektrische Raumschiff' ('The Electric Spaceship') was about electric propulsion, predicting its future role in propelling spaceships to distant targets. This book was like a bible for an entire generation of space enthusiasts. It brought electric propulsion simultaneously into the minds of science fiction writers and scientists, among them Oberth's brilliant student, Wernher von Braun, who started to dream about developing a rocket that could travel to other planets. When von Braun was brought to the United States as part of Operation Paperclip in order to continue his work on the V-2 rocket at Fort Bliss, Texas, he asked his assistant Ernst Stuhlinger to review Oberth's research on electric propulsion: "Professor Oberth has been right with so many of his early proposals; I wouldn't be a bit surprised if we flew to Mars electrically" [4].

Stuhlinger immersed himself in electric propulsion theory, and at the Fifth International Astronautical Congress in Vienna in 1954 he presented a paper entitled *Possibilities of Electrical Space Ship Propulsion*, in which he conceived the first Mars expedition using solar-electric propulsion [5]. The spacecraft design he proposed, which he nicknamed the 'Sun Ship', was comprised of three major parts: the crew/payload compartment at the ship's centre; a multi-unit solar-electric power system; and an electric propulsion system composed of a cluster of 2,000 ion thrusters using cesium or rubidium as propellant. He calculated

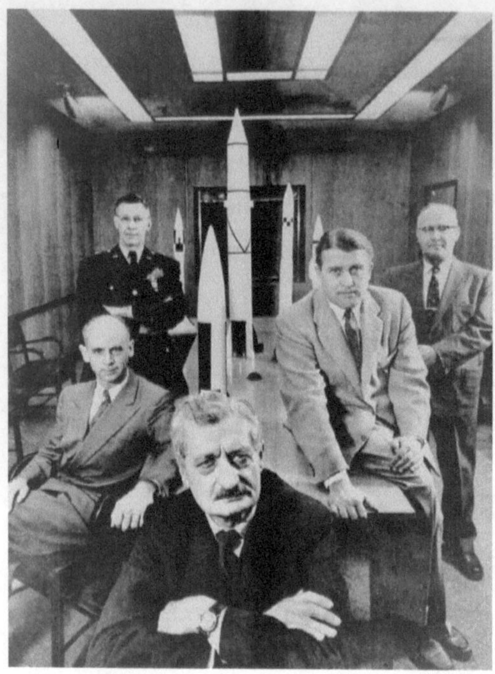

Space travel pioneers: Ernst Stuhlinger (seated, left) poses with Hermann Oberth (centre) and Wernher von Braun (seated right); behind them are US General Holger Toftoy and Robert Lusser. Image: NASA Marshall Space Flight Center.

that the total mass of the Sun Ship would be just 280 tonnes instead of the 820 tonnes necessary for a chemical propulsion spaceship for the same Mars mission.

In 1964 Stuhlinger published the first systematic analysis of electric propulsion systems – *Ion Propulsion for Space Flight* – while the physics of electric propulsion thrusters was first described comprehensively in a book by Robert Jahn in 1968 [6]. Only seven years after the dawn of the Space Age, the first demonstration of electric propulsion came in 1964, in the form of an ion engine onboard the SERT-1 spacecraft.

Throughout most of the twentieth century electric propulsion was considered the technology of the future for spacecraft propulsion. However, its complexity and the long development needed to demonstrate the lifetime required by electric propulsion missions (several thousands of hours) have delayed its use as standard propulsion system for commercial and scientific space applications. Nowadays, with literally hundreds of electric thrusters operating in orbit on telecom satellites and for primary propulsion in deep-space scientific missions such as NASA's Dawn mission to the Asteroid Belt, the future for electric propulsion has finally arrived.

ION THRUSTERS

Electric thrusters are generally described in terms of the acceleration method used to produce the thrust. These methods can be easily separated into three categories: electrothermal (resistojets and arcjets), electrostatic (ion thrusters, Hall thrusters and field emission electric propulsion or FEEP) and electromagnetic (pulsed plasma thrusters or PPT, magnetoplasmadynamic thrusters or MPD, and the variable specific impulse magnetoplasma rocket, known as VASIMR).

Ion thrusters use a variety of plasma generation techniques to ionise a large fraction of the propellant. These thrusters then utilise biased grids (from a few kilovolts to more than 10 kilovolts) to electrostatically extract ions from the plasma and accelerate them to high velocity. Ion thrusters can provide very high efficiencies (from 60 to more than 80 percent) and specific impulses (from 2,000 to over 10,000 seconds) compared to other electric thruster types; hence, they are the best candidate for interstellar precursor missions [7].

However, ion thrusters carry a fundamental price: the power imparted to the exhaust increases with the square of its velocity while the thrust increases only linearly. Table One shows a list of ion thrusters with their main characteristics and their technology readiness level (the maximum level, nine, corresponds to a flight-proven technology); notice how the specific power, which is the electric power per thrust unit, rapidly increases with increasing specific impulse (see the graph on page 186). This drawback is particularly severe for interstellar precursor missions, which require very high specific impulses in order to reduce the propellant mass to acceptable values (see Table Three on page 203). Unfortunately, the power source mass rapidly increases with increasing specific impulse, cancelling the advantage of a reduced amount of propellant. Hence, an advanced EP system for interstellar precursor missions cannot be realised without a suitable power source with very low specific mass, expressed as mass per unit of electrical power (much less than 50 kilograms per kilowatts electrical).

The 20-kW Nuclear Electric Xenon Ion thruster System (NEXIS) developed at NASA's Jet Propulsion Laboratory for the Jupiter Icy Moons Orbiter (JIMO), later cancelled. Image: NASA.

The cancellation of NASA's NERVA Project in 1972 signalled the onset of a three decade period during which NASA effectively sidelined nuclear spacecraft propulsion. Then in 2003 came NASA's Project Prometheus, with the aim of developing nuclear-powered systems for long-duration space missions. The first planned mission to come out of Prometheus was going to be JIMO, the Jupiter Icy Moons Orbiter, which would have explored the Galilean satellites Europa, Ganymede and Callisto and remotely probed the oceans believed to exist beneath their icy crusts. JIMO would be powered by a small fission reactor, which would convert 100 kilowatts of heat from the reactor into electrical power for the spacecraft's advanced ion propulsion engine using either the HiPEP (High Power Electric Propulsion] system or NEXIS (Nuclear Electric Xenon Ion System).

To reduce the risk of nuclear accidents, JIMO's reactor would have only been powered up once the vessel had moved well beyond Earth's orbit, avoiding the controversy that radioisotope thermoelectric generators (RTGs) have caused on prior missions, most notably the Cassini mission to Saturn.

Table 1: characteristics of ion thrusters

Engine	Specific impulse (s)	Required power (kW)	Thrust (mN)	Specific power (kW/N)	Verified lifetime	TRL	Mission
NSTAR	3,300	2.3	92 max	25	30,000h	9	Dawn, DS1
T5	3,200	0.48	18	27	5,000h	9	GOCE
RIT-10	3,810	0.59	21	28	23,000h	9	ARTEMIS
RIT-22	4,760	5.8	175	33		7	IHP probe
T6	4,700	5	210 max	23.8		7	Bepi Colombo
HEMPT	2,500	1.4	45	30	10,000h	7	SmallGEO
NEXT	4,100	7	236 max	30	48,000h	7	NASA Discovery
NEXIS	7,000	20	440	45	2,000h	5	JIMO
HiPEP	6,000–9,600	25–50	460–670	55–75	2,000h	5	JIMO
VASIMR	5,000	200	6,000	33		5	Tin Tin
FEEP	3,000–10,000	0.001–0.1	1μm– 1mN	50–90	4000h	5	Tin TIn
DS4G	28,000	240	1,500	160		1	TAU

Alas, the drive to complete the construction of the International Space Station meant that Prometheus lost funding in 2005, killing the JIMO mission with it. One of the criticisms of Project Prometheus was that the nuclear technology was considered too ambitious, with an optimistic specific mass of 25 kilograms per kilowatts electrical.

SOLAR ELECTRIC PROPULSION (SEP)

Typically, electric propulsion systems have so far used solar panels as their power source. The first spacecraft to demonstrate this form of propulsion with solar arrays was NASA's Deep Space One, launched in 1998. It thrusted for 678 days – a record only recently beaten by Dawn – using just 74 kilograms of xenon as it accelerated up to a final velocity of 4.3 kilometres every second. The solar arrays were pioneering a new technology, the Solar Concentrator Array with Refractive Linear Element Technology, or SCARLET for short. The array focused sunlight onto solar cells using Fresnel lenses made of silicone and was capable of generating 2.5 kilowatts at one astronomical unit from the Sun, which powered the craft's NSTAR electrostatic ion thruster.

The aforementioned Dawn mission also uses three NSTAR engines. Launched in 2007 and designed to study the two largest bodies in the Asteroid Belt, Ceres and Vesta, Dawn fires one engine at a time and is powered by a solar array capable of producing 10 kilowatts at one astronomical unit. In order to have reached Vesta in 2011, Dawn had used up 275 kilograms of xenon propellant, and another 110 kilograms will be utilised to reach Ceres in the summer of 2015. Despite the relatively small amount of xenon, Dawn's delta-V (change in velocity) has been in excess of 10 kilometres per second, which is greater than any previous mission has been able to accomplish with onboard propellant (those that have achieved greater velocity changes have done so with gravity assists). However, it is

The required electric power per thrust unit rapidly increases with the specific impulse (dots are measured or predicted values from Table One on the previous page).

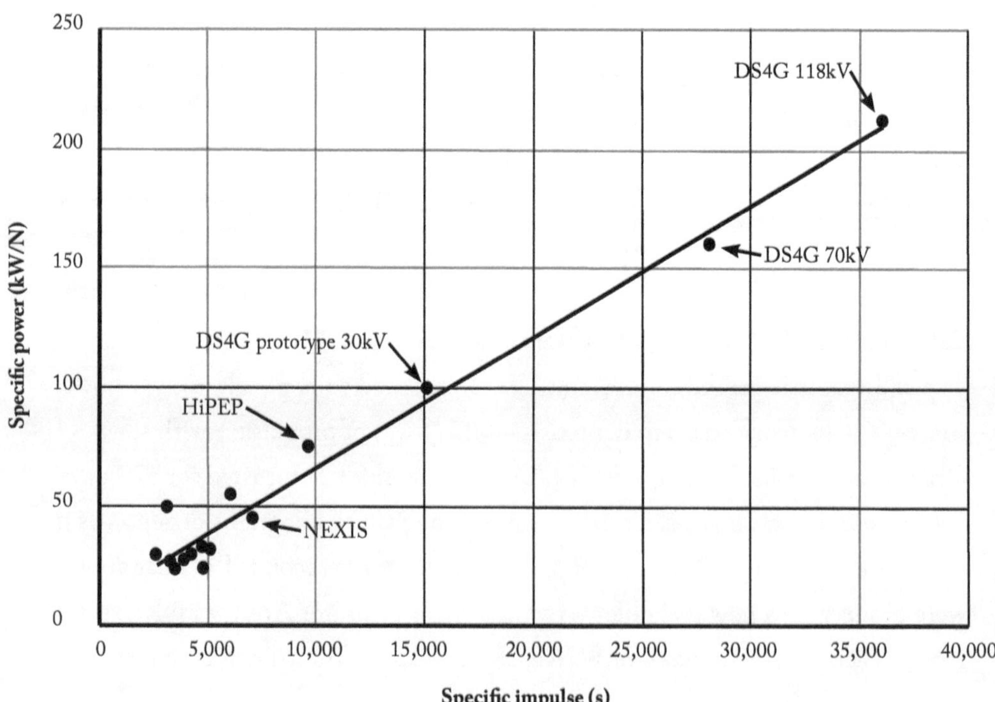

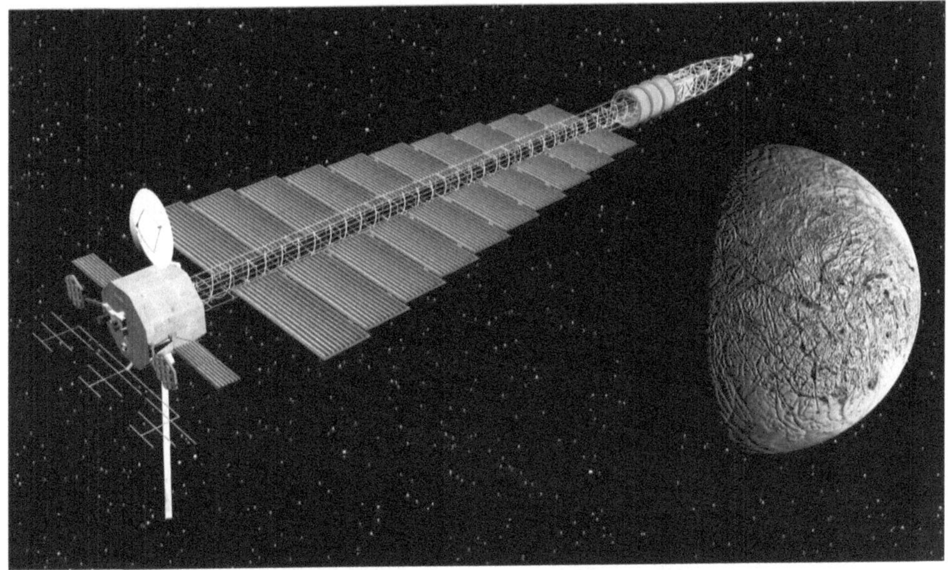

An artist's concept of the Jupiter Icy Moons Orbiter (JIMO) with a 100-kW nuclear fission reactor on top, a huge radiator in the middle and the ion thrusters with the payload behind. Image: NASA.

still not enough to be able to drive even the most simple of interstellar precursor missions.

RADIOISOTOPE ELECTRIC PROPULSION (REP)

REP systems are low-thrust, ion propulsion units based on multi-hundred watt radioisotope thermoelectric generators (RTGs) and ion thrusters.

RTGs have been the main power source for US deep space missions over 50 years. They use plutonium-238 as an electricity source; its radioactive decay heat is converted to electricity through static thermoelectric elements (solid-state thermocouples), with no moving parts. RTGs are safe, reliable and maintenance-free and can provide heat or electricity for decades as the isotope plutonium 238 has a half-life of 86.4 years.

RTGs have powered several iconic space missions, including Apollo, Pioneer, Viking, Voyager, Galileo, Ulysses, New Horizons and the aforementioned Cassini. The Voyager spacecraft have already operated for over 35 years and are expected to continue operating, powered by their RTGs, until the early 2020s.

High power-to-mass radioactive power sources are the key enabling technology of REP systems for extra-solar missions. Currently flying RTGs have power-to-mass ratios that are too small for this type of mission. Possibilities for advanced RTGs include 'second generation' Stirling Radioisotope Generators (SRGs) and high-temperature, high-efficiency RTGs. Both approaches have pluses and minuses and both need further development. SRGs produce about four times as much electric power from their plutonium fuel as RTGs do. There are issues with electromagnetic interference of instruments and the limited lifetime of the generator as a result of wear to the moving parts. Advanced high-temperature RTGs using skutterudite (a cobalt-arsenide mineral) converters show a long lifetime with no moving parts but they have inherently lower conversion efficiencies.

NUCLEAR ELECTRIC PROPULSION

The demonstration of solar electric propulsion on the Deep Space One mission has paved the way for applications of advanced electric propulsion on more demanding future missions such as outer planet orbiters and interstellar precursor flights. However, the most ambitious of these missions are required to produce a change in velocity ranging from 40 to over 100 kilometres per second. To accomplish these missions with reasonable initial masses and tolerable trip times requires advanced nuclear electric propulsion (NEP) systems capable of processing from 100 to 500 kW of power at specific impulses ranging from 5,000 seconds to over 14,000 seconds.

NEP systems convert nuclear thermal energy into electrical energy, by way of three important sub-systems:

- a controlled fission reactor core to produce heat
- a cooling loop that removes the heat from the core
- a power conversion subsystem that receives the heat from the cooling loop and converts a portion of it into electrical power.

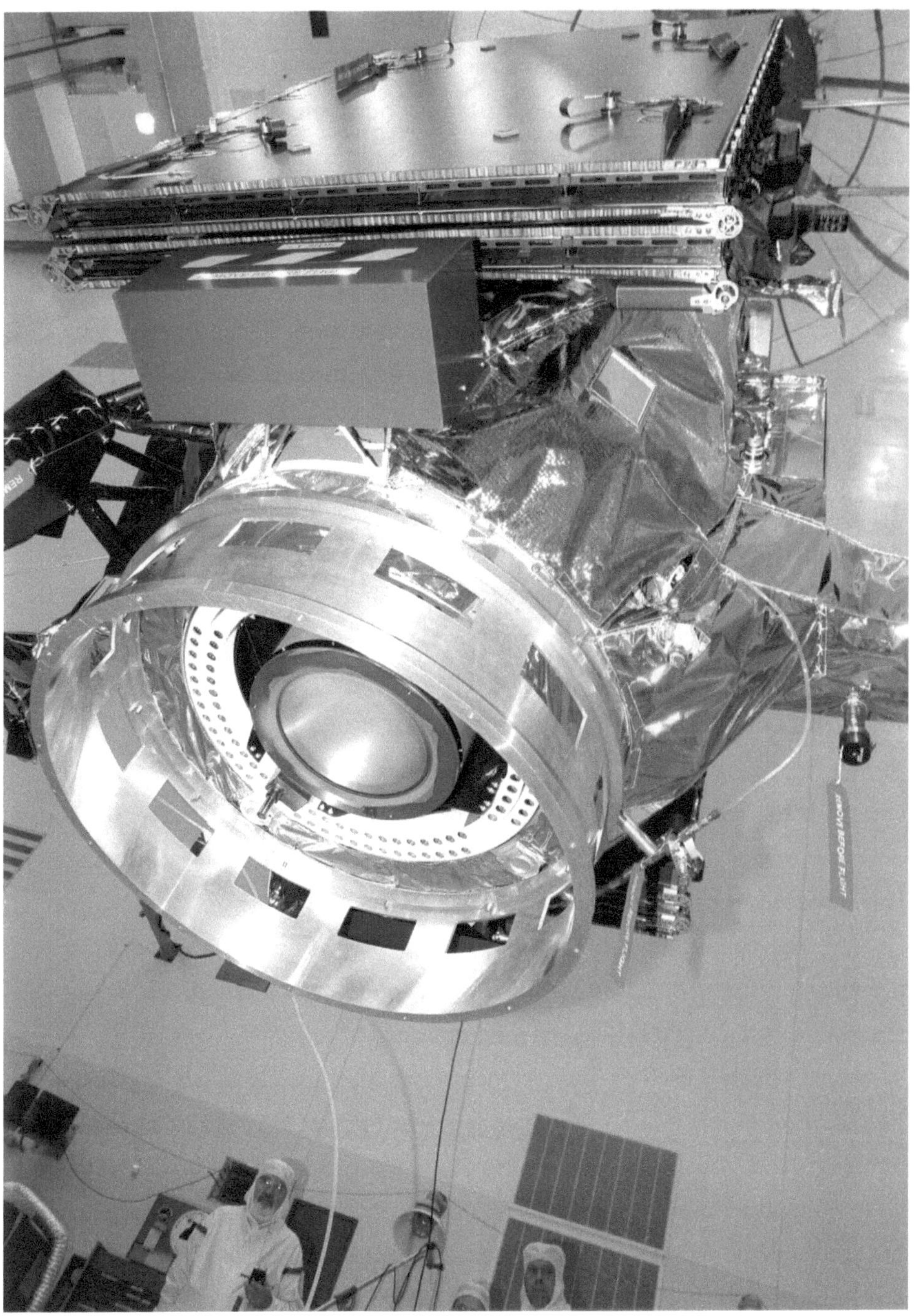

The Deep Space One spacecraft in the clean room prior to launch. Image: NASA.

The key elements of an advanced NEP system include a compact reactor core, a gas turbine or Stirling engine used as an electric generator, a compact heat rejection system such as heat pipes, a power conditioning and distribution system, and a high power propulsion system based on ion thrusters.

The current NEP pacesetter is NASA's Safe Affordable Fission Engine, the SAFE-400, which utilises a 400 kilowatt thermal reactor, fuelled by 381 uranium nitride pins clad in rhenium, to produce 100 kilowatts of electricity. It is also small-scale: the reactor is only about 50 centimetres tall, 30 centimetres across and weighing 512 kilograms. NEPs are also safe – they are launched 'cold' and only switched on once in space.

The problem with appropriating NEP for interstellar precursor missions out to distances greater than several hundred astronomical units, is that the specific mass must be much smaller than 50 kilograms per kilowatt electrical. Included in this mass must be the radiator system to expel waste heat, but so far no NEP system has been built that achieves this.

BEAMED ELECTRIC PROPULSION

Rather than carry its power source onboard, beam-powered propulsion uses energy that is beamed, either using microwaves or lasers, to the spacecraft from a remote power plant. One of the great advantages of beamed energy propulsion is that it reduces the mass of your spacecraft significantly.

Lasers can send power to photovoltaic panels coupled with an electric propulsion system, for Laser Electric Propulsion (LEP). Alternatively, a microwave beam could be used to send power to a rectifying antenna, or 'rectenna', which is used to convert microwaves into DC electricity for Microwave Electric Propulsion (MEP). Microwave beamed power has been demonstrated on the ground – the rectenna's inventor, William Brown, flew a model helicopter by beaming microwaves at it, while in 1975 scientists led by Brown and Richard Dickinson demonstrated at NASA's Jet Propulsion Laboratory the ability to transmit tens of kilowatts of power over a distance greater than 1.6 kilometres, or one mile, with 54 percent efficiency.

A beamed microwave power transmission system consists of a DC current to microwave convertor, a beam-forming antenna, free space transmission, and reception and reconversion to DC. Brown [8] estimates that the maximum possible overall DC-to-DC transmission efficiency can be as high as 76 percent. Brown states that beamed microwave power represents a technological breakthrough for a spacecraft propelled by an electric propulsion system, because of the low specific mass of the rectenna (one kilogram per kilowatt), which would allow very challenging extra-solar missions including the exploration of the Oort Cloud, at distances greater than 1,000 AU, in less than 50 years.

Near-future EP Interstellar Precursor Missions

In the coming decades, it is hoped that a progression of small body missions will push the exploration horizon outward from the Asteroid Belt, to the comets and asteroids (Centaurs and Trojans) in proximity to the giant planets (5–30 AU), the trans-Neptunian objects and plutoids (30–100 AU), the Kuiper Belt Objects (100–1000 AU) and, finally, the Oort Cloud (greater than 1,000 AU). These missions will demonstrate most of the robotic technologies needed to explore a star system. Two key attributes for interstellar robotic explorers have already been partially mastered: longevity and storability. Voyagers 1 and 2 were launched in 1977, have exited the heliosphere into the interstellar medium and are expected to operate at reduced capability until the early-to-mid 2020s, or more than forty years, limited only by attitude-thruster propellant and radioisotope electric power. All indications are that if resupplied with consumables, these vehicles could operate for several more decades.

A third key attribute is a suitable propulsion system that can provide the high/ultra-high delta-Vs needed by any interstellar precursor mission (see table three). As mentioned previously, the state-of-the-art top propulsion performance is represented by the Dawn spacecraft's SEP system with a delta-V of 10 kilometres per second. This propulsion technology can enable the first real interstellar precursor mission to explore the local interstellar medium LISM, which can be reached by travelling towards the nose of the heliosphere for more than 200 AU.

THE LISM PROBE

Voyager 1 is the first human-made object to venture into interstellar space; it crossed the heliopause, the boundary separating solar and galactic plasmas, in August 2012 at 120 AU from the Sun, 35 years after its launch from Cape Canaveral in 1977. The spacecraft is now flying through the bow shock, a crescent-shaped shockwave formed by the Sun as it moves through interstellar gas. However, today its measurement capabilities are very limited, and in five to ten years the onboard power will be too low for the probe to operate any scientific instrument further; at that time Voyager 1 will still be less than 150 AU from the Sun. The minimum required distance to reach the unperturbed 'virgin' interstellar medium is expected to be at least 200 AU. A trip time of 25 years, well within the professional lifetime of a scientist or engineer, is the target mission duration for a real LISM probe equipped with modern scientific instruments. Further distances (~1,000 AU) for longer times (~50 years) are preferred but pose significantly more demands on both propulsion and spacecraft.

There are several fundamental reasons to perform a mission to investigate the outer regions of the heliosphere and the LISM as soon as possible; I would like to mention just one of them.

The heliosphere play a critical role in shielding out the majority of galactic cosmic radiation from the local interstellar medium. Galactic cosmic rays, in turn, strongly affect life on our planet. Cosmic rays are high-energy charged particles that bombard Earth from above the atmosphere. Several thousand pass through a person's body every minute. These can cause biological damage but also cause mutations that accelerate evolution.

The majority of cosmic rays present in interstellar space are blocked by the outer heliosphere, presumably via a strong magnetic barrier that forms the inner heliosheath, where the solar wind slows prior to being deflected by the interstellar flow.

Large changes in the LISM have dramatic effects on the heliosphere and the radiation environment of the Solar System. For example, a typical enhancement in the density of the local interstellar medium by a factor of ten causes the entire heliosphere to shrink to about a quarter of its current size, and increases the fluxes of cosmic rays at Earth by a factor of between two and six. Such large changes in the LISM have certainly occurred in the past and will occur again in the future. Furthermore, observations of this global interaction are essential for understanding the radiation environment that must be traversed by astronauts for long missions to distant destinations, such as Mars. Future manned space travel will rely heavily on a better understanding of the LISM's influence over the heliosphere and the potential short- and long-term changes to the radiation environment [9].

The challenge of reaching a solar distance of 200 AU within 25 years needs the realisation of mean heliocentric flight velocities of 38 kilometres per second, corresponding to eight astronomical units per year. This requires a very advanced and sophisticated propulsion system. Furthermore, the payload requirements are challenging as well:

- A high-gain antenna
- A low-mass payload focused on fields and particle measurements
- Mass budget of less than 50 kilograms
- Power budget of less than 40W

As of today, there are three options considered capable of performing the mission:

- Ballistic-REP
- SEP-REP
- Solar sail

BALLISTIC-REP OPTION

This is the NASA's Innovative Interstellar Explorer (I²E) concept. According to NASA's Ralph McNutt [9], the top-level mission requirements for the I²E mission are:

• launch the spacecraft to have an asymptotic trajectory within a 20-degree cone of the 'heliospheric nose;'

• provide data from 10 to 200 AU;

• arrive at 200 AU as fast as possible, at least within 25 years;

• consider all possible missions that launch between 2010 and 2050;

• use existing launch hardware;

• use no 'in-space' assembly;

• launch to escape velocity;

• keep new hardware and technology to a minimum;

• provide accepted 'adequate' margins.

A propulsion solution based only on a ballistic launch is not feasible at all; to reach 9.5 AU per year (45 kilometres per second, almost three times Voyager 1's current velocity) with only a full ballistic launch from Earth would require a hyperbolic excess energy C3 (where C3 is the energy requirement of an interplanetary mission that needs to achieve an orbital velocity above escape velocity) of 1,016 km²/s², an unachievable initial launch energy per mass with any foreseeable rocket technology. Hence, launch remains only one component of an interstellar probe mission solution.

According to McNutt, maximum propulsion capability uses an Ares V launcher, two-engine Centaur upper stages, a close fly-by of Jupiter and extended electrical propulsion at low thrust using radioisotope electric propulsion.

Launching the I²E spacecraft with an Ares V coupled with a Centaur upper stage would obtain a C3 of 270 km²/s² and a corresponding asymptotic speed from the Solar System of 19 kilometres per second, or four astronomical units per year.

For the REP spacecraft McNutt's team adopted a standard configuration with a dry mass of 790 kilograms and 440 kilograms of xenon for two 1.0 kilowatt ion engines (one firing and one as a cold spare), yielding a total launch mass of 1,230 kilograms for the REP system. The thrust time is almost 15 years

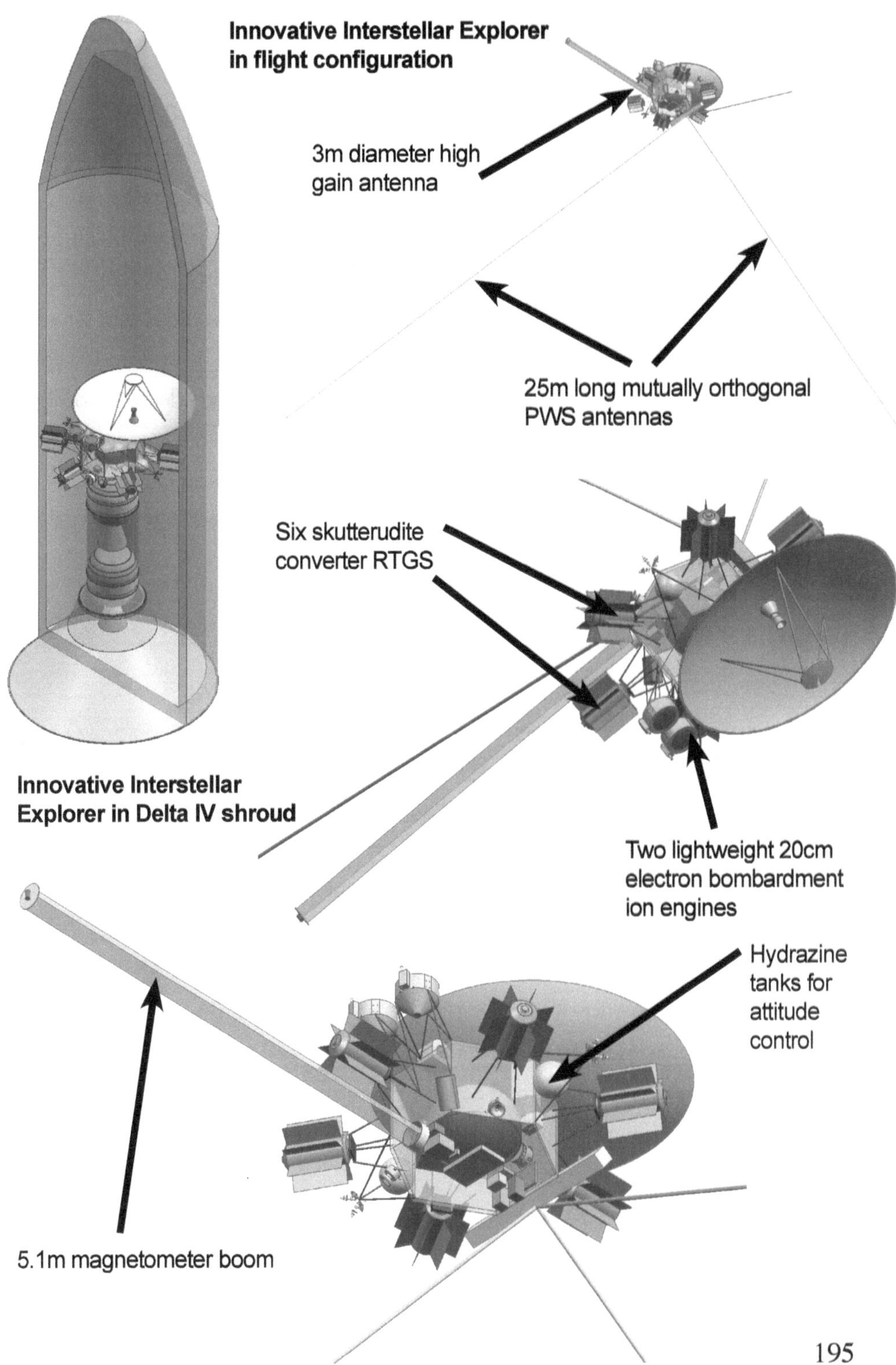

Innovative Interstellar Explorer
in flight configuration
3m diameter high
gain antenna
25m long mutually orthogonal
PWS antennas
Six skutterudite
converter RTGS
Innovative Interstellar
Explorer in Delta IV shroud
Two lightweight 20cm
electron bombardment
ion engines
Hydrazine
tanks for
attitude
control
5.1m magnetometer boom

(challenging but feasible combining the lifetime of two advanced ion thrusters) providing a delta-V of 14.8 kilometres per second.

With a final asymptotic speed, including a Jupiter fly-by, of 10.2 AU per year and with a hypothetic launch in 2015, the Innovative Interstellar Probe would reach 200AU in 2038, some 22 years after launch. Further, the spacecraft would reach 300AU some 32 years after launch and 1,000AU – well into the undisturbed local interstellar medium – in 2116, just over a century after launch. Although 1.1 plutonium-238 half-lives would have elapsed, plenty of power would still be present to run the spacecraft and downlink. Whether a spacecraft could be built to last for such a long time is questionable. However, the Voyagers have exceeded their design lifetimes of five years by over a factor of six; they are expected to operate at reduced capability until about 2020–25, more than forty years from their launch in 1977. As mentioned before, all indications are that, if resupplied with consumables, these vehicles could operate for several more decades.

SEP-REP OPTION

Professor Horst Loeb of the University of Giessen [11] has proposed a combination of solar electric propulsion (SEP) with radioisotope electric propulsion (REP) in order to reach the goal of 200 AU within 25 years.

The SEP option consists of first flying into the inner Solar System, closer to the Sun, in order to use the increased solar radiation flux and gravity and to take enough momentum. In this case the SEP stage provides the energy that would otherwise be provided by a large ballistic booster, as is the case with NASA's I²E concept. Six RIT-22 ion thrusters would compose the SEP stage, running with five kilovolts of positive high voltage corresponding to a specific impulse of 7,377 seconds. The propellant storage and feed system, as well as the electronic components and thermal control parts, are mounted within the bus structure. The REP-stage, which contains the LISM probe, is mounted on top of the SEP stage.

After launch, the SEP-thrust is used to lower the perihelion distance to within Earth's orbit but no closer than 0.7AU by thrusting in a direction opposite to the direction of flight. Then, near perihelion when the solar panels are providing maximum power, the probe accelerates with maximum thrust. The SEP-stage's propellant will run out after 831 days at a distance of 3.05AU from the Sun, resulting in a velocity relative to the Sun of 30.5 kilometres per second.

The REP stage is effectively the LISM probe. Onboard will be around a dozen instruments, mounted on a two-metre long boom, that will have a complete field-of-view all around them thanks to the probe spinning at a rate of three rotations per minute, which will also provide axis stabilisation. The REP stage will also contain four RTGs, capable of delivering 648 watts at the start of the mission.

Smaller thrusters than the RIT-22 model are required for the REP stage, with the RIT-10 thruster the best choice, which has previously been used on European Space Agency missions. Four of these thrusters will be fixed to the REP stage. With a beam voltage of 1.5 kV, a specific impulse of 3810 seconds and a thrust of 21 milli-Newtons, they will run one after the other consuming 158 kilograms of xenon in total. They will accelerate the REP stage for ten years continuously (except for when data is being returned to Earth by the two-metre diameter high-gain antenna mounted on the rear of the spacecraft), providing a delta-V of eight kilometres per second.

The total mass of the REP stage will be about 500 kilograms and tests have shown the the REP/SEP combination results in a specific impulse of 3,810 seconds.

Important in reducing the flight time will be the choice of launcher. A capable launcher like the Ariane 5, which has a C3 of 45.1 km^2/s^2, could provide a hyperbolic excess velocity of 6.7 kilometres per second and an achievable flight time of 27.5 years to 200AU. In addition, a gravity assist from the giant planet Jupiter could speed the flight-time up to 23.7 years, although this manoeuvre will reduce flexibility in the mission, such as the direction the probe needs to go to reach the heliopause's nose.

SOLAR SAIL OPTION

In 2007 the European Space Agency performed a technology reference study for a realistic mission to explore the boundary of the heliosphere [12]. The

Interstellar Heliopause Probe (IHP) is based on a solar sail propulsion concept. The study has selected a characteristic acceleration (maximum acceleration of a solar sail at one astronomical unit from the sun) of 1mm per second squared; this corresponds to a sail with an efficiency of 0.85 and a size of 246 × 246 metres squared for an in-flight spacecraft mass of 467 kilograms (including 20 percent system margin). Although this acceleration may seem small, a continuously applied acceleration of one millimetre per second squared would result in a delta-V of around 30 kilometres per second per year.

Two fly-bys of the Sun are carried out in order to gain delta-V before proceeding to the outer Solar System. For thermal reasons, the minimum distance from the Sun has been set to 0.25 AU. The sail will be jettisoned at five astronomical units after which the probe will travel to the heliopause, subject to a reduced deceleration caused by the Sun's gravitational force. The velocity at five astronomical units will be 10.4AU per year and the travel time to 200 AU will be 27 years.

An acceleration of three millimetres per second squared would reduce the travel time to just 16 years, as proposed by Richard Wallace [13]. Such accelerations require either a much larger sail or more time for technological development in order to learn how to reduce the mass of the spacecraft or sail system. Hence a 27 year transfer, based on less demanding but nevertheless challenging technology, has been selected as the baseline for the IHP mission concept study.

The IHP mission concept clearly requires many new and improved technologies with a current Technology Readiness Level (TRL) of just two or three (a flight-proven technology has a TRL of 9).

According to McNutt [14], on paper a solar sail looks like the best approach, with a small launch vehicle, a high speed out of the Solar System and no worrying about a Jupiter gravity-assist manoeuvre as no advantage can be gained from it. However, while Japan's IKAROS and NASA's NanoSail-D missions have demonstrated sail deployment, which is a significant technical step, this technology is very far from the characteristics of a sail required for an interstellar probe.

Indeed, to get the required performance, the solar sail will have to get as close to the Sun as approximately 0.25 AU and then reflect the full solar pressure at that distance. A severe problem is that there is no way of actually testing this critical part of the mission without actually doing the mission.

NASA's I^{2}E concept, which sports a large launch vehicle, an upper stage, a Jupiter gravity assist as well as REP looks more complex, yet the required technological advances to provide the required performance are closer at hand. Such an implementation trades more cost for less implementation risk. The next window for a Jupiter gravity assist opens in 2024.

Finally, a combination of SEP plus REP is a competitive option to only solar sailing or ballistic-REP. Using SEP requires a smaller launcher and, when compared to solar sail propulsion, it has a more distant approach to the Sun of around 0.7 AU. Hence, the SEP plus REP option could be the best solution for an interstellar heliopause probe.

These preliminary concepts do indicate that such a mission is possible and that the scientific return would be immense. This mission is mandatory in order to finally put our nose out of the maternal heliosphere shield that has protected our planet from the deadly cosmic rays for aeons and discover what is really waiting for us in the immense void between the stars.

PROJECT TIN TIN

Currently there are no interstellar exploration missions planned or in progress. Even the two Voyager spacecraft are not heading towards any particular star. The closest approach will be when Voyager 1 passes within 1.6 light years of the star Gliese 445 in about 40,000 years.

Andreas Tziolas [15] has started a mission profile and spacecraft design feasibility program named Project Tin Tin, with the main objective of launching the first ever interstellar spacecraft to Alpha Centauri. This project wants to demonstrate that an interstellar journey to our nearest star is feasible within 25,000 years using current technology and reduced costs, cutting Voyager's theoretical 75,000-year travel time by a factor of two-thirds, with reasonable room for improvements. Tziolas presents an idealised calculation demonstrating that the essential capability for an interstellar nanosat mission exists today, using cost-effective technologies that are either flight-proven or with a high technology readiness level. The spacecraft mass budget is 10 kilograms, thus remaining within the formal range of a 3U CubeSat. In order to reach Alpha Centauri in 25,000 years the spacecraft velocity at burnout must be 50.8 kilometres per second, assuming a hyperbolic orbit with negligible initial velocity and ignoring the Sun's gravitational drag. However, Alpha Centauri currently has a radial velocity of 25.1 kilometres per second towards the Sun; in the year 29,700 AD it will be 'just' 3.3 light years away before then starting to move away from the Solar System. If we take into account this radial velocity, the spacecraft's final velocity becomes 25.7 kilometres per second. If half of the spacecraft mass is allocated to propellant (five kilograms), the corresponding specific impulse is 3,800 seconds, a reasonable value for several electric propulsion technologies. If we chose a burnout time of 20 years we get a mass flow of 7.9^{-9} kilograms per second and a thrust of about 0.3 milli-Newtons.

Critical to the performance of all electric propulsion systems is the specific power, i.e. the electrical power output per unit mass (in watts per kilogram) that the spacecraft power source is capable of supplying to the engine. As Project Tin Tin's thruster will operate at increasing distances from the Sun, solar panels are not suitable for this mission; instead, a nuclear battery has been selected as the power source. Using the power source performance characteristics of the recently developed Advanced Stirling Radioisotope Generators (ASRG), which are said to reach around 10 watts per kilogram at maturity, Tziolas has calculated a minimum mass of 0.8 kilograms for Project Tin Tin's miniaturised power source. The final mass budget is shown in the following table:

TABLE 2: PROJECT TIN TIN'S MASS BUDGET

Component	Mass (kg)
Propellant	5.00
ASRG power	0.78
Total ion engine assembly	1.00
Instrumenrs and payload	1.00
Spacecraft structure	1.00
Remaining mass out of 10kg	1.22

A strong candidate for the propulsion system of the interstellar Tin Tin mission is the Indium Field Emission Electric Propulsion (FEEP) ion thruster, under development at FOTEC Forschungs-und Technologietransfer GmbH, in Austria. This technology offers several unique features: demonstrated operation at very high specific impulse (greater than 10,000 seconds), the most efficient way of carrying propellant (solid state), ionisation and acceleration taking place in one step with the same electric field, and very low thermal losses as the emitter electrode is kept just some degrees above 157 degrees Celsius (which is the melting point of Indium).

The core of the Indium FEEP thruster consists of a liquid metal ion source with a sharp tungsten needle protruding out of a propellant reservoir tank. This reservoir is heated to above 156.6 degrees Celsius to melt the Indium. If a sufficiently high electric field is applied between the needle and an extractor electrode, a so-called Taylor Cone is formed and ions are directly pulled out of the liquid metal surface at the tip of the needle. These ions are accelerated out by the same electric field that created them. Typical emitter voltages are 6–10 kilovolts for currents up to a few hundred micro-amps, which correspond to a thrust of a few micro-Newtons per needle.

The FOTEC liquid metal ion source technology has been developed in Austria over the last three decades; it is the only technology of its kind ever to be operated under space conditions. First successfully tested onboard the Russian Mir space station in 1991, Indium liquid metal ion sources have logged more than 15,000 hours of in-space operational hours in various applications (mass spectrometry, space-charge compensation) for space missions like GEOTAIL, EQUATOR-S, Cluster-II, DoubleStar and Rosetta. These liquid metal ion sources are suitable as ion thrusters as well, as shown in 2006 by an endurance test of 5,000 hours with a cluster of four Indium sources [16].

The last version of the Indium FEEP is characterised by the replacement of the solid tungsten needles with porous tungsten ones through a process named 'micro powder injection molding', in which a mixture of tungsten powder and binding polymers is injected into an appropriately shaped mold; this in order to further improve the reliability of this technology.

This new device is a multi-emitter array using a common extraction electrode, designed for operational voltages up to 15 kilovolts and using 28 needles arranged in a circle. For obvious reasons, this device is named the 'crown emitter', and it is capable of generating a thrust from sub micro-Newtons up to one milli-Newton [17].

The Indium FEEP crown emitter is a valid candidate for the Tin Tin mission. In principle one crown emitter can provide the required thrust of 0.3 milli-Newtons (indeed this is its nominal thrust level) with a specific impulse of 3,800 seconds or higher. However, because of the very challenging required lifetime of 20 years, it is necessary to use a cluster of these micro-thrusters, firing one at a time. The lifetime of a crown emitter has not been assessed yet; an endurance test is already planned in the framework of an ESA R&D project. If we consider that the longest demonstrated lifetime of an ion thruster is the one from the NASA's NEXT thruster, which operated for more than five years, then at least four crown emitters are needed to perform the mission. Furthermore, the specific power of this technology – electrical power per thrust unit – is greater than 50 watts per milli-Newton, which is higher than traditional ion thrusters. This means that more mass has to be allocated for the ASRG (at least double), or a RTG with a higher electrical power output per unit mass (watts per kilogram) has to be developed.

Another propulsion technology that could provide supplementary thrust is the solar sail. While the useful solar flux is limited to well before Jupiter orbit at five astronomical units, the inclusion of an ultra-light solar sail could provide additional thrust while doubling as a high-gain antenna after judicious choice of metallisation of the sail surface. A dual use sail of this type would have to be rigidly set and not spin stabilized, because the sail structure will fold onto the spacecraft when the main high specific impulse thruster is engaged.

DISTANT-FUTURE ELECTRIC PROPULSION INTERSTELLAR PRECURSOR MISSIONS

An interstellar precursor mission beyond the heliosphere needs an advanced electric propulsion system with very high specific impulses. In fact, as the required velocity increment (delta-V) increases, the propellant mass becomes greater, resulting in a heavier spacecraft and a longer mission duration. In order to keep the flight time compatible with the working life of the scientists and engineers involved (a maximum of 30 years), the specific impulse must significantly increase, thus minimising the propellant consumption. Table Three shows how the typical requirements of electric propulsion systems become more and more demanding as the missions extend out from nearby planets to interstellar travel. It is clear that for a real interstellar precursor mission aiming for the Oort Cloud (greater than 1,000AU), specific impulses greater than 30,000 seconds are required, which is 5–10 times the specific impulse demonstrated by the current ion thrusters (see Table 1 on page 185). How is it possible to achieve these ultra-high specific impulses? David Fearn [18] has suggested that with a combination of a reduced atomic mass and very high extraction voltages an advanced gridded ion thruster can have a specific impulse as high as 150,000 seconds.

TABLE 3: ELECTRIC PROPULSION REQUIREMENTS

The typical requirements (delta-V and specific impulses) of electric propulsion systems for a variety of missions with a reasonable flight time (maximum 30 years) [18].

Distance from Sun (AU)	Mission	Delta-V	Minimum specific impulse
0–1.5	Nearby planet	5–7km/s	3,000s
5–40	Outer planets	10–15km/s	3,000–5,000s
100–1,000	Heliopause	100km/s	10,0000–30,000s
10,000	Oort Cloud	1,000km/s	100,000s
Interstellar	Alpha Centauri	30,000km/s	3,000,000s

ULTRA-HIGH SPECIFIC IMPULSE ION THRUSTERS: THE DUAL-STAGE FOUR-GRID ION THRUSTER

The innovative Dual-stage Four-Grid Ion Thruster has been proposed by Fearn [18] in order to extend gridded ion thruster performance to very high specific impulses, thus enabling interstellar precursor missions. Four grids are used instead of the usual three-grid arrangement in order to separate the ion extraction and acceleration processes (done simultaneously in current systems). This enables very high ion beam potentials to be put on the grids in the acceleration stage, thereby significantly increasing exhaust velocity, specific impulse, power density and thrust density. Fearn calculates that, using a beam potential of 70 kilovolts and a propellant with a mean atomic mass of 4.5 AMU (atomic mass unit) – namely compounds of hydrogen with carbon and nitrogen – a specific impulse as high as 150,000 seconds is achievable. However, the corresponding input power per thrust unit is ultra-high too, 900 kilowatts per newton. This means that the corresponding power source must have a specific mass less than one kilogram per kilowatt in order to keep the power source mass to a reasonable value.

In order to demonstrate the feasibility of this new thruster concept, a small experimental laboratory prototype has been designed and constructed. The experimental campaign comprised two successful test phases that were conducted in a vacuum facility at ESTEC, ESA's Space Research and Technology Centre, during November 2005 and May 2006, with the aim of demonstrating the practical feasibility of the four-grid concept, verifying the high performance predicted by the analytical and simulation models and investigating critical design issues and technological challenges. Total accelerating potentials of up to 30 kilovolts were demonstrated. Narrow beam divergences of the order of two to four degrees were also achieved. The specific impulse reached values as high as 14,000–15,000 seconds and total efficiencies of 70 percent and thrust over 5 milli-newtons were obtained. The specific power was between 90 and 110 watts per milli-newton. The values achieved represent an improvement by several times on the current state-of-the-art, whilst maintaining very small direct ion impingement of the beam on the grids [18].

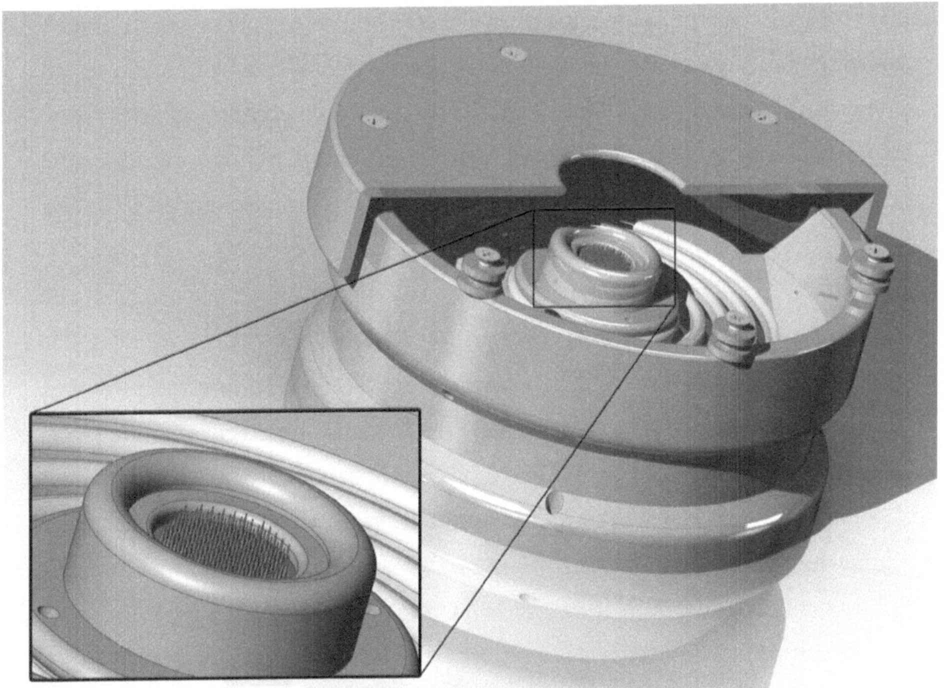

A 3D rendering of the Ultra-FEEP proof-of-concept thruster with a zoomed view showing the emitter assembly based on a 122-needle array. Assuming that every needle will emit 50 micro-amps (mean value) at 120 kilovolts, a thrust of three millinewtons could be generated with a specific impulse up to 30,000 seconds. Image: FOTEC.

ULTRA-FEEP

An improved indium FEEP thruster design with much higher beam potentials, up to more than 100 kilovolts, could lead to values of specific impulse higher than 30,000 seconds [19]. This novel design consists of a two-dimensional array of porous tungsten needles filled with indium, shaped in such a way that a homogenous electric field is present across the whole surface of the array thus providing the same operational conditions for every needle in the array (see diagram, above).

The previously proposed design could be scaled up to meet the requirements of ambitious interstellar precursor missions such as a probe built for Oort Cloud exploration. Preliminary simulation shows that a direct ten-times scaling of the proof-of-concept module together with the emitter should be feasible and would allow operational voltages in excess of one mega (million) volts and thrusts of 0.5 newtons. A cluster of ten Ultra-FEEP thrusters would permit a total thrust of five Newtons. Such a propulsion system will require an input power of four megawatts that could be supplied only by an ultra-light nuclear reactor with specific mass less than one kilogram per kilowatt.

ADVANCED NUCLEAR POWER GENERATION SYSTEMS

Between 1960 and 1992, more than 30 reactors have been used in space. However, their power levels were limited to less than 10 kilowatts. Those reactors were using sodium-potassium as a coolant and low efficiency thermoelectric or thermionic technologies for conversion. Specific masses made great progress during this period of time but their values remained very high: 570 kilograms per kilowatt for Romashka in 1967, 400 kg/kW for Buk in 1977 and 180 kg/kW for TOPAZ in 1987.

Today, the specific mass can be expected to be in the range of 30 to 40 kg/kW at the 100 kilowatt power level. Such a specific mass for this power could be achievable using thermoelectric conversion but with lithium coolant instead of the sodium–potassium in order to increase efficiency by increasing temperature. Such a system could be developed in a relatively short term. However, these state-of-the-art specific mass values are still too high for ambitious interstellar precursor missions.

Recently, a CNES/AREVA study on advanced space nuclear power generation systems has shown that a five megawatt nuclear reactor using a high-temperature Brayton conversion cycle could achieve a specific mass of 15 kg/kW [20]; this technology could be available in the short-to-medium term. A high-temperature Rankine conversion cycle could allow further mass reduction (potentially as low as five kilograms per kilowatt for the total system). However, it must be noted that feasibility of a space propulsion nuclear reactor operating in a zero or very low gravity environment with Rankine conversion is still under question. The French study suggests that this technology could be developed within 20 years.

A further reduction of specific mass (less than three kilograms per kilowatt) would necessarily need an even higher core temperature, which questions the feasibility of a solid core. Vapour core reactors have thus been considered, but their feasibility in less than two decades is questionable.

TAU

The Thousand Astronomical Unit mission was an interstellar precursor mission concept, studied by JPL scientists in the late 1980s with the potential for enabling an unmanned probe to reach a distance of 1,000 astronomical units (0.016 light years) within a 50-year trip time [21]. TAU would be powered by a nuclear reactor in the one megawatt class with a specific mass of 12.5 kilograms per kilowatt and a full-power operating (thrusting) time of 10 years. The ion thrusters required would have a specific impulse of 12,500 seconds, a thrust of 6.8 newtons, an input power of 490 kilowatts and a burn time (per thruster) of two years. As two thrusters would have to fire simultaneously to provide the total thrust of 13.6 newtons and, taking into account a 20 percent redundancy, a cluster of 12 thrusters is considered suitable to perform the mission (see Kelvin Long's chapter for more details on this challenging mission).

Table four shows the main figures of the baseline TAU mission (1987) resulting in a significant initial mass of 62 metric tons (65 percent of it is just propellant). By keeping the thrust constant and increasing the specific impulse in order to reduce the propellant mass, the initial mass can be reduced down to 52 metric tons; a further increase in specific impulse does not bring any advantage because of the contemporary increase in the mass of the nuclear power source as a result of the higher thruster input power per unit of thrust. The key to significantly reduce the TAU spacecraft's initial mass is the power source specific mass; by considering a fission reactor with a specific mass of five kilograms per kilowatt, it is possible to reduce the initial mass down to less than 38 metric tonnes at specific impulses as high as 28,000 seconds. Table four compares the baseline TAU with an advanced TAU concept based on foreseeable developments in ion thrusters and nuclear reactors in the medium-to-long term, as outlined by Fearn [22] and Nicolas Berend [20]. Advanced ion thrusters with a specific impulse of 28,000 seconds based on the Dual Stage Four Grid configuration will need at least 2.2 megawatts of input power to keep the total thrust constant at 13.6 newtons, as their specific power will be roughly double that of the baseline TAU thrusters. However, a next-generation 2.2-megawatt fission reactor with a specific mass of five kilograms per kilowatt will weigh 'just' 11 metric tonnes. The advanced TAU initial mass would be less than 38 metric tonnes; after 10 years of thrusting, the final speed will reach more than 150 kilometres per second, or 32 AU per year. The travel time to 1,000 AU would be just 37 years, which is 13 years less than the old TAU.

TABLE 4: THE TAU MISSION

The original TAU mission compared to an advanced TAU concept based on next-generation gridded ion thrusters.

	TAU (1987)	Advanced TAU (2030)
Ion propulsion system dry mass	4,160kg	4,160kg
Propellant (xenon)	40,000kg	17,860kg
Nuclear fission reactor	12,500kg	10,800kg
	(12.5kg/kWe)	(5 kg/kWe)
Payload	5,000kg	5,000kg
TAU initial mass	61,660kg	37,820kg
Ion thruster specific impulse	12,500s	28,000s
Total thrust (two engines)	13.6N	13.6N
Burn time	10 years	10 years
Burnout speed	112km/s	153km/s
	(23.5AU/yr)	(32.3AU/yr)
Time to 200AU	15 years	12 years
Time to 500AU	27 years	21 years
Time to 1,000AU	50 years	37 years
Time to 10,000AU	430 years	316 years

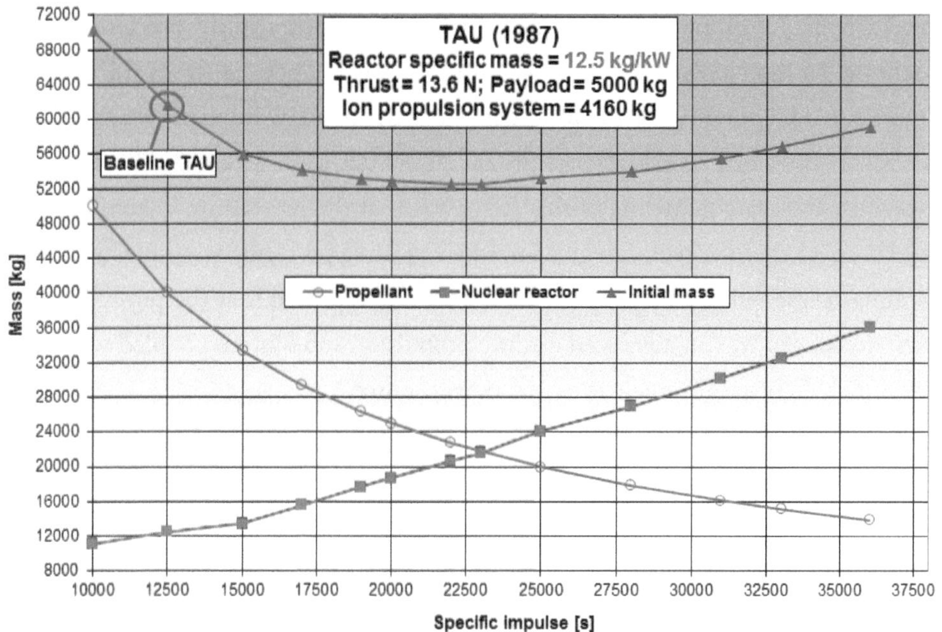

Above: baseline TAU. propellant mass, nuclear power source mass and S/C initial mass as a function of specific impulse.

Below: advanced TAU. propellant mass, nuclear power source mass and S/C initial mass as a function of specific impulse.

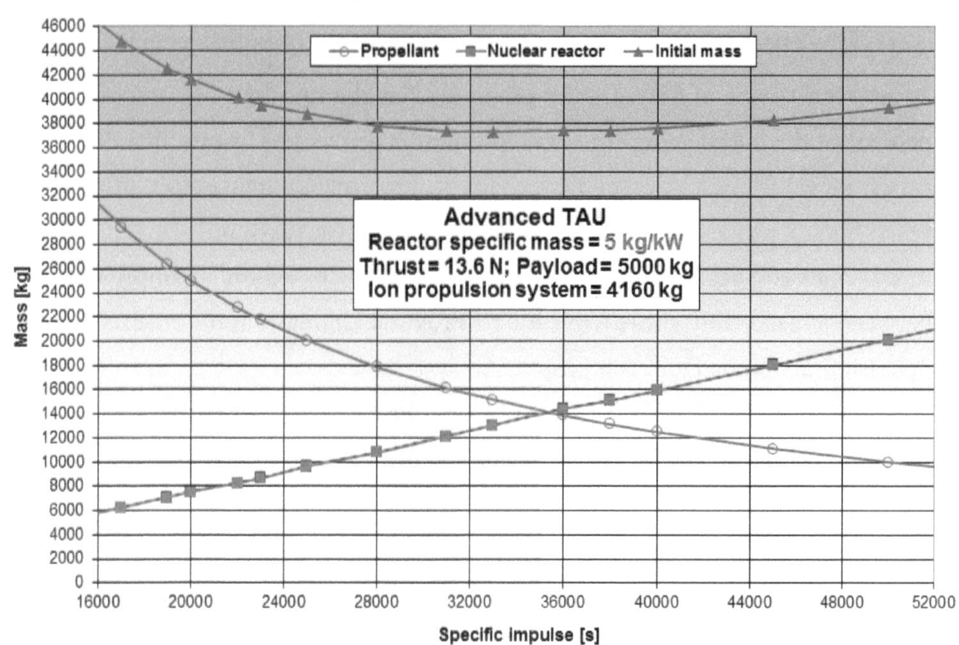

In 1985 the Jet Propulsion Laboratory (JPL) performed a study directed towards defining an interstellar propulsion system that could support a fly-by mission to Proxima Centauri with a trip time in the order of a few hundred years, corresponding to a spacecraft velocity of 0.01–0.02c (3,000–6,000 kilometres per second). The main guideline of this study was that the propulsion system had to be based on familiar technologies for which reliable performance projections and scaling relationships existed. Graeme Aston [23] reported that the propulsion system chosen was an advanced electric propulsion sub-system coupled with an extremely lightweight nuclear fission power source.

The fission reactor technology chosen was the rotary bed reactor, which promised to have very low specific masses at multi-megawatt power levels. As regards the ion thruster, a simple rule of thumb for efficient travel suggests that the exhaust velocity be of the same order as the required mission velocity. This implies a specific impulse of several hundred thousand seconds to achieve the required terminal velocity, with beam potentials in the order of several megavolts. Many different star-probe performance scenarios have been investigated, leading to a near-optimum mission solution with an initial mass of 427 tonnes, an electric propulsion system providing a thrust of 500 newtons with a specific impulse of 400,000 seconds and an input power of one gigawatt. After a burn-time of 65 years the probe reaches the terminal velocity of 1.22 percent of the speed of light (3,660 kilometres per second, more than 200 times the velocity of Voyager 1), arriving to Proxima Centauri after 'just' 389 years. Such a probe could reach the heliopause in five years and the Oort Cloud in less than 100 years. This optimal configuration is based on a compromise between a reasonably short trip time to Proxima Centauri and non-excessive operating demands from the propulsion system.

How realistic is this JPL study? David Fearn [1] reported that specific impulses as high as 150,000 seconds are possible with improved gridded ion thrusters. Furthermore, there are neutral-beam injection gridded ion sources for nuclear fusion research that have already demonstrated specific impulses of the order of 350,000 seconds. Hence, a 400,000 second ion thruster is in principle feasible, however, the associated engineering issues are really challenging, especially if coupled with a required lifetime of 65 years; the longest demonstrated lifetime of a standard ion thruster is about 5.5 years (see Table One).

As regards the fission reactor, there are recent studies for multi-megawatt reactors with a specific mass down to on kilogram per electrical kilowatt, but we are still very far from the star-probe reactor suggested by Aston with a very optimistic specific mass of 0.003 kilograms per thermal kilowatt. However, the NERVA solid core reactor built in 1965 showed a specific mass of 0.007 kilograms per thermal kilowatt [22], not so far from Aston's challenging requirement.

VISIONARY INTERSTELLAR MISSIONS: LASER-POWERED INTERSTELLAR PROBE

Propulsion of an interstellar probe by a laser-pushed light-sail has been proposed by Robert Forward and others. A difficulty of this approach is that, while the specific impulse is effectively infinite, the energy efficiency is very low, thus a light-sail requires extremely high power. Since the limiting factor in the construction of a laser-propelled interstellar probe is the amount of laser power required, it is useful to consider alternatives that increase the energy efficiency. An alternative is the laser-powered rocket, where the laser is converted into electrical energy, which is used to power an electric propulsion system.

Geoffrey Landis [24] showed that solar cells illuminated under monochromatic (laser) illumination achieve far higher efficiency than under solar illumination; efficiencies up to 75 percent are possible with a wavelength of 550 nanometers.

Comparing Robert Forward's fly-by mission of a laser-pushed light-sail and a mission velocity of 11 percent of the speed of light with the same probe propelled by a laser-electric thruster, Landis showed that the laser power can be almost halved, from 65 gigawatts to 38 gigawatts.

Furthermore, unlike a light sail, a laser-energised rocket could be powered by a beam from ahead of the array as easily as by one behind the array. This can reduce the required power to 'just' 18 gigawatts.

A major issue of this interesting concept is that the optimal specific impulse of the laser-powered electric propulsion system is about two million seconds, a value clearly outside the bounds of current technology.

CONCLUSIONS

It is common opinion that every interstellar precursor probe that can be built now or in the future will, sooner or later, be overtaken by more advanced probes; hence, better not waste money now and just wait for the more advanced probes. However, Christopher Columbus did not wait for the realisation of a fast battleship in order to start his incredible voyage towards the unknown, crossing the mighty and scary Atlantic Ocean; similarly, we shall not wait for the realisation of a hyper-fast Starship Enterprise in order to begin exploring the interstellar void that is waiting for us outside the Solar System. The time to pave the way towards future starships is now; as shown in this chapter, electric propulsion systems can be extended in the near term in order to launch very challenging interstellar precursor missions. A 200AU heliopause probe with a burnout speed of 10 AU per year (three times Voyager 1's speed) can be assembled and launched in less than five years; the same is valid for the Project Tin Tin cubesat mission that can be the first human-made artifact to be sent to Alpha Centauri. A medium-term development programme focused on ultra-high specific impulse ion thrusters and ultra-light fission reactors can give us the possibility of launching an advanced TAU mission (to 1,000 AU) with a burnout speed of 30 AU per year within the next 20–25 years.

The discoveries that these early missions will make will greatly increase the interest in interstellar exploration, thus greatly increasing the probability of proper funding for the realisation of faster interstellar probes.

In 1961 USA President Kennedy said: "We choose to go to the Moon in this decade and do the other things not because they are easy but because they are hard… Because that challenge is one we are willing to accept and unwilling to postpone."

Interstellar exploration is an extraordinary challenge we should be willing to accept and unwilling to postpone.

REFERENCES

[1] Fearn, D; 'The Potential Future Capabilities of Gridded Ion Thrusters', Third International Conference on Spacecraft Propulsion, Cannes, 10–13 October 2000.

[2] Goddard, R H; *The Green Notebooks Volume 1*, The Dr Robert H Goddard Collection at the Clark University Archives, Clark University, Worchester, Massachusetts.

[3] Choueiri, E Y; 'A Critical History of Electric Propulsion: The First 50 Years (1906–1956)', *Journal of Propulsion and Power*, 20, pp193–203 (2004).

[4] Stuhlinger, E; *Ion Propulsion for Space Flight*, McGraw-Hill (1964).

[5] Stuhlinger, E; 'Possibilities of Electrical Spaceship Propulsion', Fifth International Astronautical Congress, pages 100–119, Innsbruck (1954).

[6] Jahn, R G; *Physics of Electric Propulsion*, McGraw-Hill (1968).

[7] Goebel, D M and I Katz; *Fundamentals of Electric Propulsion: Ion and Hall Thrusters*, JPL/Wiley Press (2008).

[8] Brown, W; 'Beamed Microwave Power Transmission and its Application to Space', *Proceedings of the IEEE*, 40, 6 (June 1992).

[9] Wimmer–Schweingruber, R F; 'Interstellar Heliospheric Probe/Heliospheric Boundary Explorer Mission – A Mission to the Outermost Boundaries of the Solar System", *Experimental Astronomy*, 24, pp9–46 (2009).

[10] McNutt, R L; 'Enabling Interstellar Probes', *Acta Astronautica*, 68, pp790–801 (2011).

[11] Loeb, H W et al; 'Solar/Radioisotope Electric Propulsion Combination for a Mission to the Heliopause', 32nd International Electric Propulsion Conference, IEPC-2011-052, Wiesbaden, Germany, 11–15 September 2011.

[12] Lyngvi, A E et al; 'Study Overview of the Interstellar Heliopause Probe', ESA Technology Reference Study, SCI-A/2006/114/IHP, issue 3, rev 4 (2007).

[13] Wallace, R A et al; 'Interstellar Probe Mission/System Concept', *Proceedings of the IEEE Aerospace Conference*, 7, 385–396 (2000).

[14] McNutt, R L; 'Update on Innovative Interstellar Explorer', *Centauri Dreams* [online]. Available at www.centauri-dreams.org/?p=21050 (last accessed 5 August 2014).

[15] Tziolas, A, et al; 'Project Tin Tin – Interstellar Nano-Mission to Alpha Centauri', 63rd International Astronautical Conference, Naples, Italy (2012).

[16] Genovese, A, et al; '5,000h Endurance Test of an Indium FEEP 2 × 2 Cluster", AIAA Joint Propulsion Conference, AIAA-2006-4827 (2006)

[17] Vasiljevich, I, et al; 'Porous Tungsten Crown Multi-emitter Testing Programme Using Three Different Grain Sizes and Sintering Procedures', IEPC-2011-065, 32nd International Electric Propulsion Conference, Wiesbaden, Germany, 11–15 September 2011.

[18] Bramanti, C, Fearn, D et al; 'The Innovative Dual-Stage 4-Grid Ion Thruster Concept – Theory and First Experimental Results', IAC-06-C4.4.7 (2006).

[19] Genovese, A, et al; 'Innovative Ultra-FEEP Thrusters for Interstellar Precursor Missions', Starship Congress, Dallas, Texas (2013).

[20] Berend, N, et al; "How fast can we go to Mars using high power electric propulsion?", 48th AIAA Joint Propulsion Conference, AIAA-2012-3889, Atlanta, Georgia (2012).

[21] Nock, K T; 'TAU: A Mission to a Thousand Astronomical Units', 19th International Electric Propulsion Conference, AIAA 87-1049, Colorado Springs, Colorado, 11–13 May 1987.

[22] Fearn, D; 'Technologies to Enable Near-Term Interstellar Precursor Missions: Is 400 AU Accessible?', *Journal of the British Interplanetary Society*, 61, pp279–283 (2008).

[23] Aston, G; 'Electric propulsion: A Far-reaching Technology', 18th International Electric Propulsion Conference, AIAA 85-2028, Alexandria, Virginia, 30 September to 2 October 1985.

[24] Landis, G; 'Laser-powered Interstellar Probe', Conference on Practical Robotic Interstellar Flight, New York University, 29 August to 1 September 1994.

CHAPTER 10

POWER FOR THE STARS:
BOOT-STRAPPING OUR WAY TO ALPHA CENTAURI

ADAM CROWL

The stars are, at minimum, thousands of times further away than the planets. Presently flights to the planets take a year or more, thus to reach the stars in years, rather than millennia, requires speeds thousands of times faster. Present day chemical fuels have sufficient energy content to propel us to perhaps 15 kilometres per second, via rockets. To reach 15,000 kilometres per second, one-twentieth the speed of light, requires fuels with an energy content at least a million times greater than what mere chemistry can achieve. Fortunately, thanks to Einstein's theory of relativity, we know that matter itself contains billions of times more energy than chemical bonds.

Philosopher Olaf Stapledon imagined some of the first realistic starships in fiction in his multi-billion year history of the cosmos, *Star Maker* (1937). Intelligent species at first achieve interplanetary travel via 'normal fuels' – presumably chemical rockets. They then develop interstellar travel after the invention of 'sub-atomic energy' – the total conversion of mass into energy. After they develop artificial planets – what we now call 'space colonies' – they combine the two and launch forth to explore the Galaxy. Some of the faster artificial planets achieve half the speed of light, which would require them to arrive with one third the mass they launched with. Stapledon's starships propelled by beams of sub-atomic radiation are known nowadays as 'photon rockets.'

Photon rockets have featured in astronautical discussions almost since the discovery that light has momentum and can produce push [1]. Stapledon's advanced alien civilisations learnt the trick of turning raw matter into pure energy, then directing it in a useful beam, but his description was rather short on details for serious engineering purposes. As a means of propulsion, pure photon thrust is very power intensive, requiring 300 megawatts of energy for every newton of thrust. Power is defined as the rate at which energy is used/released per unit time – one watt is one joule of energy used per second. Humanity presently uses about 20 trillion watts from all sources, other than food. Thus a thousand tonne starship accelerating at 10 metres per second squared (just over one Earth gravity or 'gee') would require a photon power of at least three quadrillion watts, roughly 150 times the total power output of human civilisation.

With such immense power levels the difficulty of fitting a space vehicle with a photon rocket or drive has led many to abandon the task as hopeless. More modest rocket concepts have been proposed over the last 100 years of astronautics, using nuclear and thermonuclear reactions to achieve smaller fractions of the speed of light [2]. However there is a minimum power requirement to reach other star systems including the nearest, alpha Centauri, in a given amount of time. In astronautical engineering a useful concept is the specific power, which is the amount of power produced, or used, per unit mass of the space vehicle. This is generally called alpha and for a 100-year mission to alpha Centauri, just 4.36 light-years away, alpha needs to be approximately 90 kilowatts per kilogram of starship. Alpha is proportional to the square of the distance and the inverse cube of the time - in other words to go twice as far, one increases alpha by 2 squared i.e. 4. To travel to alpha Centauri in 50 years (half 100 years) that means alpha must increase by two-cubed, or eight-fold, to 720 kilowatts per kilogram. For a 1,000 tonne starship that is a minimum power of 720 gigajoules of kinetic energy being added to the starship by the engine every second. Because of the inefficiencies in the engine the total amount of energy being handled might be 10 times greater. If a tiny fraction of that energy is absorbed as waste heat then the vehicle will get very hot, very quickly. To control waste heat, space vehicles need heavy radiators to emit the heat into space, which makes engineering a starship with a high alpha very difficult indeed.

For speeds below about a fifth the speed of light, rockets using nuclear fission or fusion are probably sufficient to the task, although immensely challenging to design with a sufficiently high-alpha for a minimum-time journey [2]. For higher speeds a starship needs to use very energetic particles and an exhaust increasingly composed of pure photons. Matter–antimatter reactions are the simplest way we know of turning matter into photons, but the very high-energy photons produced are very difficult to point in a useful direction – they tend to fly from the reacting particles in random directions. Being very high energy they are also difficult to reflect and direct; gamma rays tend to pass through most forms of matter that we can engineer.

THE LASER-SAIL

Instead of carrying the photon-making machines, why don't we send the photons to the starship? This idea was explored in fiction for decades before a means of projecting photons across immense distances in tight beams was found. One way was invented in 1960 – the laser. A starship would then be an immense reflective 'sail' that rode on a tight beam of photons. This concept, the laser-sail, was first seriously proposed by physicist Robert Forward in 1962 [3]. Other physicists studied the concept through the 1960s and 1970s [4, 5], usually considering X-ray lasers as the preferred means of sending a tight laser beam over the large distances needed to accelerate a laser-sail.

To accelerate to close to the speed of light, denoted as 'c' by physicists, requires a large acceleration distance: at one Earth gee a laser-sail would need to accelerate over 0.15 light years (about 1.4 trillion kilometres) to reach 0.5c. The distance can be shortened by increasing the acceleration, but the light intensity quickly becomes greater than known materials can stand. This was one difficulty identified in the laser-sail discussions of the 1970s. Another problem at that time was that no one could conceive of a plausible means for slowing the sail down. P C Norem [6] conceived of a way of turning using the magnetic field of the Galaxy, then slowing the vehicle down with the same laser than accelerated it, but modern measurements of the magnetic field have shown this to be impracticably slow.

As inventive in the 1980s as he was in the 1960s, Robert Forward produced a series of new laser-sail concepts over the period 1982–1985 [7]. His first break-through was the use of an immense Fresnel Zone Lens to increase the useful range of optical frequencies of laser light. Prior to this suggestion, laser propulsion advocates had suggested X-ray lasers would be needed to beat the diffraction limit on useful laser range. Because of diffraction, which causes the laser wave-front to spread out, the range of optical lasers would be limited to just astronomical units of distance. Useful range was inversely proportional to the wavelength of the laser-light, so very short X-ray wavelengths would be needed to reach across the vast light years of distance until Forward's lens allowed the more tractable optical wavelengths to be used.

Perhaps more significantly Forward demonstrated a means of braking laser-sails to a halt at their destination with his invention of the 'stage-sail' – a laser-sail composed of successively smaller sails to be pushed via light reflected from the larger sail surrounding it. In Forward's analysis, this allowed for the exciting prospect of reaching nearby stars, like epsilon Eridani, at 0.5c and then returning the vehicle to the Solar System. Laser-sails could thus be used for fast round-trip missions.

However, to send laser light usefully concentrated across ten light years or more, the beam would need to be focused by an immensely large Fresnel Zone Lens – about 50,000 kilometres in diameter. The lens itself, which shapes the wave-front of the beam via alternating rings of light and dark, need only be made of thin space-capable material, but such an immensity would still mass half a million tonnes. The laser stage-sail itself would mass 75,000 tonnes and be 1,000 kilometres in diameter, requiring a pushing power of many thousands of terawatts of laser energy. So where would those trillions of watts of laser energy come from? Forward suggested a vast array of energy collectors and laser generators in orbits close to the Sun.

The discovery in 1986 of high temperature superconductors allowed astronautical engineers to contemplate use of superconductivity for creating low-mass magnets on a grand scale. One such application was the magnetic sail. Jonathan Vos Post, Dana Andrews and Robert Zubrin had discovered [8] the long-sought braking system for laser-sails from interstellar speeds. Coupled to a laser-sail, a magnetic-sail would allow deceleration from relativistic speed in a couple of decades or less, without the problems of focusing laser light across light-years in order to brake as required by Forward's stage-sail design. In the late 1980s NASA scientist Geoffrey Landis [9] studied high temperature laser-sail materials that would allow much shorter acceleration track-ways to be used, reducing the scale of the focusing optics to more manageable sizes. In one analysis, by Zubrin and Andrews [10], a focusing mirror and magnetic sail-equipped laser-sail as 'small' as 50 kilometres across would achieve the same performance as Forward's 1,000-kilometre Fresnel Lens/light-sail combination for a mission to epsilon Eridani. The laser power would be about 5,000 terawatts – just 250 times present day energy production levels.

PELLET PROPULSION

Alternatively the lasers could be entirely done away with and the magnetic-sail could be pushed via a particle or pellet beam. Charged particle beams tend to be susceptible to deflection by the ambient magnetic fields around them, which limits their useful beam range. A beam composed of heavier particles would deflect less, if at all. One of the earliest concepts using pellets was Clifford Singer's 'interstellar pellet-launcher'.

Developed at roughly the same time as Forward's stage-sail, Singer [11] proposed using small pellets to project momentum to a starship. His concept used a very large magnetic accelerator, spread out over a very large track-way, to propel small particles to relativistic speeds and then push against the magnetic field of an interstellar vehicle. The advantage of the pellets was that they were too big for the feeble magnetic fields of interstellar space or random impacts of interstellar gas and cosmic rays to deflect from their path to the starship. Being so numerous, the loss of a few pellets to random dust collisions would not impact the system's performance significantly.

Then, in the 1990s, as nanotechnology was becoming a part of the conceptual landscape, a new pellet-propulsion concept emerged, championed by

USAF researcher Gerald D Nordley. Instead of 'dumb pellets', externally steered and shot towards a starship, Nordley [12] proposed nanotechnological 'smart-pellets' that could maintain themselves in the beam path to a starship via a simple artificial intelligence and tiny single-atom rockets for course correction.

A more recent concept, developed by Jordin Kare [13], would combine the best of both laser-sails and pellet propulsion - the 'laser micro-sail beam' - or 'sail-beam' concept. Instead of pushing a large starship with a laser and building an immense pellet accelerator array, Kare proposed to use a smaller laser to accelerate micro-sails at immense accelerations to then fly towards the starship, be blasted into ions by a laser on the starship and those ions would then push against the starship's magnetic field. The micro-sails would be designed to reflect most of the laser light, but otherwise transmit the light they don't reflect and only absorb a tiny fraction. This would allow an acceleration approaching three million gee, so the micro-sails could accelerate to 36,000 kilometres per second (0.12 c) in just 1.2 seconds.

All these concepts require readily available supplies of energy for the laser or pellet systems. In space one power source dominates all others: the Sun. Earth absorbs, every second, about 122.2 petawatts (122.2×10^{15} watts) from the Sun. Thus using the energy that falls on Earth alone we could power starships, but covering the Earth in solar collectors is impractical. In 1968 a physicist named Peter Glaser [14] proposed building immense solar arrays in space called 'solar power satellites', or power-sats. Glaser imagined that power-sats would beam energy back to Earth via microwave beams or lasers. Built via regular techniques, as imagined in the 1970s, every component would be launched into orbit via rocket and each power-sat assembled piece-by-piece via teams of astronauts at a potential cost of trillions of dollars to supply, perhaps, a gigawatt of power to a ground receiver. For starships something much, much cheaper is needed.

The intensity of sunlight decreases with the inverse square of the distance from the Sun. At Earth's orbital distance about 1,400 watts of sunlight power falls on every square metre. At the orbit of Mercury the intensity of sunlight increases ten-fold. At just 0.1 astronomical units (AU; one AU is the average

Earth–Sun distance, 149.6 million kilometres) the intensity is 100-fold higher. Thus to get more useful power for the same mass of equipment the power-sats need to be moved closer to the Sun. High temperature photovoltaic materials can be made to operate in such conditions, but at higher temperatures converting raw sunlight into useful electrical energy can use highly efficient thermoelectronic conversion [15].

Thus a 1,000 tonne power-sat that produces a gigawatt (a thousandth of a terawatt) at Earth's orbit could produce 100 gigawatts at 0.1AU. To produce, for example, the 5,000 terawatts for a laser-sail starship to fly to epsilon Eridani, about 50,000 power-sats would be needed. Launching the required 50 million tonnes of power-sat into orbit does not seem feasible.

No astronautical designer has ever seriously proposed launching so much material from Earth, but translating sufficient traditional industrial capacity into a more conducive location in the Solar System, like Mercury or the Moon, seems an equally Herculean task. Current world solar-power manufacturing capacity is only a gigawatt or so, thus many thousands of years of manufacturing would be required using present day techniques. Instead of traditional manufacturing technology, relocated to space, something new is needed.

Gerald Nordley [16] has suggested that we utilise a self-replicating factory to build more replicators and produce power-sats, deriving materials from the asteroids in space. Two questions arise: (1) what difference does self-replication make? And (2), can it really be done?

To answer the first, we will use Gerald Nordley's replication rate: a self-replicating factory (a SeRF) that can make a copy of itself and a power-sat, situated inside the orbit of Mercury and generating a 100 gigawatts, in one year. Thus, at the end of the first year, we have two SeRFs and a power-sat. At the end of two years we have four SeRFs and two more power-sats. At the end of ten years there are 1,024 SeRFs and 1,023 power-sats, thus more than a 100 terawatts (10^{14}, or a hundred trillion, watts) of power-generating capacity.

At the end of year twenty there will be 1,048,576 SeRFs and 1,048,575 power-sats producing over a 100 petawatts (a thousand trillion watts). Thus in just twenty years enough power to propel 20 laser-sail starships to epsilon Eridani at half light-speed. Earth intercepts from the Sun about 175 petawatts of energy, thus in 21 years more usable energy will be available than is received by the Earth. Nikolai Kardashev, a Russian astronomer, proposed a scale for measuring civilisations based on their energy usage. In the Kardashev Scale a Type I Civilisation utilises the energy equivalent to that received by its home planet from its star. A Type II Civilisation utilises the total energy produced by its star, while a Type III utilises the total energy produced by its home Galaxy.

At the end of year thirty there will be over a billion SeRFs and 100 exawatts (10^{20} watts) being produced by over a billion power-sats, enough to launch 20,000 laser-sail starships at half light-speed per year. To reach Kardashev Type II, mathematically speaking, the SeRFs would require just 52 years of self-replication to fully envelope the Sun in an energy gathering shell. At 1,000 tonnes per SeRF and power-sat, however, we may use up the available mass floating free in space in the form of asteroids. To merely power thousands of starships will require a tiny fraction of what is available. Clearly self-replicating technologies can make an immense difference in a relatively short span of time.

In answering question two we must step back to the late 1970s when President Jimmy Carter directed NASA to study what would be needed to build a self-replicating factory on the Moon, with the final report being delivered in 1980, edited by Robert Freitas and William Gilbreath [17, 18]. For a seed mass of 100 tonnes, using robotic and manufacturing technology just ahead of 1970s technology, the study considered all the aspects of how to gather, process and manufacture an automated factory's components on the Moon. An important aspect that determined the level of difficulty was parts/component closure, i.e. what proportion of parts could be manufactured on site from lunar resources and what had to be imported from Earth. Closure of 90–95 percvent was considered straight-forward. More challenging would be 100 percent closure, with the lunar factory being totally autonomous of inputs from Earth. Chiefly, the difficulty of producing computer chips meant, in the early days, that they would be easier to produce on Earth. Eventually even this could be moved to the Moon.

A more recent study from 2011 by NASA researchers Philip Metzger,

Anthony Muscatello, Robert Mueller and James Mantovani [19], re-examined the concept, updating it with modern 3-D additive manufacturing techniques, concluding that the seed mass could be considerably reduced. Instead of immediate 100 percent closure of parts and components, the system would evolve through several generations of improving robotics and industrial techniques, adapting to the local resources. For a seed mass of 41 tonnes the system could produce 100,000 'robonauts' in 20 years to sustain industry on the Moon, eventually expanding to the Asteroid Belt and beyond. Human beings would remain a part of the process, with tele-operation, i.e. advanced remote control, of the robonauts as a key part on the road to increasing autonomy of the robotic systems. Humans would train the automation and remain key to the whole system's evolution as it expands.

Once a space-based industrial infrastructure is well underway, with the corresponding experience of manufacturing and prospecting in space, then a more focused effort on developing self-replicating power-sat factories can begin. Gerald Nordley [16] has plotted the following development timetable over the next century. Initially the resources of the asteroids will be prospected, as the company Planetary Resources are already planning to do. Next step will be refining materials in situ, with tele-operated machinery at first, but also by developing ever more capable robotics as activities move beyond speed-of-light control-range. In Nordley's starship architecture an early stage is a high-speed flyby of alpha Centauri (and other target star systems) to confirm the necessary resources are available. Then a robotic base is launched that will send out a braking trail of slow-pellets for the following starship to brake against. As the braking trail pellets are much slower than the accelerating pellets, the robotic base does not need to produce power-sats for decades to build-up to full power. Once the braking trail is confirmed to be in place, the first crewed starship to alpha Centauri will depart, arriving in about six years.

Key to powering that first starship, as we have seen, will be developing the Solar System's resources. Not only does bootstrapping space industry make starships possible, immense potential wealth is opened up for the human species and bulk industry is relocated into space. For this potential to become actual the broader public will need to be involved every step of the way. Prospecting thousands of asteroids will need thousands of citizen scientists analysing the data. Controlling thousands of tele-operated robonauts on the Moon and beyond will need thousands of people on Earth and eventually in space as we move further afield and the light-speed communication limit makes tele-operation difficult. Co-ordinating and maintaining the immense power-sat arrays that will power the starships will also need human supervision.

What is the limit? Consider for a moment the power of the Sun, some 400 trillion trillion watts, which is a total power sufficient to propel 80 billion laser-sails to half light-speed. Harnessing a tiny fraction of that allows everyone to go to the stars.

REFERENCES

[1] M M Michaelis and Andrew Forbes; 'Laser Propulsion: A Review', South African Journal of Science, 102, pp289–295 (2006).

[2] L D Jaffe and D F Spencer; 'Feasibility of Interstellar Travel', NASA Report: JPL-TR-32-233 (1963).

[3] R L Forward; 'Pluto – The Gateway to the Stars,' Missiles and Rockets, Vol 10, pp. 26–28 (1962).

[4] George Marx; 'Interstellar Vehicle Propelled by Terrestrial Laser Beam,' Nature, Vol 211, pp22–23, (1966).

[5] W E Moeckel; 'Comparison of Advanced Propulsion Concepts for Deep Space Exploration,' J Spacecraft and Rockets, 9, pp863–868 (1972).

[6] P C Norem; 'Interstellar Travel, A Round Trip Propulsion System with Relativistic Velocity Capabilities,' American Astronautical Society, Paper No 69–388, (1969).

[7] R L Forward; 'Roundtrip Interstellar Travel Using Laser-Pushed Light-Sails,' J Spacecraft and Rockets, Vol 21, No 2, pp187–195 (1984).

[8] D F Andrews and R M Zubrin; 'Magnetic Sails and Interstellar Travel,' IAA Paper 88–553, presented at the 39th IAF Congress, Bangalore, India (1988).

[9] G A Landis; 'Optics and Materials Considerations for a Laser-Propelled

Light-Sail,' Paper No. IAA-89-664, presented at the 40th IAF Congress, Malaga, Spain (1989).

[10] D G Andrews and R M Zubrin; 'Use of Magnetic Sails for Advanced Exploration Missions', in Vision-21: Space Travel for the Next Millennium; NASA Lewis Research Center, pp202–210, (1990).

[11] C E Singer; 'Interstellar Propulsion Using a Pellet Stream for Momentum Transfer,' Journal of the British Interplanetary Society, Vol 33, pp107–115, (1980).

[12] G D Nordley; 'Interstellar Probes Propelled by Self-Steering Momentum Transfer Particles,' Paper IAA-01-IAA.4.1.05, 52nd International Astronautical Congress, (2001).

[13] J T Kare; 'SailBeam: Space Propulsion by Macroscopic Sail-Type Projectiles,' Space Technology and Applications International Forum-2001, M S El-Genk, ed., AIP Conference Proceedings Vol 552, 2001, pp402–406.

[14] P E Glaser; 'Power from the Sun: Its Future', Science Magazine 162, 857–861 (1968).

[15] S Meir, C Stephanos, T H Geballe and J Mannhart; 'Highly-Efficient Thermoelectronic Conversion of Solar Energy and Heat into Electric Power,' arXiv:1301.3505v1 [cond-mat.mtrl-sci]

[16] G D Nordley and A J Crowl; 'Mass–Beam Propulsion: An Overview,' presented at the 100 Year Starship Symposium, Orlando, Florida (2011).

[17] NASA Conference Publication 2255 (1982), based on the Advanced Automation for Space Missions NASA/ASEE summer study Held at the University of Santa Clara in Santa Clara, California, from 23 June to 29 August 1980.

[18] R A Freitas Jr and W Zachary; 'A Self-Replicating, Growing Lunar Factory', paper delivered at the fifth Princeton/AIAA/SSI Conference on Space Manufacturing, 18–21 May 1981.

[19] P Metzger, A Muscatello, R Mueller, and J Mantovani; 'Affordable, Rapid Bootstrapping of the Space Industry and Solar System Civilisation.' J Aerosp Eng 26, Special Issue: In Situ Resource Utilisation, 18–29, (2013).

CHAPTER 11

WARP DRIVES AND FASTER-THAN-LIGHT TRAVEL

TIFFANY FRIERSON

According to Albert Einstein's Special Theory of Relativity, nothing can go faster than the speed of light through space. However, our nearest star, Proxima Centauri, is 4.3 light years away. That means that if a spacecraft is travelling at the speed of light, then it would take 4.3 years to reach Proxima Centauri from Earth, and that's one way. What is the problem? Well, according to Special Relativity, nothing with mass can travel at the speed of light. As an object with mass approaches light speed, its mass increases, so that at the speed of light, the object's mass would be infinite! Even if one could travel near or at the speed of light, while the crew would not experience much, if any, time passing during a mission to more distant stars, observers on Earth would measure the elapsed time to be hundreds or thousands of years. This is a result of the time dilation effect of relativity. So, if we can't travel at the speed of light, is there a way that we could reach our nearest star within a human lifetime?

Consider the fastest spacecraft we have launched, Voyager 1. It blasted off on 5 September 1977 [1], originally as a scientific mission to the outer planets of our Solar System. Voyager 1 is travelling out of the Solar System at a velocity of about 1/18,000th the speed of light. If it were headed in the direction of Proxima Centauri, it would take 73,000 years to get there. Using a new type of propulsion system, say nuclear pulse or a matter/antimatter annihilation system, we could reach Proxima Centauri in about 80 years. That, however, is still only one-way. An astronaut could reach the star within his/her lifetime, but he/she could never return to Earth. It is also desirable to lessen the travel time for human astronauts as much as possible. Long-term exposure to radiation in space could be very harmful to a human being's health. Also, for a mission of around 100 years, planning would be needed to deal with human conception and birth in a low-gravity environment, the effects of low gravity on bodily functions, food rations and other issues.

Is there some kind of loophole in the laws of physics that could get us to our nearest star in less than a human lifetime? What if we could travel faster than the speed of light, so that an astronaut could venture to the alpha Centauri system and maybe even be able to come home on the same day? Is this possible?

According to Albert Einstein's General Theory of Relativity, it is possible! For several decades, scientists have been working with the field equations of the General Theory of Relativity to see if there is any way to actually modify space-time to allow for faster-than-light travel through the Universe. To do this, one specifies a configuration of space-time, then plugs that configuration into Einstein's field equations to analyse the energy requirements and other properties of the said configuration.

There are two main classes of faster-than-light metrics permitted by General Relativity: wormholes and warp drive. In this chapter, we will take a look at the warp drive metric, while Remo Garattini will tackle wormholes in chapter 12.

WARP DRIVE

Many people's first introduction to warp drive came from an exposure to science fiction television shows and books. One very popular example is *Star Trek*, where the starship Enterprise zips across the Galaxy using warp engines. According to *Star Trek* lore, Zephram Cochrane is the scientist credited with the design of the "modern warp drive." [2] Working with a team of engineers in 2061, he successfully designed a working prototype of his warp drive engine that allowed a spacecraft to alternate between light and sub-light speeds. This alternation prevented the infinite energy expenditure problem, which states that an object with mass would require infinite energy to reach the speed of light. The early 'Continuum Distortion Propulsion' engine was successful and enabled Cochrane and his team to relocate to a colony in the alpha Centauri system with a transit time of four years.

The Cochrane warp drive system worked by creating layers of warp field energy. The force of each layer on its next outer layer is controlled and the cumulative effect drives the vehicle forward. The higher the warp factor, the more layers are required. The warp coils are located in the two nacelles, the cylindrical structures in the aft of the ship. The warp coils are fired in sequence from fore to aft. The frequency of the firing of the warp coils determines the number of layers of warp field energy. A matter/antimatter reaction is used (with moderation from fictional dilithium crystals) to produce a massive amount of energy, in the form of plasma, to be sent to the nacelles by way of the power transfer conduits.

Star Trek and other science fiction novels, television shows and movies have even captured the imagination of many scientists, who have turned to physics in an attempt to bring fiction to life. Miguel Alcubierre is one of these scientists. In 1994, while a graduate student at Cardiff University, he published the first serious scientific study of the theory behind the warp drive metric. Published in the *Classical and Quantum Gravity* journal of the Institute of Physics, Alcubierre's paper, entitled *The Warp Drive: Hyper-fast Travel Within General Relativity* [3], outlined how the warp drive space metric could be achieved using General Relativity.

The warp drive is a 'bubble' of space-time. It allows the simultaneous contraction of space-time in front of the spacecraft and the expansion of space-time behind it. The contraction of space-time pulls two points in space towards each other, while the expansion of space-time behind the spacecraft pushes it away from its departure point. The advantage of the warp drive is that there is no violation of Special Relativity or General Relativity. Objects are constrained to move in space-time at the speed of light, but space-time itself does not have a speed limit. Space-time may move at many times the speed of light.

This idea was inspired by Alan Guth's inflationary universe theory. Guth formulated his paper on the inflationary universe in 1980 as a solution to the problems of the then-current theories of the big bang [4]. One problem that he developed a solution to was the 'horizon problem'. Light and the information it carries has a finite speed limit. It takes a long time for light to reach one part of the Universe from a distant point. Yet the Universe everywhere has the same physical properties, so how did distant parts of the Universe reach this equilibrium even though these points are causally disconnected (information and light from one point in the Universe not having had chance to reached these other points)? Guth's answer is that during the first moments following the Big Bang, space-time must have expanded at an extremely rapid rate, driven by negative pressure. Then the distant points of space that seem to be causally disconnected were, in fact, causally connected at one point in time. If one could harness the process that caused space-time to expand so rapidly, then we could use that to travel to distant parts of the Universe at a speed faster than light.

Imagine two stars: our own star Sol and Proxima Centauri. Then imagine a spacecraft starting out on a voyage from Earth. A conventional propulsion system would carry the spacecraft from Earth to some point between Sol and Proxima Centauri, coming to a complete stop. At the activation of the warp drive, the simultaneous contraction of space-time in the front of the craft and expansion behind the craft will push the spacecraft to a point close to Proxima Centauri at an apparent velocity that is faster than that of light relative to a distant observer. During this phase, the spacecraft's acceleration within the actual warp bubble is zero. This means that the astronauts will not suffer any effects of fast acceleration (called the G-forces). After the warp phase, the conventional propulsion system would be restarted to carry the spacecraft the remaining distance to Proxima Centauri. An outside (non-accelerating) observer, as well as the astronaut, would observe the spacecraft as going many times the speed of light during the warp drive phase. What the outside observer would see would be the spacecraft 'surfing' on a 'wave' of space-time. This (in theory) allows the possibility of a mission to our nearest star at faster-than-light speeds without breaking any of the laws of physics!

Since Alcubierre's paper, there have been many papers published perfecting the warp drive concept. One very interesting idea is to use String/M-Theory to find another way to manipulate space-time to enable the warp drive. This idea uses Quantum Field Theory, instead of purely General Relativity, to formulate the warp drive.

In 2008 Richard Obousy and Gerald Cleaver of Baylor University came up with the novel idea of manipulating one of the extra dimensions of M-Theory to push the spacecraft to its destination [5]. One popular formulation of M-Theory is that there are ten spatial dimensions and one time dimension. Where are the extra seven dimensions? Well, according to M-Theory these extra dimensions are curled up into a very compact space no more than 10^{-35} metres in diameter. The cosmological constant of the Universe describes its expansion (although no one knows exactly what is causing space to expand). According to Obousy and Cleaver, one could manipulate the cosmological constant of the compacted space, causing it to contract. According to General Relativity, the contraction of the compact space will cause the non-compact dimensions (our familiar three dimensions) to simultaneously expand. This simultaneous contraction of the compact space and expansion of the other three dimensions will cause the familiar warp drive effect, enabling the spacecraft, according to an external observer and the spacecraft occupants, to travel with an apparent speed that is faster than that of light. However, String/M-Theory has not been experimentally verified, so it is unknown whether extra spatial dimensions even exist. Nevertheless this new warp drive idea seems promising and should be further expanded upon, but the physical theory behind it must mature and be verified before serious consideration of bringing it into physical reality can occur.

So, it is known that warp drive is theoretically possible. Why is humanity not already using this technology to travel around the Galaxy? One of the answers is that there are more practical problems in making warp drive a reality. After solving the general relativistic field equations to come up with the warp drive metric, further analysis shows that positive energy, the type of energy that is encountered everyday on Earth, is not sufficient to produce the warp bubble. Not even vast amounts of it. What is needed is exotic matter.

EXOTIC MATTER

The very hard part of engineering warp drives is the fact that, according to General Relativity, exotic matter is needed to create them. What exactly is this?

Exotic matter is matter that violates the four classical Hawking–Ellis energy requirements: the Strong Energy Condition, the Weak Energy Condition, the Null Energy Condition and the Dominant Energy Condition. The Strong Energy Condition states that the observed energy and pressure of matter combined must be positive or at the very least zero. The Weak Energy Condition states that the energy density of matter must be observed to be non-negative. The Null Energy Condition is implied by the Strong Energy Condition and states that for a null (light-like) vector the energy observed must be greater than or equal to zero. Finally the Dominant Energy Condition states that the flow of energy must not exceed the speed of light. All of these energy conditions would seem to dictate that exotic matter with a negative energy density could not exist. However, each of the energy conditions has been experimentally verified to be false [6]. For example, the Strong Energy Condition was necessarily violated in the inflationary period of the beginning of the Universe. Also, semi-classical quantum effects and quantum field theory effects are shown to violate all of the energy conditions. Quantum Theory and Quantum Field Theory have excellent agreement with experiment, so the fact that the processes described by them violate the energy conditions means that nature itself violates these conditions on a regular basis. So violations of the energy conditions do not preclude the existence of exotic matter, although at present we have no idea what form this matter would take, where we could find it or how we could create large quantities of it.

Negative energy density is the specific form of the energy condition violations that is needed to form the warp bubble. What is meant by a negative energy density is that the configuration of matter fields that would generate the warp bubble would have an energy density that is less than its pressures. The energy density and flux could also be algebraically negative [7]. Quantum Field Theory allows the existence of local regions of negative energy densities, stating that the vacuum has a ground state of fluctuations of varying frequencies. If one could lower the fluctuations of the quantum vacuum, then one could generate a negative energy density from the quantum vacuum. Negative energy has been produced in the lab, albeit at small quantities [primarily via the Casimir effect; see below]. How much negative energy does it take to generate a warp bubble? The answer is: a lot! Obviously, the faster the velocity of the warp bubble, the more negative energy needed. For a warp velocity of the speed of light, -3.03×10^{50} joules is needed [6]. For a warp velocity that is 100 times the speed of light, -3.03×10^{54} joules is required.

There have been formulations that decrease the amount of negative energy needed to produce a warp bubble. Harold White, of NASA's Eagleworks Laboratory, found that if the thickness of the warp bubble is increased, then the negative energy requirements decrease as well [8]. The increase in bubble thickness would result in a smaller space in the middle of the bubble for the spacecraft, although that is not a serious issue. The reduced real estate results in a drastic decrease in the negative energy requirement. Also, White found that the increased bubble thickness would result in a topological change in the warp bubble, such that the warp topology becomes imprecise. However, it seems that imprecise warp bubbles are easier to engineer than more precise ones. However, even with White's mathematical formulation, the negative energy requirement cannot be made to disappear. We must find some way to engineer more negative energy than has been detected to date in the laboratory.

NEGATIVE ENERGY TECHNOLOGIES

One way to achieve a negative energy density is to squeeze the quantum vacuum. According to Quantum Field Theory, empty space is not 'empty', but consists of virtual particles and vacuum wave fluctuations that give rise to the 'zero point fluctuations'. These fluctuations are actually an implication of Heisenberg's Uncertainty Principle, which states that complementary variables (such as energy and time) cannot simultaneously be measured precisely. This means that virtual particles can pop in and out of existence, changing the energy of the quantum vacuum in a very small amount of time. However, it is known that one can squeeze the state of a quantum field (such as the vacuum) by increasing the accuracy of one quantum observable at the expense of its conjugate (whose variance is increased).

A squeezed electromagnetic field has actually been realised in the lab [7]. An electromagnetic field is another example of a quantum field, which necessarily has ground state fluctuations. In a study using a nonlinear optics technique, the fluctuations of one observable of a laser beam was moved into the other observable. This resulted in the first observable's fluctuations being 'squeezed' at the expense of the second. One could, in principle, have a squeezing device with the quantum vacuum as the 'input'. Squeezing the zero point fluctuations of the vacuum would result in an energy density that is less than the ground state energy of the vacuum. This would result in a negative energy density. If one could scale up this technology, one in principle could generate enough negative energy to power a warp drive engine.

Another technology that generates negative energy is that of the Casimir plates. Two perfectly-conducting parallel plates are brought together such that a very small distance stands between them. At this very small distance, the allowed frequencies of the quantum vacuum fluctuations in between the plates are less than those outside of the plates, which results in a 'focused' quantum vacuum. The effect that arises from this configuration, the Casimir effect, causes the plates to become attracted to one another. In order to keep the quantum vacuum focused between the plates, an opposing force separating the two plates must be applied, to keep the quantum vacuum between the plates focused. However, to generate the amount of negative energy needed for generating a warp bubble, the plate separation would need to be much smaller than the electron Compton wavelength, or much less than 2.46×10^{-12} metres. At this distance, the two plates would cease to be perfectly conducting, so that the Casimir effect would stop.

An accelerating, conducting and perfectly reflecting mirror could be used to generate the dynamical Casimir effect. Increasing the acceleration of a moving mirror would cause a negative energy flux to pass in front of the mirror. In 2011, this effect was observed in the laboratory. However, this effect is known to be exceedingly small. Thus, it is woefully inadequate for generating a warp bubble.

The idea that mirrors could also be used to generate negative energy can be expanded for actual focusing. The Casimir effect can be viewed as the reflection of vacuum fluctuations by a reflecting mirror. Larry Ford and N F Svaiter came up with the idea to use parabolic mirrors to generate negative energy [9, 10, 11]. Every incoming ray of light that is parallel to a parabolic mirror's axis becomes focused at the mirror's focal point. If one could somehow input the quantum vacuum instead of light, parallel to a parabolic mirror, then the vacuum could also be focused by the mirror. The focusing of the quantum vacuum results in a reduction of the quantum vacuum fluctuations, which results in a negative energy density. Ford and Svaiter used a geometric optics approximation to calculate the energy density near the focus of a parabolic mirror. They found, using the geometric optics approximation, that a light ray incident to the mirror can be made into two reflected rays. The interference between these two rays is the dominant contribution to the scalar field (such as a light field) near the focal point of the mirror. This effect is actually larger far from the focus of the mirror. For a perfectly conducting parabolic cylindrical mirror, the negative energy density near the focus of the parabolic mirror can reach negative infinity the closer to zero a light ray incident to the mirror is to the mirror's axis.

The idea of using parabolic mirrors to generate negative energy is actually a scalable one. It is possible, in theory, to use a large array of perfectly conducting, small parabolic cylindrical mirrors to generate a large amount of negative energy. As the negative energy requirements for generating a warp drive are quite large, this scalable technology could be the answer to the problem of generating enough negative energy to power a future warp drive. The array of parabolic cylindrical mirrors could be deployed right outside of the spacecraft. On input of the quantum vacuum, the vacuum fluctuations would be focused by the large amount of parabolic cylindrical mirrors to generate a large amount of negative energy in the configuration needed to create the warp bubble. After the warp phase is over, the mirror array could be de-aligned and covered to cease the production of negative energy and deactivate the warp drive.

Although the parabolic cylindrical mirror array could theoretically generate the large amounts of negative energy needed to power a warp drive, this idea is still theoretical. It has not been tested in the laboratory to verify the results in

Ford and Svaiter's study. However, as this idea is currently the best chance there is to make faster-than-light concepts a reality (including warp drives), there is a justification for including the concept in this chapter. Hopefully, current or future scientists will take this concept and study it in the lab, or extend the theory surrounding this idea. Or, someone could find another scalable idea for generating negative energy.

CONCLUSIONS

As discussed in this chapter, it is known that warp drive engineering is theoretically possible, although technically challenging [12, 13]. Much research has already been done on the theory behind it, but much also needs to be done before the technology can actually be brought into existence. Take the concern about negative energy for example. Negative energy is absolutely essential to bringing about the existence of warp drives. Being able to accumulate a large amount of negative energy is actually a very, very big step that will have to be taken before knowledge about warp drive technology can really mature. Research into detecting negative energy in the laboratory is still in its infancy. It will take time before research in this area can bring about the progress needed to start actually building the technology. Also, research about navigation and guidance will need to be undertaken. It turns out that a difficulty with warp drive theory is that the spacecraft, in its warp bubble, is causally disconnected with the outside of the bubble. That means that light and information cannot pass to or from the spacecraft crew. How does one navigate through the Universe with this fact in mind?

Being that humanity has not even gone back to the Moon, it is very easy to wonder why any effort at all should be put towards futuristic ideas like warp drives. It is very important to research and progress towards the actualisation of warp drives because they would open up the entire Universe to human exploration. Exploring beyond the Solar System is very important to long-term human survival for many reasons. The Sun's lifetime is not infinite. There will come a time, five billion years hence, when it will expand into a red giant, consuming Earth in the process, before dimming down. Before then, the Sun will grow hot enough to render life on Earth impossible within the next 1.5 billion years. This may be a very long time from now, but it is important to start the research now, perfect the technology and be ready for such a cataclysmic event. More urgently, many dangers to the survival of human civilisation exist. Examples include galactic gamma-ray bursts, large and deadly asteroids, or even human error as we grapple towards being a more technologically advanced species. Earth does not exist in its own Universe. Events that happen within our cosmic neighborhood have a direct effect on all life on Earth.

One may also argue for the continuation of warp drive and other breakthrough propulsion studies for purely scientific reasons. The human race has always been curious about its surroundings. Exploration has been of interest to us almost since we have existed. Imagine being able to explore not only our Solar System and the nearest stars, but the entire Universe. The benefits this would bring to human knowledge are absolutely enormous and almost inconceivable. The only way to achieve such a capability is by studying and then realising faster-than-light space travel.

Acknowledgments

The author would like to thank the Initiative for Interstellar Studies, especially Kelvin F Long, for the opportunity to be a part of this project. Also, the author would like to thank Dr Eric Davis for his valuable advice and mentorship. Additional thanks go to Icarus Interstellar for the valuable opportunities that it has afforded the author.

References

[1] NASA; 'Voyager – The Interstellar Mission', [online] available at http://voyager.jpl.nasa.gov/index.html. Last accessed 4 June 2014.
[2] Sternbach, R and M Okuda; *Star Trek: The Next Generation Technical Manual*, Pocket Books, 1991.
[3] Alcubierre, Miguel; 'The Warp Drive: Hyper-Fast Travel Within General Relativity', *Classical and Quantum Gravity*, 11-5, L73–L77, 1994.

[4] Guth, Alan; 'Inflationary Universe: A Possible Solution to the Horizon and Flatness Problems', *Physical Review D*, 23, 2, 1981.

[5] Obousy, Richard and Gerald Cleaver; 'Warp Drive: A New Approach," *Journal of the British Interplanetary Society*, 61, 364–369, 2008.

[6] Visser, Matt, and Carlos Barcelo; 'Energy Conditions and Their Cosmological Implications', Plenary talk given at Cosmo99, Trieste, Sept/Oct 1999.

[7] Davis, E W; 'Faster Than Light Approaches in General Relativity', *Frontiers of Propulsion Science*, Progress in Astronautics and Aeronautics, Volume 227, 471–507, AIAA, 2009.

[8] White, Harold; 'Warp Field Mechanics 101', presented at the 100 Year Starship Symposium, Orlando, 2011.

[9] Ford, L H, and N F Svaiter; 'Focusing Vacuum Fluctuations', *Physical Review A*, 62, 062105, 2000.

[10] Ford, L H, and N F Svaiter; 'Focusing Vacuum Fluctuations II', *Physical Review A*, 66, 062106, 2002.

[11] Ford, L H, and N F Svaiter; 'Fluid Analog Model for Boundary Effects in Field Theory', *Physical Review D*, 80, 065034, 2009.

[12] Long, K F; 'The Status of the Warp Drive', *Journal of the British Interplanetary Society*, 61, 347–352, 2008.

[13] Davis, E W; 'Faster Than Light Space Warps: Status and Next Steps', *Journal of the British Interplanetary Society*, 66, 68–84, 2013.

Chapter 12

Interstellar Travel and Traversable Wormholes

Remo Garattini

Crossing deep space to explore the Universe is a big challenge that has had a beginning with the advent of rocket jet propulsion. In antiquity, the availability of black powder (gunpowder) to propel projectiles was a precursor to experiments with rocket-propelled fire arrows. We had to wait for World War II and Wernher von Braun to assist in the birth of modern rocketry with the production of the V-2 rocket. After World War II rocket technology saw huge improvements; however, such improvements are nought compared to the developments in propulsion required to overcome the enormous distances between the stars. Indeed, it is these huge distances that represent the major obstacle to eventual interstellar travel.

As far as we know, from Albert Einstein's Special Theory of Relativity, the speed of light represents a barrier that cannot be surpassed. Therefore even the most efficient rocket, for example a photon rocket, could not be able to transport a human in a reasonable amount of time. In addition, a massive rocket with living people on board will be subjected to relativistic effects like time dilations that render the voyage very problematic, especially if the crew wanted to come back home. Indeed, elapsed time on Earth would be large compared to the elapsed time on the ship. Therefore, the first reaction to such a challenge would be a total frustration caused by our incapability to move beyond the Earth–Moon distance in a reasonable amount of time. One could invoke some exotic proposals like tachyon particles, from the Greek ταχυς (tachys, 'swift, quick, fast, rapid'), which are hypothetical particles that move faster than light. Unfortunately, or fortunately depending on your point of view, such particles have never been detected. From one side I say unfortunately because tachyons could hold the secret to crossing deep space in a reasonable amount of time with respect to a human lifetime. On the other side, I say fortunately because their existence could create paradoxes like causality violation. However, because of the huge distances involved, interstellar travel cannot be examined using just Special Relativity, but instead should be examined with the help of the General Theory of Relativity. To this end General Relativity offers an amazing solution to the problem: traversable wormholes. However, before talking about traversable wormholes, it is useful to introduce Einstein's field equations. In particular, we shall qualitatively examine one solution, known as the Schwarzschild solution, which is the prototype of a traversable wormhole.

THE SCHWARZSCHILD SOLUTION

Einstein's field equations are deduced by imposing that not only are inertial observers equivalent to each other, which is the domain of Special Relativity, but even that accelerated observers are equivalent. The result is that, in some conditions, space–time acquires the geometrical property of being curved. However, how space–time can be curved is a question that can be answered only by solving Einstein's field equations. This set of equations regulates the gravitational behavior of every particle with energy. It does not matter if a particle is massive or not, it will obey Einstein's field equations: light is the best example. However, in what way is the gravitational behavior of a particle regulated by these equations? It is usual to say that energy curves space–time by means of a gravitational field and the curvature of space-time tells energy how to move. For example, if we try to solve Einstein's field equations without matter fields, we find that the solution represents a flat space–time, known as Minkowski space–time and the geometry looks like a table surface. It is for this reason that Minkowski space is also known as flat space. The key point is encoded into a mathematical object that represents the difference between the amount of space crossed by light in a time interval dt squared, namely c^2dt^2, and the square distance crossed by a particle dl^2. For brevity this mathematical object is indicated by ds^2. For example, in cartesian coordinates, one gets $dl^2 = dx^2 + dy^2 + dz^2$, namely Pythagoras' theorem in three dimensions. Then the ds^2 becomes $ds^2 = c^2dt^2 - dl^2$. It is interesting to note that there exists one time variable and three space variables, globally a description of space–time. For light $ds^2 = 0$, which implies that $c^2dt^2 = dl^2$, as it should be. When $ds^2 > 0$, causality is preserved and the particle crosses an amount of space smaller than the one crossed by the light. These particles are termed bradyons, from the Greek βραδνζ (bradys, meaning 'slow'). On the other hand when $ds^2 < 0$, we are in the presence of tachyons, which are the particles travelling faster than light. The key feature of Minkowski space–time is to have in front of the temporal and spatial part only constant coefficients like the speed of light, given by the letter 'c'. For the set of coordinates (t, x, y, z), the coefficients of the Minkowski space–time are $(c^2, -1, -1, -1)$. To enter the realm of General Relativity we have to substitute the constant coefficients with coefficients that are functions of the coordinates. Of course these functions must satisfy Einstein's field equations to describe General Relativity phenomena. There exists a simple solution, known as the Schwarzschild solution, which has the property of being spherically symmetrical [1]. This means that the set of coordinates change from (t, x, y, z) to (t, r, q, f), where r is the radial distance, being the square root of $(x^2 + y^2 + z^2)$; q is the azimuthal angle and f is the polar angle. The Schwarzschild solution has an intriguing feature: it possesses a region of space–time described by the Schwarzschild radius r_s that is known as an 'event horizon', below which light signals cannot escape because of the strong gravitational forces. In this situation all contact with the external world is lost, justifying the name black hole assigned to the Schwarzschild solution.

Another interesting feature of the Schwarzschild solution is that for large distances from the event horizon, space–time becomes flat. However, this is not the whole story. Indeed, the Schwarzschild solution describes a black hole very well, but this is only a part of the complete solution that also includes a white hole.

A white hole is a region of space–time that cannot be entered from the

A picture of an eternal black hole. The horizontal direction is space and the vertical direction is time. This representation has only two dimensions.

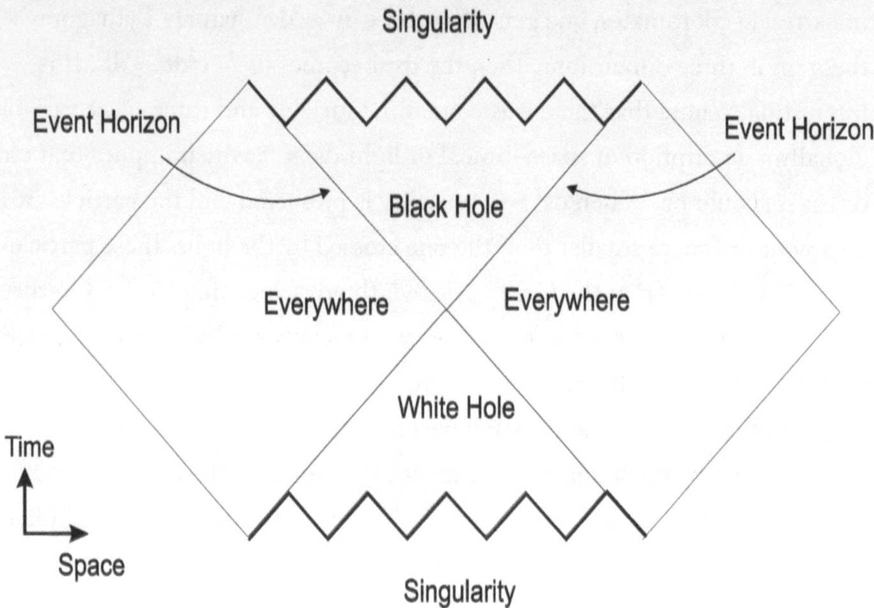

outside, but from which matter and light have the ability to escape. In this sense it is the reverse of a black hole, which can be entered from the outside but from which nothing, including light, has the ability to escape. Matching a white hole with a black hole creates an eternal black hole, namely a black hole that has always existed. Note that the word 'eternal' refers to the point of view of an external observer; a more realistic black hole that forms at some particular time from a collapsing star would require a different metric. In an eternal black hole, if we consider a hyper-surface of constant time and we try to represent the curvature of space at that time with the help of an embedding diagram, we will find that such a diagram looks like a tube connecting the two exterior regions, known as an Einstein–Rosen bridge or Schwarzschild wormhole.

Note that in 1916, Ludwig Flamm [2] recognised that the Schwarzschild solution of Einstein's field equations represents a wormhole. Then in 1935, Albert Einstein and Nathan Rosen [3] showed that implicit in General Relativity formalism is a curved-space structure that can join two

A picture of an eternal black hole. Σ is a hypersurface with a constant time. S_0 represents the wormhole.

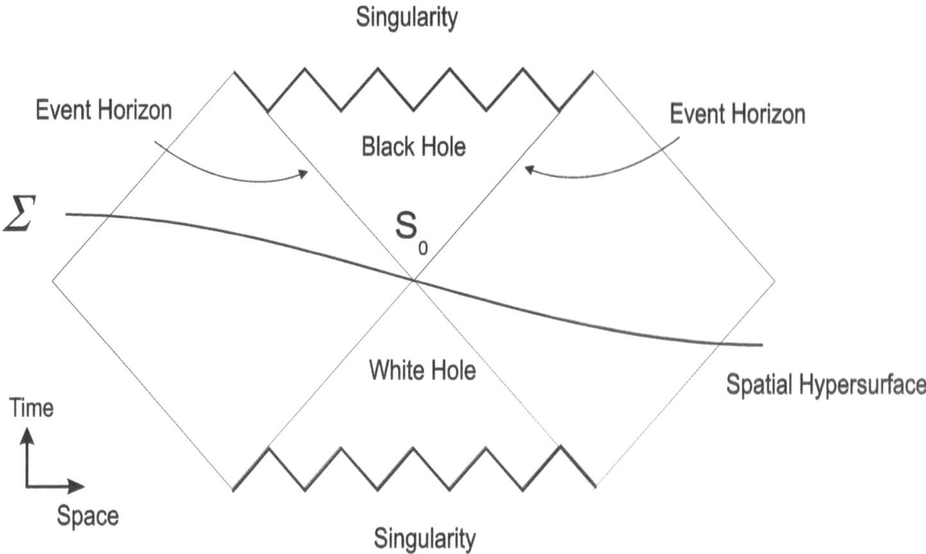

distant regions of space–time through a curved, tunnel-like, spatial shortcut. The purpose of Einstein and Rosen's work was not to promote faster-than-light or inter-universe travel, but to attempt to explain fundamental particles like electrons as space tunnels threaded by electric lines of force. However, from the point of view of whether it is traversable, the Schwarzschild wormhole is unfortunately highly unstable and any tentative attempt to cross it causes it to be disrupted. It is for this reason that in 1988, in the *American Journal of Physics*, Michael Morris and Kip Thorne published a paper entitled 'Wormholes in Space–Times and Their Use For Interstellar Travel: A Tool for Teaching General Relativity' [4, 5], where they sought to understand not only whether General Relativity and traversable wormholes are compatible, but more importantly whether they are stable with respect to matter crossing their boundary.

TRAVERSABLE WORMHOLES AND THE PROBLEM OF NEGATIVE ENERGY

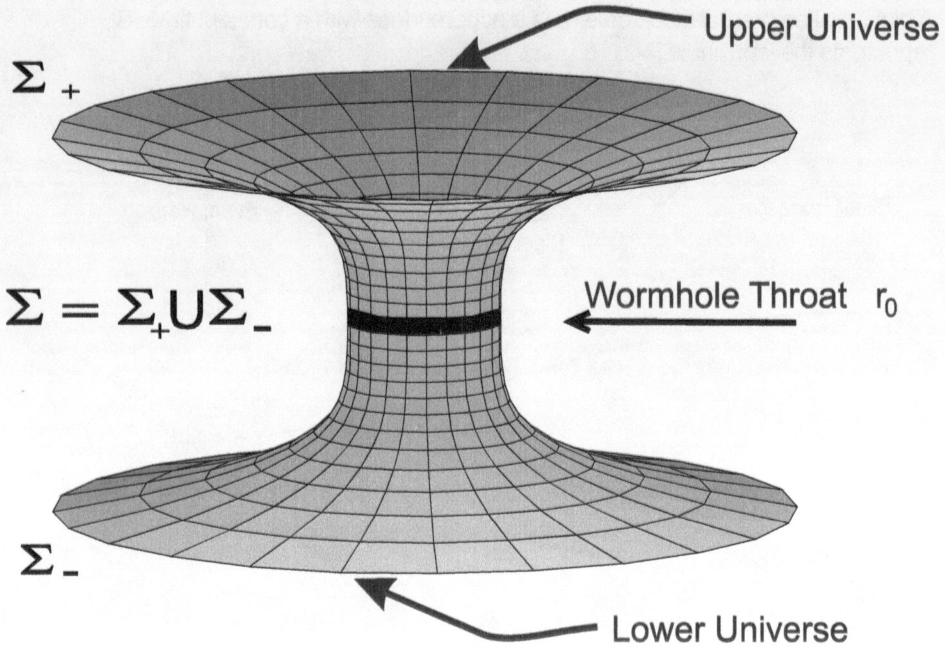

Wormhole represenation for the Schwarzschild metric. The bifurcation surface S_0 divides the hypersurface S+ (upper Universe) and S− (lower Universe).

Depending on where the space-like hyper-surface is chosen, the Einstein–Rosen bridge can either connect two black hole event horizons in each universe (with points in the interior of the bridge being part of the black hole region of space–time), or two white hole event horizons in each universe (with points in the interior of the bridge being part of the white hole region). It is impossible to use the bridge to cross from one universe to the other, however, because it is impossible to enter a white hole event horizon from the outside and anyone entering a black hole horizon from either universe will inevitably hit the black hole singularity.

However, traversable wormholes are born to overcome these problems. The main difference between a Schwarzschild wormhole and a traversable wormhole is that the function responsible for the creation of a black/white hole is defined to avoid the horizon formation and is substituted by a new function termed the 'redshift function', while the spatial part containing information about the radial

Light rays focusing

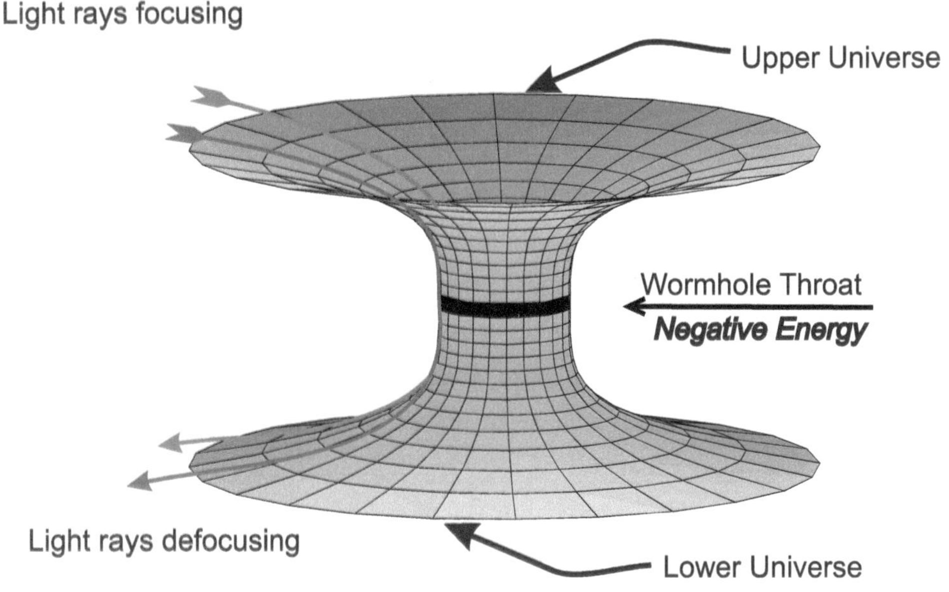

A picture of light rays entering into a wormhole from the upper Universe and exiting from the lower Universe. The negative energy focuses the light rays before they enter the wormhole and de-focuses them while exiting from the lower Universe.

coordinate is substituted by the 'shape function' and is defined in such a way as to generalise the description of the Schwarzschild wormhole. Of course these new functions must satisfy Einstein's field equations.

To be traversable the wormhole must have some features. First of all, the existence of a minimum radius, or in other words the wormhole throat. Then, it ought to allow signals, in the form of light rays, to pass through it. Light rays entering one mouth of a wormhole are converging, but to emerge from the other mouth they must defocus. In other words, they must go from converging to diverging somewhere in between.

Moreover, if wormhole travel is to be at all convenient for the traveller then we require that:

i) the entire trip should take less than or of the order of one year as measured both by the traveller and by people living far from the wormhole;

ii) the acceleration felt by the traveller must not exceed one Earth gravity, g, equal to 9.8 metres per second;

iii) the tidal accelerations between various parts of the traveller's body must not exceed *g*.

When we put these constraints into Einstein's field equations, we find that in the proximity of the wormhole throat the mass density must be negative. Therefore to exist, traversable wormholes must violate the null energy conditions, which means that the matter threading the wormhole's throat has to be 'exotic'. For example, the defocusing of the light rays requires negative energy. Whereas the curvature of space produced by the attractive gravitational field of ordinary matter acts like a converging lens, negative energy acts like a diverging lens. It is interesting to note that even in the warp drive space–time of Alcubierre [6], where a bubble with no gravitational forces inside is allowed to have a superluminal velocity, exotic matter is required. However, the Alcubierre warp drive has two additional and quite severe problems: first, as proved by Barceló et al [7] using a semi-classical approach, the Alcubierre drive at faster-than-light velocities is impossible, mostly because extremely high temperatures caused by Hawking radiation would destroy anything inside the bubble at superluminal velocities and lead to instability of the bubble itself. These problems do not arise if the bubble velocity is kept sub-luminal, but exotic matter is still necessary for the drive to work. Secondly, there are damaging effects at the destination, namely when an Alcubierre-driven ship decelerates from superluminal speed and the particles that its bubble has gathered are released in energetic outbursts; in the case of forward-facing particles, they would be energetic enough to destroy anyone at the destination directly in front of the ship [8].

Regardless, in summary, both traversable wormholes and warp drive need exotic matter [9]. However, we have to draw the reader's attention to the point that negative energy should not be confused with antimatter, which has positive energy. When an electron and its antiparticle, a positron, collide, they annihilate. The end products are gamma rays, which carry positive energy. If anti-particles were composed of negative energy then such an interaction would result in a final energy of zero. Therefore, since ordinary matter satisfies the usual positive energy conditions, it is likely that wormholes must belong to the realm of semi-classical or perhaps a possible quantum theory of the gravitational field. Nevertheless, for the warp drive, if we take seriously the result of [6], then for the moment we have to abandon our pursuit of this line of research. I say momentarily, because the result in [6] is obtained in a particular framework that does not account for the effect induced by the gravitational field itself.

Either way, we are left with traversable wormholes. Even in this case we can approach this problem semi-classically. This is also justified by the fact that a complete theory of quantum gravity has yet to come. The energy produced by the Casimir effect on a fixed background has the correct properties to be a substitute for exotic matter: indeed, it is known that, for different physical systems, Casimir energy is negative.

Before going on, let me recall what the Casimir effect actually is. Predicted by Hendrik B G Casimir [10, 11, 12, 13] it was experimentally confirmed at the Philips Research Labs [14]. The Casimir effect is induced when the presence of electrical conductors distorts the zero point energy of the quantum electrodynamics vacuum. The device producing the Casimir effect involves the following: two parallel conducting surfaces, in a vacuum environment, which attract one another by a very weak force that varies inversely as the fourth power of the distance between them. This kind of energy is a pure quantum effect; no real particles are involved, only virtual ones [15, 16]. This result is based on a calculation involving the difference between the energy density (in general the stress–energy tensor) computed in the presence and in the absence of the plates with the same boundary conditions. It is interesting to note that the energy density computed in the presence of the plates and the same one computed in the absence of the plates is divergent. This is because each contribution coming from the summation over all possible resonance frequencies of the cavities is divergent and devoid of physical meaning, but the difference between them with and without the plates is well defined. However, the appearance of negative density is not a prerogative of Quantum Electro Dynamics.

Additionally, in quantum optics there exists the possibility of finding negative energy. Indeed, some special states of fields in which destructive quantum interference suppresses the vacuum fluctuations can be created. These particular states are known as squeezed vacuum states involving negative energy. For example, one can create these particular energy states with a laser beam. More precisely, they are associated with regions of alternating positive and negative energy. The total energy averaged over all space remains positive; squeezing the vacuum creates negative energy in one place at the price of extra positive energy elsewhere [17, 18, 19, 20, 21, 22].

Intense laser light upon the metal plates induces the material to create pairs of light quanta, photons. These photons alternately enhance and suppress the vacuum fluctuations, leading to regions of positive and negative energy, respectively. To summarise, the concept of negative energy is not purely academic; some of its effects have even been produced in the laboratory. They arise from Heisenberg's uncertainty principle, which requires that the energy density of any electric, magnetic or other field fluctuates randomly. Even when the energy density is zero on average, as in a vacuum, it fluctuates. However, although quantum theory allows the existence of negative energy, it also appears to place strong restrictions on its magnitude and duration. These restrictions are known under the term of quantum inequalities [23, 24, 25] and these inequalities have some similarities with the uncertainty principle. They say that a beam of negative energy cannot be arbitrarily intense for an arbitrarily long time. The permissible magnitude of the negative energy is inversely related to its temporal or spatial extent. An intense pulse of negative energy can last for a short time; a weak pulse can last longer. Furthermore, an initial negative energy pulse must be followed by a larger pulse of positive energy. The larger the magnitude of the negative energy, the nearer its positive energy counterpart must be. These restrictions are independent of the details of how the negative energy is produced. One can think of negative energy as an energy deficit that must be compensated for, just as a debt is negative money that has to be repaid. However, it appears that it is not sufficient to compensate the negative energy density to pay off the debt; one has to overcompensate with an interest in positive energy density.

CONCLUSIONS AND OUTLOOK

If the traditional approach to interstellar travel is plagued by a series of tremendous limitations, such as the barrier of the speed of light, one is tempted to invoke the use of other methods that differ from ordinary rocket propulsion. In this chapter we have proposed to use traversable wormholes that are hypothetical astrophysical objects able to connect two distant regions of the same Universe or two different universes. However, this perspective is also plagued by one fundamental point: the detection of a wormhole and whether it is traversable. Notwithstanding these practical and theoretical problems, studying taversable wormholes is feasible. When John Wheeler in 1957 [26, 27, 28, 29] conceived the term 'space–time foam', he conjectured that space–time is composed of tunnels connecting bubbles that are in fast transformation: these tunnels are nothing but wormholes.

However, Wheeler wormholes are of the Planck size, which in numbers means they are $1.61619926 \times 10^{-35}$ metres. Therefore it is quite impossible to use such tiny objects as real tunnels. However, Wheeler wormholes are conceptually important because they can be considered as building blocks for a quantum theory of the gravitational field. Therefore if, on theoretical grounds, we need to consider wormholes as if they were fundamental objects like particles, then on the other hand we would like to be able to enlarge one of them from the quantum foam. As discussed in the previous sections, to make a wormhole traversable we need to use exotic matter or negative energy and one method is the Casimir effect. However, when one tries to solve Einstein's field equations with some matter fields acting as a gravitational source, one discovers that the ability to traverse through a wormhole is present in principle but not in practice, given that the wormhole remains at the Planck scale. In a series of papers I proposed, in collaboration with Francisco Lobo, to use the gravitational field itself as a source of negative energy in such a way as to realise a gravitational Casimir effect. We have christened such wormholes 'self-sustaining traversable wormholes' because they are sustained by their own gravitational quantum fluctuations. The good news is that the sustainability is possible. The bad news is that we have found in various examples that it appears possible to traverse it in principle but not in practice. Therefore, even if we have found a method that mimics exotic matter, we are very far from being able to use such traversable wormholes for human travel. However, once it is established that traversable wormholes can be stabilised by their own quantum fluctuations and therefore one can claim that they exist, the next step is to find a mechanism able to enlarge some of them to human-sized or a ship-sized, which represents the real challenge for interstellar travel.

REFERENCES

[1] Schwarzschild, K; *Proceedings of the Prussian Academy of Sciences*, 424 (1916).

[2] Flamm, L; *Physikalische Zeitschrift*, 17, 448 (1916)

[3] Einstein, A and N Rosen; *Physical Review*, 48, 73 (1935).

[4] Morris, M S and K S Thorne; 'Wormholes in Space–times and Their Use For Interstellar Travel: A Tool for Teaching General Relativity,' *American Journal of Physics*, 56, 395 (1998).

[5] Morris, M S, K S Thorne and U Yurtsever; 'Wormholes, Time Machines and the Weak Energy Condition,' *Physical Review Letters*, 61, 1446 (1988).

[6] Alcubierre, M; 'The Warp Drive: Hyper-fast Travel Within General Relativity,' *Classical and Quantum Gravity*, 11, 5: L73–L77 (1994).

[7] Barceló, C, S Finazzi and S Liberati; 'Semi-classicial Instability of Dynamical Warp Drives,' *Physical Review D*, 79 124017 (2009).

[8] McMonigal, B, G F Lewis and P O'Byrne; 'The Alcubierre Warp Drive: On the Matter of Matter', arXiv: 1202.5708 (2012).

[9] Lobo, F S N; 'Exotic solutions in General Relativity: Traversable Wormholes and Warp Drive Space–Times,' arXiv:0710.4474 (2007).

[10] Casimir, H B G; *Proceedings of Klijke Nederlandse Akademie van Wetenschappen*, 51, 793 (1948).

[10] Lamoreaux, S K; 'Demonstration of the Casimir Force in the 0.6 to 6µm Range', *Physical Review Letters*, 78, 5 (1997).

[11] Mohideen, U and A Roy; 'Precisions Measurement of the Casimir Force From 0.1 to 0.9µm,' *Physical Review Letters*, 81, 4549 (1998).

[12] Roy, A and U Mohideen; 'Demonstration of the Nontrivial Boundary Dependence of the Casimir Force,' *Physical Review Letters*, 82, 4380 (1999).

[13] Roy, A, C Y Lin and U Mohideen; 'Improved Precisions Measurement of the Casimir Force,' *Physical Review D*, 60, 111101 (1999).

[14] DeWitt, B S; 'Quantum Field Theory in Curved Space–Time,' *Physical Reports*, 19, 295 (1975).

[15] Visser, M; *Lorentzian Wormholes: From Einstein to Hawking*, AIP Press, New York (1995).

[16] Slusher, R E and B Yurke; 'Squeezed Light,' *Scientific American*, 258, 50–56, May 1988.

[17] Ford, L H; 'Quantum Coherence Effects and the Second Law of Thermodynamics,' *Proceedings of the Royal Society A*, 364, 227 (1978)

[18] Ford, L H; 'Constraints on Negative Energy Fluxes,' *Physical Review D*, 43, 3972 (1991).

[19] Ford, L H and T A Roman; 'Averaged Energy Conditions and Quantum Inequalities,' *Physical Review D*, 51, 4277 (1995), [arXiv:gr-qc/9410043].

[20] Ford, L H and T A Roman; 'Quantum Field Theory Constrains Traversable Wormhole Geomtetries,' *Physical Review D*, 53, 5496 (1996) [arXiv:gr-qc/9510071].

[21] Ford, L H and T A Roman; 'The Quantum Interest Conjecture,' *Physical Review D*, 60, 104018 (1999) [arXiv:gr-qc/9901074].

[22] Pfenning, M J and L H Ford; 'Quantum Inequalities on the Energy Density in Static Robertson–Walker Space-Times,' *Physical Review D*, 55, 4813 (1997) [arXiv:gr-qc/9608005].

[23] Wheeler, J A; 'On the Nature of Quantum Geometrodynamics,' *Annals of Physics*, 2, 604 (1957).

[24] Wheeler, J A; 'Geons,' *Physical Review*, 97, 511–536 (1955).

[25] Wheeler, J A; *Geometrodynamics*, Academic Press, New York (1962).

[26] Garattini, R; 'Self-Sustained Traversable Wormholes,' *Classical and Quantum Gravity*, 22, 1105 (2005) [arXiv:gr-qc/0501105].

[27] Garattini, R; 'Self-Sustained Traversable Wormholes and the Equatio of

State,' *Classical and Quantum Gravity*, 24, 1189 (2007) [arXiv:gr-qc/0701019].

[28] Garattini, R and F S N Lobo; 'Self-Sustained Phantom Wormholes in Semi-Classical Gravity,' *Classical and Quantum Gravity*, 24, 2401 [arXiv:gr-qc/0701020].

[29] Garattini, R and F S N Lobo; 'Self-Sustained Traversable Wormholes in Noncommutative Geometry,' *Physical Letters B*, 671, 146 (2009) [arXiv:0811.0919 [gr-qc]].

Chapter 13

INTERSTELLAR COMMUNICATION:
RADIO FREQUENCY AND OPTICAL LINK ESTABLISHMENT

DIVYA SHANKAR

The prime purpose of sending a probe to another star is to collect data about that star, any planets around it and the interstellar medium, and to send this data back to Earth for interpretation and analysis. This presents some difficult technical challenges with several potential technology choices for implementing the communication system. As the distance between the transmitting probe and the receiving segment on Earth increases, a more efficient link needs to be established to maintain the quality of the communication. The development of a communications system is a significant factor relevant to the design and implementation of a starship.

Over the past five decades many spacecraft have been sent into space to explore the planets and return the data collected back to receiving stations on Earth. Although these spacecraft are interplanetary spacecraft, they can serve as an example for long distance communication system design. The study of interplanetary communication provides a baseline for us to consider interstellar communication.

Voyager 1 and 2

The Voyager probes are the most distant human-built objects ever sent into space and Voyager 1 in particular is also the fastest. Built to explore the outer Solar System, Voyager 1 is now travelling at a speed of 17.1 kilometres per second (3.6 AU per year) while Voyager 2 is no slouch either, moving at 15.4 kilometres per second (3.2 AU per year). After exploring the giant planets Jupiter, Saturn, Uranus and Neptune, their mission was extended as they moved into the outer boundaries of the Solar System to study the dust and particle regime far from the Sun and, eventually, enter the Kuiper Belt between around 30 and 50AU from the Sun. Voyager 1 has now crossed the heliopause and is now making its first tentative steps into the interstellar medium.

The Voyager communication system is a radio frequency system in the S-band (2 to 4 GHz; wavelength 1.5m to 7.5cm) and X-band (7 to 11.2 GHz; wavelength 3.75cm to 2.5cm) frequency range. Each Voyager probe has both high-gain (providing a narrow and highly targeted beam appropriate for communicating across vast distance, directing more of the transmitted power towards the receiving station) and low-gain (a wide beam that can be more easily detected, but is less efficient) antennas for the two bands. A very high gain 3.7-metre diameter parabolic dish is used for both 13cm (S-band) operation and 3.6cm (X-band) operation. The data rate is variable according to the distance. During launch, the low-gain antenna is used in S-Band and the data is convolutionally coded – a form of error-correction – and communicated at 1,200 bits per second (bps).

During the first few months after launch, the data rate was increased to 2,600 bps using the S-band high-gain antenna only. During the planetary cruise, the X-band low gain antenna is convolutionally coded using error correction techniques such as Golay and Reed–Solomon coding (a systematic means of creating code that can detect and fix random errors).

When the spacecraft were returning the data collected during its planetary fly-bys, that data was transmitted using the low-gain S-band and high-gain X-band antennas at 7.2 kilobits per second and 115.2 kilobits per second respectively. Now far away in the interstellar phase of their missions, the Voyagers send back the data using the high-gain X-band antenna with a 160 kilobits per second data rate [1].

The Voyager communication system also follows the store and forward technique of communication architecture. It has a recorder to store the data when the spacecraft is unable to obtain a direct link with Earth, for example when Earth is on the other side of the Sun. This recorder can store up to 62,500 kilobytes of data – remember this is 1970s technology and hence the low storage capacity. When the spacecraft comes into direct contact with Earth, it establishes a link and sends the stored data. It takes about 17 hours for the data to reach the Earth from the Voyager 1's current location. As Voyager 1 has now entered the unexplored space of the interstellar medium, it serves as our best example for designing a communication system for a true interstellar spacecraft.

PIONEER 10 AND 11

Pioneer 10 was launched on 2 March 1972 and was the first spacecraft to enter the Asteroid Belt and encounter Jupiter. In addition, it was the first spacecraft to achieve Solar System escape velocity. Communication was lost with the probe on 23 January 2003, at a distance of 80AU from Earth, as a result of power constraints.

As a successor to Pioneer 10, Pioneer 11 was launched on 6 April 1973 to study the Asteroid Belt, the environment around Jupiter and Saturn, the solar wind, cosmic rays and eventually the far reaches of the Solar System and the heliosphere. Pioneer 11 fell silent in 1995 as its power drained.

Their communication system included a 2.74-metre diameter parabolic dish for the high-gain antenna. Each transceiver is 8 W and received data across the S-band at 2.11 GHz for the uplink from Earth, and 2.29 GHz for the downlink to Earth with the Deep Space Network – formed from a trio of receiving stations in California, Spain and Australia – tracking the signal. The data transmission rate at launch was 256 bits per second, with the rate degrading by about 1.27 millibytes per second for each day of the mission. A data storage unit was included on the probes to record up to 6,144 bytes of information gathered by Pioneer's instruments. The digital telemetry unit was used to prepare the collected data in one of thirteen possible formats before transmitting it back to Earth [2].

NEW HORIZONS

This spacecraft, the first mission to Pluto, was launched by NASA in the 2006 to reach the dwarf planet in July 2015, travelling at a velocity relative to the Sun of 16.26 kilometres per second. New Horizons' communications system is operating at X-band frequency and includes two broad-beam, low-gain antennas on opposite sides of the spacecraft for near-Earth communications, as well as a 30cm diameter medium-gain dish antenna and a large 2.1-metre diameter high-gain dish antenna for communicating at larger distances. During normal operations, the spacecraft communicates with Earth through its 2.1-meter wide high-gain antenna. A low energy beacon signal indicating the status of the spacecraft's numerous subsystems is used when the spacecraft is unable to send the payload data [3]. Data rates depend on the power used to send the data, the size of the antenna on the ground, and spacecraft distance, with 38 kilobytes per second when New Horizons flew past Jupiter in 2007 and just one kilobyte per second expected at Pluto.

THE GROUND SEGMENT

The Deep Space Network (DSN) forms the terrestrial end point on Earth for communication with long-range spacecraft. The signals from the spacecraft will be very weak because of the huge distances involved. Large antennas with high gain and very good sensitivity of the receivers are required in order to establish a communication link with the spacecraft. To achieve this, the DSN is the largest and most sensitive spacecraft communications network on Earth. It consists of three deep space communications complexes located approximately 120 degrees of longitude apart around the world: at Goldstone, California; near Madrid, Spain; and near Canberra, Australia. This equidistant placement means that there is always at one receiving station pointed at any given spacecraft. Each station consists of at least four receiving terminals including one 34-metre (112 ft) diameter high efficiency antenna, one or more 34-metre (112 ft) beam waveguide antennas, one 26-metre (85 ft) antenna and one 70-metre (230 ft) antenna.

ANALYSIS OF INTERSTELLAR COMMUNICATION LINKS

The first step in designing a communication network is to perform a link budget analysis. The size of the antennas to be used, the transmitter output power, receiver sensitivity, power amplifier requirements, bit error rate and the link margin are determined by the link budget calculation. A link budget is a simple addition and subtraction of gains and losses within a radio frequency link. This calculation will give us end-to-end system performance.

In a wireless communication system, the link margin, measured in decibels (dB) is the difference between the receiver's sensitivity (i.e., the received power at which the receiver will stop working) and the actual received power. The link margin should be greater than 0dB for the link to work. A target value should be approximately 10dB for a low cost system, 6dB for a professional system and 3dB for a deep space system.

Radio Link between Main Vehicle and the Ground Segment

All the interplanetary probes communicate at radio frequencies. A microwave beam need not be pointed as accurately as a more tightly focused optical beam. If we consider the target star as alpha Centauri, which is 4.3 light years away from us, the link budget is calculated for a link margin of 3dB to establish a communication link with the spacecraft in deep space. This is done on the basis of achieving a maximum data rate with a minimal antenna size, considering space and power constraints onboard the spacecraft.

The link analysis shows that using a microwave system would require transmitting and receiving antennas that are too large to be practical. The calculation for the transmitter antennas from the spacecraft at alpha Centauri with 20kbps data rate comes up with a 94.36-metre diameter transmitter antenna with a gain of 90dBi (dBi is decibels-isotropic, the unit of power gain) in the Ka-band frequency of 32GHz. The power required for this transmission is two megawatts, which is almost 20 times the power consumption of the International Space Station. The path loss for the Ka-band at 4.3 light years distance is 394.8dB and other miscellaneous losses are considered to be 4dB. Thus, the dominating factor in the communication link is the path loss. The calculations show that the ground segment on Earth is required to have a higher sensitivity, which leads to an antenna diameter of two kilometres. The values used for the calculation can be found in the Table One in the appendix of this chapter.

Radio Link between the Spacecraft and Sub-Probes

Table One in the appendix shows the impracticality of radio frequency communication between a spacecraft and the ground segment at a distance of 4.3 light years. Let us check for the suitability of radio frequency communication for the main probe and any sub-probes that it may launch to explore the target system, considering sub-probes to be at a distance of 1,000AU from the mothership. With the same frequency of 32GHz and the data rate reduced to 10kbps, the transmitting and receiving antenna diameter should be 28 metres with a gain of 80dBi and one kilowatt of transmitting power to get a link margin of 3.56dB for the communication.

OPTICAL LINK

'Free-space' optical communication – i.e. the transmission of optical (laser) light through space – is the technology that can be used when radio frequency communication is impractical. Because communication beams spread as the square of the distance between transmitter and receiver, the advantages of using an optical communication system include much lower diffraction-limited beam divergence, much smaller and lighter transmitting and receiving antennas and the ability to support high data rates. Optical communication involves the real challenge of estimating the sizes, weights, powers and reliabilities of many of the required optical communication components and subsystem technologies along with modulation and coding efficiencies. Table Three in the appendix shows the parameters considered for optical communication. The link budget is calculated considering Nd:YAG (neodymium-doped yttrium aluminium garnet) laser operating at a wavelength of 0.532 microns [4]. The transmitter antenna gain is 144.9dB for a transmitting aperture of three-metres diameter with an obscuration diameter of 0.2-metres. The transmitter laser power is 20W and the receiver aperture diameter required is two metres with a 155.2dBi gain. The link margin obtained is 3.9dB which is a good margin for deep space communication.

OPTICAL LINK WITH THE SUB-PROBES

Optical link analysis is carried out to check the communication between the mothership and the sub-probes. The link requires 11.4 signal photons per pulse to achieve a symbol error rate of 10^{-3}. This error rate is for the information that is not coded. The error rate can be reduced arbitrarily by adding error correction coding, albeit at the cost of useful bandwidth [4]. As per the analysis given in the Table Three in appendix, 28.3 signal photon events are received per second (i.e. 28.3 signal photon events per 10 nanosecond pulse), then link margin of 3.9 dB [5] is achieved.

The larger uncertainties in parameter values can be expected for optical system as the technology is still not being used by any deep space probe and the practical implementation varies from the theoretical value.

THE OPTICAL COMMUNICATIONS SYSTEM

The basic communication architecture involves a source that will send the information, a transmission channel that will carry the information and the destination that will receive the information. An optical communication system contains a transmitter that will convert the message into an optical signal, a channel that carries the optical signal and a photo detector to act as a receiver. Free-space optical communication involves laser and LEDs as the main optical sources. For short distances, LEDs can be used. For large distance communication lasers, especially in infrared that can penetrate interstellar dust, will be very useful. The biggest advantage of optical communication is that an extremely narrow beam can be used. In addition to the transmitter, channel and the receiver, a free-space optical communication system includes a telescope with lenses and mirrors that shape the transmitted laser beam and focus the received signal onto the photo detector. The optical system consists of a front aperture that might be a reflecting or a refracting telescope, optics that include lenses, mirrors, beam-splitters and filters, fast-steering mirrors and acquisition and tracking detectors.

A communication system that consumes less power, has better output power and an efficient coding technique, and a ground segment with very high sensitivity are essential to establishing a good communication link. A tracking technique is essential to maintain proper placement of the transmitters and receivers at the source and destination to establish an error-free communication. The tracking system at the transmitter ensures that a narrow beam is pointed at the receiver with minimal residual jitter. Meanwhile at the receiver end, the tracking system ensures a tight focus on a relatively small detector. The spacecraft optical system transmits an optical signal and receives the beacon signal from the ground. The transmission and reception can be in a single channel or two separate channels, depending on the architecture of the spacecraft's communication system. The beacon signal is used for acquisition and tracking and also to uplink command data from Earth or from other spacecraft [6].

SUB-SYSTEM TECHNOLOGIES

The basic sub-system technologies of free-space optical communication includes a laser transmitter, spacecraft telescopes, acquisition, tracking and pointing, detectors, filters and optical coding.

The main element in the transmitter is the optical source. There are many types of lasers which are being used. The aforementioned Nd:YAG laser is the best optical source for transmission known so far; it is a crystal that is used as a lasing medium for solid-state lasers. The power efficiency of a Nd:YAG laser is 0.5 percent. However, power generation onboard the spacecraft is very difficult and the conversion efficiencies are too low for the use of a Nd:YAG laser on deep space missions. A series of experiments conducted at NASA's Jet Propulsion Laboratory showed that combining many semiconductor laser diode elements together in a phase-locked array increases the transmitter output power and also the steering of the beam from a laser diode array. Hence semiconductor lasers are much more power efficient than conventional solid state Nd:YAG lasers [7].

SPACECRAFT TELESCOPES

Telescopes onboard the spacecraft act as an 'antenna' for optical signals. They align the laser radiation to a collimated beam and directs it to receiver. Optical telescopes can come in three types, namely refractors, reflectors and catadioptric telescopes. A refracting telescope uses lenses, a reflecting telescope uses mirrors and a catadioptric telescope uses both lenses and mirrors. The structure needs to be thermally stable, in order to avoid image distortions. Furthermore, the structure needs to be lightweight, in order to save mass.

ACQUISITION, TRACKING AND POINTING

The optical beam from the spacecraft must be precisely pointed and it should be diverged towards the receiving target since the attitude of the spacecraft keeps changing. There will be relative movement between the spacecraft and receiving segment on Earth. The transmitted beam from a spacecraft must be pointed ahead of the apparent location of the receiver to compensate for cross velocities between the host spacecraft and the reception location. To accomplish this, the beacon signal is tracked and received by the receiving segment on Earth in order to calibrate the attitude orientation of the transmitting aperture of the spacecraft. The beacon signal is also used to measure the vibration components of the host spacecraft. Correction of both the telescope line-of-sight error, as well as compensation for vibration disturbances, is then accomplished using one or more fine-steering mirrors in the optical path of the telescope. This compensation scheme can also be used to implement the needed point-ahead angle calculated from mission trajectory and planetary orbital predictions. Star-trackers can also be used as acquisition and tracking beacon [6].

DETECTORS

Photodetectors, also known as photodiodes, serve as the optical receiver in free-space optical communication. The photodetector converts the incoming light into current. An avalanche photodiode (APD) is a highly sensitive semiconductor device that makes use of the photoelectric effect to convert light energy into electricity. APDs provide a built-in first stage of gain through avalanche multiplication. As a consequence of this multiplication effect the output will have high variance. To use it to detect single photons, APDs should be used in 'Geiger Mode', which is when the APDs are biased beyond the avalanche breakdown voltage so that the gains are high enough to detect a single photon that has been received. Since the breakdown voltage is high, the gain will be very high and the current will be generated in milli-amps and hence it can be counted. Therefore APDs have high sensitivity [6].

FILTERS

Filters are the components used to reject or pass the desired wavelength of signals. At the receiving end, narrowband filters are used before the detectors in the optical communication system. Throughput of the filters must be high to avoid significant loss to the desired signal. Interference filters, which are multi-layer dielectric filters, are usually used [6].

OPTICAL CODING

Modulation technique best suited to optical communication is pulse-position modulation. It uses a time interval that is divided into a number of possible pulse locations, but only a single pulse is placed in one of the possible positions. The position of that pulse is determined by the information that is to be transmitted. An optical coding experiment conducted at JPL, using Reed-Solomon codes in concatenation with convolution codes, showed that Reed-Solomon codes were very efficient in correcting the bursts of errors that would result when the Iturbi decoder that was decoding the convolution code made an erroneous branch path decision. Since the loss of a single pulse-position modulation optical pulse detection caused a similar burst of errors (loss of a pulse meant that the data bits associated with that pulse were chosen at random), it appeared that such codes were well matched to the pulse-position modulation channel [6]. The higher the order of the pulse-position modulation, the better the performance, provided the modulation is used with a matching-alphabet Reed-Solomon code. The pulse-position modulation order has linear influence on the code. If the order was increased, then the code becomes complex and if the pulse-position modulation order was decreased, performance was reduced. The codes called accumulator codes are based on product-coding techniques where simpler codes are combined and then jointly decoded [8].

REFERENCES

[1] Roger Ludwig and Jim Taylor; 'Voyager Telecommunications' JPL DESCANSO Design and Performance Summary Series, March 2002.

[2] NASA; 'The Pioneer Missions' (online) www.nasa.gov/centers/ames/missions/

archive/pioneer.html (last accessed 28 October 2014)

[3] NASA New Horizons Launch Press Kit (January 2006)

[4] J R Lesh, C J Ruggier and R J Cesarone; 'Space Communication Technologies for Interstellar Mission.' *Journal of The British Interplanetary Society*, 49, pp7–14 (1996).

[5] Pat Galea; 'Optical Lasers for Interstellar Communication' Phase 3 Study of M9 Communication and Telemetry, Project ICARUS (2011).

[6] Hamid Hemmati; 'Deep Space Optical Communications,' DESCANSO Book Series, External Release No.: CL 01-1804, Volume 7, Chapter 1 (June 2006).

[7] Hamid Hemmati; 'Optical System for Free-Space Laser Communications,' JPL, Dspace, Bitstream, 2014/38461/1/03-1910 (invited paper).

[8] Mehdi Rouissat, Riad Borsali and Mohammed Chick-Bled; 'Dual Amplitude-Width PPM for Free-Space Optical Systems." *International Journal of Information Technology and Computer Science,* Vol 4, 3 (April 2012).

Appendix

Table One: Communication link parameters between a spacecraft 4.3 light years away and Earth

Frequency:	Ka-band (32GHz)
Power transmitted:	2MW
Transmitting antenna diameter:	94.37m
Gain of transmitting antenna:	90 dBi
Receiving antenna diameter:	2km
Gain of receiving antenna:	100dBi
Distance between transmitter and receiver:	4.3 light years
Space (path) loss:	394.8dB
Miscellaneous loss:	4dB
Data rate:	20kbps
Bit error rate:	10^{-3}
Link margin:	3.56dB

TABLE TWO: COMMUNICATION LINK PARAMETERS FOR 1,000AU DISTANCE

Frequency:	Ka-band (32 GHz)
Power transmitted:	1KW
Transmitting antenna diameter:	28m
Gain of transmitting antenna:	80dBi
Receiving antenna diameter:	28m
Gain of receiving antenna:	80dBi
Distance between transmitter and receiver:	1,000AU
Space (path) loss:	346dB
Miscellaneous loss:	2dB
Data rate:	10kbps
Bit error rate:	10^{-3}
Link margin:	3.58dB

TABLE THREE: OPTICAL LINK BETWEEN MOTHERSHIP AND THE SUB-PROBES

Optical Link Parameters:	
Wavelength:	0.532 microns
Modulation method:	Pulse-position modulation
Slot width:	10ns
Transmitter laser power:	20W
Transmitter aperture diameter:	3m
Transmitter obscuration diameter:	0.2m
Transmitter antenna gain:	144.9dB
Receiver aperture diameter:	10m
Receiver obscuration diameter:	2m
Receiver antenna gain:	155.2dB
Distance:	4.3 light years
Space (path) loss:	478.9dB
Link margin:	3.9dB

CHAPTER 14

STARSHIP COMMUNICATION
VIA QUANTUM ENTANGLEMENT

DAVID E FIELDS

A communications system that links a starship with Earth requires physical concepts and technologies that deal with criteria far more stringent than those normally considered in local (planetary-scale) applications. Such systems usually resolve into specifications of comparatively short-term reliability, modest signal averaging, interference rejection and current technology cryptography. Extending these challenges to starship communications adds requirements of long-term reliability (over decades or centuries, including redundant systems, voting sub-assemblies, self-diagnosis and repair); algorithmic (evolving software) driven receivers that optimise signal-averaging, frequency selection and demodulation; mission-critical information extraction and transfer; SETI signal extraction and decoding; and permitting coordinated optimisation of communication. The station-to-station spatial separation may approach light years and the implied communications delay has significant technical and cultural implications.

This chapter considers the possibility of using quantum-entangled (QE) particles (photons) to provide instantaneous full duplex communication between Earth and starship. The possibility of QE communication is a game-changer, having enormous mission-critical significance and we will explore the technical, cultural and survivability implications of this technology.

INTERSTELLAR DISTANCES AND TIME SPANS

The vast distances that interstellar missions will have to journey has been discussed in other chapters of this book, which are supported by a large body of well-developed thought [1]. Communication between a starship and Earth will be critical to the success or failure of the mission. Although unmanned probe missions may be undertaken at any time, human missions require communication schemes that, through use of high-power transmitters, high-gain antennas (perhaps using the Sun as a gravitational lens [2]) and sensitive algorithmic receivers [3], show promise of bridging the spatial gap via creative upgrades of today's light speed communications [4].

These distances impose a time delay on communications, which is daunting because of the lack of coordination that will result from even light speed communications. The aforementioned spatial scale represented by light-months [5] and light-years, places a starship mission at the cutting edge of communications, command, control and culture. Coordinated decision-making in anything approaching real time will become impossible. Separations of months or years will preclude cultural congruence between populations on generation ships and prevent those populations from assisting in decision making or evaluating new discoveries.

A striking need exists for faster-than-light (FTL) starship communications. 'Need' does not ensure a solution – communications technologies will be constrained by the physics of the Universe. The well-accepted macroscopic physical theories of Special and General Relativity limit achieving material matter speeds that exceed the speed of light. It is usually assumed that the maximum speed of information transfer is also the speed of light, since photon speeds appear the greatest possible. Yet one might inquire whether information could be transmitted through the manipulation of entangled quantum states.

The reader should consider that over a century ago it was thought that the aether was necessary for the transmission of photons. No evidence for the existence of an aether has ever been found, but another question has arisen: is it possible to transmit information faster than the speed of light?

Theoretical Basis for Quantum Entanglement

Relativistic physics is widely accepted as the paradigm that describes our space-time. Newtonian physics has been recognised to be a limiting case of relativistic physics where both accelerations (gravitational or otherwise) and relative velocities are small. Quantum physics is recognised as a limiting case for describing the physical Universe when discussing temporal and spatial scales small enough to mandate the incorporation of the wave nature of matter. Either a wave function or state variable approach may be used, but either way, a probabilistic interpretation arises. In the quantum-limited perspective, the mathematical rigour of quantum mechanics is used to explain the behavior of matter, space and time. Although quantum mechanics techniques are very useful, there exists different schools of thought that interpret the underlying physics differently. The most accepted school of quantum mechanics (the Copenhagen interpretation) lists among its predictions that two (or more) particles (photons) may be generated such that they are 'entangled'. If the particles are entangled, then their quantum states – their spins, polarisations or energies – share a reciprocal nature. If spins, then one particle might be spin up and the other spin down and if energies, then the two energies must sum to a particular value. This property of reciprocity is known as quantum entanglement (QE).

The Copenhagen school of quantum mechanics predicts that if an experimenter resolves the quantum uncertainty (wave function) and measures the state variable of one entangled particle (a), then the state variable of the other entangled particle (b) is known [6]. Admittedly there do remain questions of interpretation, including how quickly the second quantum state is known and how widely separated the particles might be. If the particles with entangled properties are widely separated then, by this school of thought, information would seem to be transferred faster than the speed of light.

Some physicists are uncomfortable with this conclusion from the Copenhagen school. One was Albert Einstein, whose objections became known as the Einstein–Podolsky–Rosen paradox [7]. Einstein and his co-authors resolved the paradox by assuming that entanglement was caused by local 'hidden

Below: a thought experiment for testing quantum entanglement consists of experimenter A attempting to send data to Experimenter B by alternately ignoring, then resolving, quantum states of the photon beam from a central source of entangled photons. Experimenter B analyses the beam in an attempt to extract information encoded in entangled states (or not). Both experimenters are aware of constraints on bit rate and statistical validity.

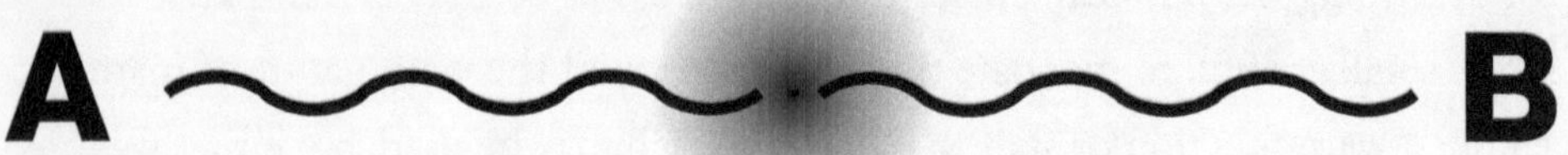

Experimenter A intermittently sometimes reads his photon beam and resolves the quantum uncertainty. Thus he encodes digital information.	The entangled photon source, midway between A and B, generates two beams of entangled photons and sends the beams in opposite directions.	Experimenter B monitors the photon beam and analyses their quantum states in an attempt to determine if experimenter A has encoded a message.

variables'. They concluded that there is no net information transfer, but that the experimenter who resolved state (a) simply knows the state of (b).

The supporters of quantum entanglement choose the interpretation of Bell's Theorem [8], which asserts that the Einstein–Podolsky–Rosen interpretation is invalid and that information may be transferred over great distances and at great speed. Bell's Theorem is significant not just for its implications for starship communications. It considers the fundamental basis for understanding causality in modern physics.

If Bell's Theorem holds and quantum entanglement may be used for FTL information transmission, then there are yet other challenges to exploiting quantum entanglement for starship communications. First, quantum mechanics is a probabilistic discipline. The statistics might be daunting; the detectors might be insensitive. Furthermore, it is difficult to resolve, amplify and recreate an entangled state without destroying it. Nevertheless, most of these challenges are technical difficulties and they might be overcome. The diagram below shows a simple experiment that could test predictions about quantum entangled states.

Transmission of information by quantum entanglement is now accepted as demonstrated by many in the physics community, although most experiments are not as 'clean' as suggested in the accompanying diagram. Transmission of information by quantum entanglement at a velocity exceeding the speed of light remains controversial. Nevertheless, the quantum entanglement theory appears valid. Quantum physicists generally agree that if a quantum entanglement experiment is truly described by quantum mechanics, then the only way to interpret the experiment is probabilistically and that information can be sent via quantum entanglement. However, if the appearance of entanglement is a result of the existence of some hidden classical variables, then although information might be sent via quantum entanglement, it would have none of the instantaneous transmission properties that are so very desirable for starship communications.

EXPERIMENTAL EVIDENCE FOR QE COMMUNICATIONS

Information transmission via entangled quantum states (carrying qubits of information) has been the grail of laboratory experiments since Bell's Theorem was published in 1964. Qubits – which are units of quantum information – may be encoded as particle (atomic, photon, or sub-atomic particle) spin, polarisation [9] or energy [10]. The first irrefutable quantum entanglement experiments with high levels of statistical validity have been especially important since they appear to show that hidden variable interpretations (assumed in the Einstein–Podolsky–Rosen paper and its extensions) are a totally inadequate representation of nature [11]. These experiments in general support the existence of quantum entangled states and most have been accepted by the larger physics community. They are demonstrably valid for communications, although FTL communications have not yet been demonstrated.

CONFIGURING A STARSHIP MISSION WITH QE COMMUNICATIONS

An essential condition of quantum entangled communications between a starship and Earth must be that the communicating stations on each simultaneously (or within some small time increment) have access to entities (particles) sharing corresponding entangled quantum states. Thus the two stations must be equally separated in time from a source that generates entangled particles. Many arrangements are possible, but the simplest may be that for a starship travelling at some time-dependent velocity v(t), there is between it and Earth a generator moving at velocity v(t)/2. This simple configuration is shown in the diagram on the opposite page.

IMPLICATIONS OF QE COMMUNICATIONS

The primary implication of quantum entangled communications is that Earth and the starship will share event space. Cultures may remain congruent. Difficulties in decision-making may not be easily resolved, but mission challenges, decision premises and resolutions may be shared in real time.

Starship Communication via quantum entanglement mandates a unique photon generator source position, for which the light-times are equal for the starship and Earth. Thus the possibility of having communications monitored by a third party is minimal. The corollary is that when the starship encounters signals generated by other extraterrestrial cultures [12], then these signals, if transmitted using quantum entanglement, will not normally be readable by the starship (if the starship is populated, then the first verifiable extraterrestrial culture already exists). However, they may be readable with delay between transmission and receipt, depending on photon coherence length (a classical concept, certainly, but with some quantum entangled analogue). This inability to communicate may be yet another frustration for SETI researchers, but at least the recording of information that is shown to be encoded via quantum entanglement may suggest where to search for other communicative sites.

CONCLUSIONS

Differences in interpretation of the results from quantum entanglement experiments remain in the physics community, leading to uncertainty about the validity of quantum entanglement for some applications, notably the possibility of FTL communications. Thus quantum entanglement concepts, while theoretically regarded as robust under current interpretations of quantum mechanics, are

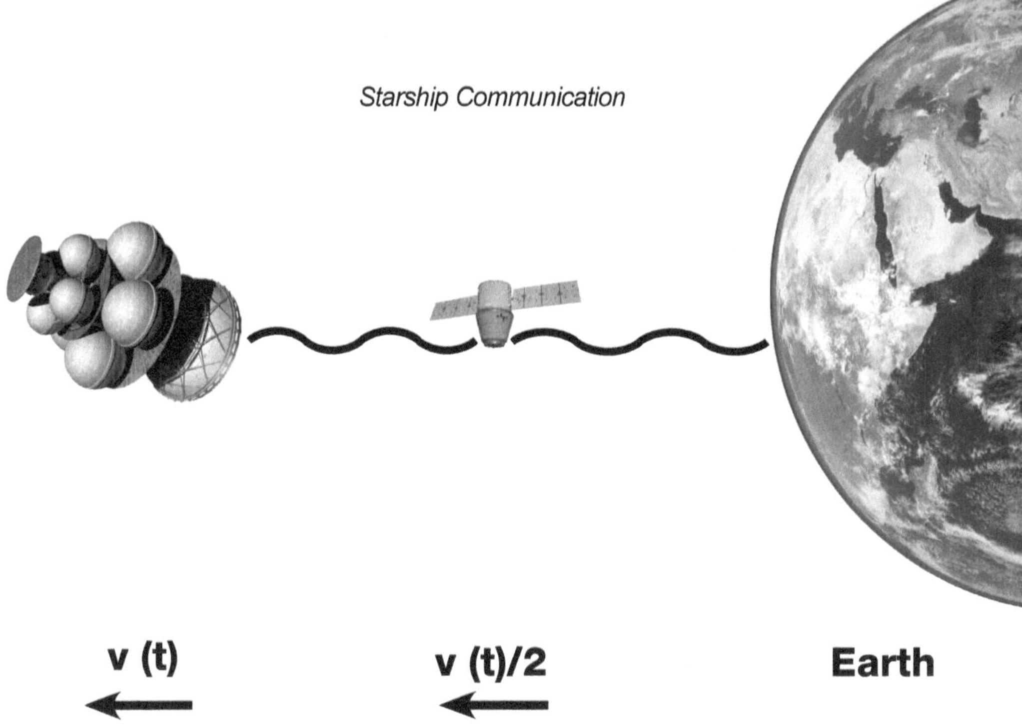

Above: a source of quantum-entangled particles, that maintaining position between a starship and earth could facilitate FTL communications. The source must maintain a position that is temporally equidistant from both communication stations. In the arrangement shown, either station (starship or earth) could transmit by resolving quantum states, and the other could receive. If the scheme is physically valid, then full duplex communication is only slightly more difficult.

not yet uncritically verifiable. Determining which quantum mechanics school of thought applies to the physical Universe and whether or not this permits FTL starship communications, are two critical questions for our culture and for planning starship missions. Technologically, we are just beginning.

Current starship design activities anticipate technologies and schemes that are not extant, but which may be available at mission initiation and these technologies will certainly have capabilities that exceed those currently available. Some of these technologies and schemes may appear as magic, yet others may not even appear as such; rather, they may simply not be conceivable from our current perspective. We are only slightly beyond the conceptual frontier of quantum entanglement.

REFERENCES

[1] Long, K F, R Obousy, A Tziolas, A Mann, R Osborne, A Presby and M Fogg; 'Project Icarus: Son of Daedalus – Flying Closer to Another Star', *Journal of the British Interplanetary Society*, 62, 11/12, pp403–416 (2009).

[2] Maccone, C; 'Sun Focus Comes First, Interstellar Comes Second', *Journal of the British Interplanetary Society*, 66, 102, pp25–37 (2013).

[3] Fields D E, P Oxley, B Vacaliuc, P Shuch, S Hogle, K Roy and R Kennedy; 'Algorithm Defined Receivers for Interstellar Travel and SETI', Tennessee Valley Interstellar Workshop, 3–6 February 2013 [online]. Available at: www.youtube.com/watch?v=fLvwq1OhEUc [last accessed 16 July 2014].

[4] Galea, P; 'Communication with World Ships – Building the Diasporanet', *Journal of the British Interplanetary Society*, 65, 6, pp180–184 (2012).

[5] Roy K, R Kennedy and D Fields; 'Colonising Brown Dwarfs and Orphan Planets', Tennessee Valley Interstellar Workshop, 3–6 February 2013 [online]. Available at: www.youtube.com/watch?v=O8r0WE_yJFY [last accessed 16 July 2014].

[6] Genovese, M; 'Research on Hidden Variable Theories: A Review of Recent Progress', *Physics Reports*, 213 (2005).

[7] Einstein A, B Podolsky and N Rosen; 'Can a Quantum-Mechanical Description of Physical Reality be Considered Complete?', *Physical Review*, 47, p777 (1935).

[8] Bell, J; 'On the Einstein–Podolsky–Rosen Paradox', *Physics*, 1, 3, pp195–200 (1964).

[9] Clauser J, M Horne, A Shimony and R Holt; 'Proposed Experiment to Test Local Hidden Variable Theories', *Physical Review Letters*, 23, pp880–884 (1969).

[10] Richards, R; 'Quantum-Entangled Photon Interferometry', *Proceedings of the SPIE*, 5531, Interferometry XII: Techniques and Analysis, 17 (2004).

[11] Aspect A, P Grangier and G Roger, 'Experimental Realisation of Einstein–Podolsky–Rosen–Bohm Gedankenexperiment: A New Violation of Bell's Inequalities', *Physical Review Letters*, 49, 2, pp91–94 (1982).

[12] Baxter, S; 'Project Icarus: Interstellar Spaceprobes and Encounters with Extraterrestrial Intelligence', *Journal of the British Interplanetary Society*, 66, 1–2, pp51–60 (2013).

[13] Yirka, Bob; 'Los Alamos Reveals Quantum Network', *Physorg* [online]. Available at: phys.org/news/2013-05-los-alamos-reveals-quantum-network. html#nwlt (last accessed 16 July 2014).

CHAPTER 15

AN INTRODUCTION TO ARTIFICIAL INTELLIGENCE AS APPLIED TO SPACE EXPLORATION

JEREMY E CLARK

From the beginning of the last century, two strong themes emerged in the world of science fiction: spaceships, as flown by Buck Rogers (1928), and artificial intelligence, from this point on referred to as AI. The spaceships took mankind on adventures while the AI in *Metropolis* (1927) played on the fears earlier expressed in *Frankenstein* (1818). Humankind attempts to play God but creates a soulless nightmare reflecting the fears of the ordinary person's life in an industrial world dominated by automation. These two themes were probably most notably brought together in *2001: A Space Odyssey* (1968). So, at least in the public imagination, these two subjects have a strong connection right from the beginning. Today that has become a very practical connection. It takes approximately five minutes for a radio signal to reach Mars when it is at its closest, five minutes for a rover to tell us it is in trouble and five minutes for people in a control centre on Earth to get a message back to Mars. If the message was "I am nine minutes away from a cliff edge," that will be the rover's last message. Clearly, our unmanned space explorers need some level of ability to think for themselves, but how can we give them this ability? What is intelligence? Is there just one kind: ours? What new problems will the intelligent machines of the future face?

As described above, AI and space travel have grown up together in the world of science fiction but what place does AI have in space travel today? There has been a public debate about the virtues of manned space exploration versus robotic space exploration. It would seem like a very depressing admission of defeat if astronauts where not to tread upon alien worlds in the future. However, even a short trip to Mars would involve exposure to high levels of radiation, confinement in a claustrophobic environment and physical weakening of the body because of the effects of low gravity plus myriad other unforeseen dangers. Clearly, exploration by robotic probes will avoid all of these problems and will be much cheaper. Given a distance of 4.2 light years to the nearest star, even using fusion propulsion with a theoretical velocity of ten percent of the speed of light, it would take a lifetime for a return trip. Not only do astronauts face huge physiological problems but it would seem far more likely that it would be the astronauts that suffered a breakdown under the psychological stress rather than a HAL-2000. Even if the missions are manned, a class of computer software known as an expert system would be invaluable. Far from Earth, expert systems could save the lives of astronauts by providing medical diagnoses and other specialist advice not otherwise available to a small community. In short, space is a dangerous place where humans will be isolated by huge distances. AI systems will be our helping hand to the stars.

This chapter will present the current uses of AI in space exploration and then provide an overview of how the science of AI developed. This draws upon two academically separate areas of knowledge: computer science and psychology. However, despite the fact that computer science and psychology would seem very far apart in their application, the development of AI cannot easily be put into one domain or the other. To confuse matters, at times the terminology differs between the two disciplines; neural networks for computer scientists and 'connectionist' models or 'parallel distribute processing' for psychologists.

To achieve any progress in this area it is first necessary to conceive of an intelligence having a physical embodiment rather than the product of some ethereal soul. The western world has tended to separate out mind from body as philosophically exposed by René Descartes in the 1600s. In 1890, the psychologist William James put the mind and its intelligence firmly into the physical organ of the brain, a place that was now accessible to science. Sixty years later, Alan Turing gave humankind a framework for the creation of thinking machines. Perhaps over-optimistically, in 1950 he wrote that by the year 2000 the world would have built the sort of intelligent machines that are common in the imaginary worlds of science fiction [1]. Today NASA's Jet Propulsion Laboratory (JPL) has an AI group that, amongst other things, has developed the AEGIS software that allows the Mars rovers Opportunity and Curiosity to roam around the surface of an alien planet with some degree of independence.

When a probe is sent out into the far reaches of space it has to manage a variety of goals and sub-goals. These are activities common to the demands of daily human life. JPL develop plans and schedules for all kinds of space exploration. However, no matter how detailed the plan, explorers will still be required to shift their attention as new vistas are investigated. Target goals must be selected and juggled with the goals and demands of flight control. Far from Earth, distant probes have also been given enough intelligence to make use of limited communication bandwidth and long delays. For example, when a probe is scanning the surface of Earth for a selected target, it needs to filter out images that have captured cloud cover or other obstacles. This is an easy task for an intelligent human but a challenge for a dumb machine. AI must have the ability to 'appreciate' the difference between an object or photograph of interest and data without value. Far away from human intervention, AI must be able to prioritise goals. It must extract 'meaningful' information from its environment and act with 'purpose'.

Before we can discuss how to create AI or give it a useful job in space, it might be helpful to establish what we mean by it and want from it. It will not help a Mars rover to tell it, "I can't tell you what a rock is but I'll know it when I see it". However, the first and most famous test of AI uses precisely that criterion. Alan Turing's (1950) test entails a person and the subject to be tested to be unable to see each other but able to communicate. The person conducting the test holds a discussion, for example by written note or telephone. If the tester is not able to tell the difference between a human subject and a machine subject based on this communication, then the machine or software is deemed to have passed the test. This would certainly seem useful as a means for astronauts to interact with the control system of a space vehicle. It also highlights one of the many functions that AI can serve, namely natural communication via speech. At the same time it feels like an unsatisfactory definition. Just because a computer is able to say that it has a favourite song, does it actually mean it? These are questions of our personal experience of reality or phenomenology. We would certainly expect an alien intelligence to have an inner world with meaning and purpose; does it matter if our intelligent machine does? I will return to this question later.

ALAN TURING AND THE BIRTH OF COMPUTING

Humankind's attempts to make calculating machines probably go back to the dawn of civilisation. For example, Stonehenge as a means of calculating phases of the solar year, or the two thousand-year-old Greek Antikythera mechanism. In more recent years (1822), Charles Babbage designed a difference engine to aid mathematical and in particular astronomical calculations. However, I would argue that the moment the world changed was in 1936 when Alan Turing published *On Computable Numbers, With An Application to the Entscheidungsproblem* [2]. This provided the theoretical bases for today's ubiquitous computers. Despite the common experience of feeling exasperated at the lack of intelligence present in these machines, such as when machines take everything literally or insist that they cannot find the printer that is so clearly (to us) plugged into the back of them, these dumb machines will be the building blocks of AI systems for the foreseeable future.

Turing devised the concept of a universal computing machine by examining the way a person undertook the process of a computing task with pen and paper. As a result, today's computers have an embryonic psychological component at their heart although current experience of them is as machines of mathematical logic. In designing a model for computing, one based on a person using pen and paper, Alan Turing highlighted implicit limitations of mathematics, today's computers and perhaps the intelligence that is currently being programmed into them. These problems may seem a little esoteric but who knows what unmanned probes might find far from humankind's guiding hand? The first limitation is the impossibility of determining whether a series of calculations or algorithm would ever terminate. This is known as the halting problem. Clearly it is easy to tell if a process will terminate simply by simulating it. If it ends after a short run you have a positive answer. If on the other hand it is still running, it may not be possible to determine if it will finish in two minutes or never. For some problems it will be impossible to give an answer. Secondly, the word Entscheidungsproblem in the paper's title refers to a question set in 1928 by the mathematician David Hilbert. This concerned whether the logic of an algorithm can be universally valid. Here too the mathematical abilities of a calculating machine have limitations. Finally, in 1929, Kurt Gödel [3] demonstrated that the whole of mathematics has fundamental limitations with his completeness theorem. This theorem stated that a formal language such as mathematics can either be complete or consistent but not both. There will always be statements that are true but which cannot be proven without adding additional rules. In a colloquial form this can be compared to the natural language logic statement, "This statement is not true." It is a science fiction cliché for a computer to blow up when presented with this type of problem.

The birth of computers and AI came hand-in-hand with the loss of certainty offered by mathematics. A hundred years ago the possibility of a predictable and calculable reality was under siege. Just prior to Turing's work, quantum mechanics had destroyed the notion that physics might offer humankind the key to a predictable clockwork Universe. As AI developed so did another area of mathematics, namely chaos theory. Perhaps the earliest example of chaos theory is the three body problem. It is impossible to calculate exactly how the orbits of three gravitational bound objects will evolve. Chaos theory states that no matter how accurately initial conditions (position, velocity etc) are measured, the tiniest uncertainty in these initial conditions will, over time, produce wildly varying results; the most clichéd of examples being the flapping of a butterfly's wings in South America leading to a hurricane in North America. However, despite the fact that mathematics might have limitations unearthed by AI, there is no reason for us not to use textbook mathematics to send astronauts into space. It may also be the case that other methodologies are discovered in the future that are less restrictive and possibly human intelligence is built upon these processes rather than simply being a very sophisticated calculating machine. In the 1990s the mathematical physicist Roger Penrose argued that human intelligence is based upon something other than Turing's model [4]. However, for the foreseeable future, any AI system sent into space will likely have at its heart a universal Turing machine. Was it inevitable that humankind should start on the path to artificial minds at the point in scientific history when human minds encountered what may be insuperable obstacles to our mental and intellectual models mapping exactly on to reality?

NEURAL NETWORKS, MACHINES START TO LEARN

Turing and John von Neumann gave us electronic problem solving machines but nature gave the animal kingdom biological problem solving machines. In the case of the simplest creature, this may just be a brain that is capable of allowing its host to swim towards sunlight and food. In 1943 McCulloch and Pitts started to model brain cells and with them nature's version of the PC [5]. The brain's information processing cells known as neurons can take an array of inputs (dendrites) that receive electrical impulses. These inputs are combined and result in one output along an axon. We could imagine simulating or even cloning one of these cells so that a multitude of analogue inputs, perhaps from the sensors of an engine, produce one analogue output that might control the thrust. However, in order to start modelling a neuron it makes sense to start by modelling the simplest type. Instead of a multitude of inputs, we will start with just two. From these inputs are four possible combinations of input (11, 10, 01, 00). It would be equally possible to refer to these inputs as true or false instead of 1 or 0. As there is only one output, this can only produce a single result of true or false. In this simple model we can imagine that the neuron could act as an AND gate. In this case 1 AND 1 equals 1. All other combinations result in a 0. This may be simpler to understand using the true or false notion. True (1) AND false (0) equals false (0). A practical example being (Are you tall?) AND (Are you short?); if you can only be either be tall or short then the answer to this very contrived question must be false. Artificial neurons based on these simple logic gates can then be

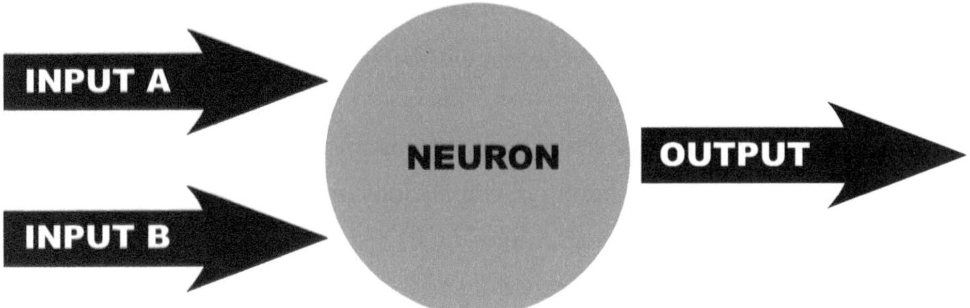

A schematic of a single neuron with two inputs and one output. The inputs can be considered to be the dendrites and the output the axon.

combined to tackle much harder problems than the question of height. This raises the question of how can this simple neuron adapt to be an AND, OR or any other type of processing unit that is capable of mapping some environmental input data into a single output or action.

To perform as an AND gate the neuron only needs to add the two inputs (11, 10, 01, 00). If the neuron produces an output greater than the threshold value of 1 then it will be an AND gate. To be an OR gate all it needs to do is have an activation threshold greater than 0. A biological neuron would be far more complicated and may include inputs that would excite an output and others that would inhibit the output. For example, to identify a circle, sensing a curved line would activate a positive interpretation of the shape as a circle, while sensing a corner would inhibit that interpretation. As a human being learns a skill, so some actions are reinforced while others will be suppressed. In 1949 the psychologist Donald Hebb provided a biological base for the adaptability of these processing units by describing how the connections between neurons were strengthened or withered as a result of use [6]. It is not difficult to use software to model a simple neuron. However, given that there are approximately half as many neurons in the human brain as stars in the Milky Way and that each one can receive somewhere between 5,000 to 15,000 inputs, it is not surprising that we cannot yet artificially simulate a human brain. Whereas in the case of a simulation of the Milky Way,

Newton's law of gravity will get us off to a good start, chaos theory notwithstanding.

AI simulations based upon neurons, known as 'artificial neural networks', consist of a variety of elements including numerous inputs to each neuron, which may be the output from another neuron. This would imply either a specific architecture of connections or alternatively all possible connections but allowing some to wither away. This withering of connections may result in a specialised architecture emerging but one determined by its training environment. AI researchers have used a variety of equations to calculate how the strength of these connections is varied and the threshold at which a neuron activates. For the threshold a simple step function could be used i.e. above a specified level is an excitable on and below some other specified level is an inhibited off, with a zero output in between. Most important to the actual learning or training is how the effect of each input can be altered by applying a weighting factor. As the weighting factor of an input is increased it will become more likely that the neuron produces an output. As that weighting factor reduces to zero that input will have less of an effect on the output of the neuron. These inputs could also act to inhibit the neuron as well as to excite it. However, the important conceptual leap is that the neuron should be able to change its behaviour as feedback or tuition causes the relative importance of an input to be altered. The bigger the error detected, the greater the adjustment. This permits machine learning and, as long as a training error is not fatal, makes for adaptability.

REASONING: CONTEXT FOR PEOPLE AND THE XOR FOR A NEURON

Earlier in this chapter the question of how to define intelligence was raised. Having proposed a method by which machines can learn and adapt it would seem appropriate that we return to this question. Sadly, Alan Turing did not live long enough to conduct his proposed experimental laboratory AI test. This test was designed for a machine emulating human intelligence while communicating. So far the analysis of brain cells highlights the adaptive nature of intelligence, not its social component. As described above this would be handy for having a chat with the ship's computer but of less use to a distant probe. In fact the machine intelligence may be more reliable because it is more general and less tailored. The cognitive psychologist Peter Wason in 1968 used an experiment to demonstrate how tailored to social thinking human reasoning is [7].

In the first test a participant is presented with four cards. On one side is a letter and on the other is a number. The reasoning task is to test the rule: if there is an A on one side then there must be a 2 on the other side. Presented with the cards B, A, 5 and 2, which cards do you need to turn over to prove the rule? I feel sure my readers will be able to solve this problem. However, it proves far harder for people to solve in this context than the identical problem in a social format. For example, consider four people with four drinks when only those above the appropriate legal age are allowed to drink. Let us call them Drinker A, B, C and D and you know only one fact about each of them. Drinker A is 50, Drinker B is 10 (well below the legal age), Drinker C has water and Drinker D has beer. Based on this knowledge, who do you have to ask to check the law is not being broken? I expect that everybody would ask Drinker B what they are drinking and Drinker D how old they are. Returning to the card problem, however, perhaps surprisingly only a small percentage of people turn over just Card A and Card 5. People tend to turn over card 2, which is like checking that the 50 year old is drinking beer. They are legally entitled to be drinking but are not required to. Similarly, card 2 can have an A on its reverse but there is nothing to stop it having other letters, whereas an A must always come with a 2 so could not come with a 5. We are excellent at reasoning in a social context but not as purely rational as we would hope AI to be.

Evolution, helped by culture and personal experience, has tailored us to our environment. This is a largely two-dimensional environment, obstacles protruding in the third dimension may have to be circumnavigated and those that plunge sharply into the third dimension, such as ravines, should be approached with care. During instrument training pilots have to learn to ignore their senses and fly by the instruments when put into an unusual attitude. AI will have the advantage that it will be tailored for the task at hand. It may fail Alan Turing's intelligence test but from the perspective of an unmanned probe there would be little need to imitate human social intelligence, whereas adapting to unforeseen conditions may be vital.

The work of McCulloch and Pitts allowed the creation of a first generation of artificial neural networks to be implemented with software. These could be used to model simple logic gates, for example an AND gate would have two inputs. If these inputs were both active then the neural network produced an active output. (Input A AND Input B = Output '+1') otherwise (No Output = '-1'). Perhaps not very sophisticated but the neural circuits could be combined to perform much more complicated tasks. The most important element being that the neural network could learn or be trained to have certain behaviour. A computer is simply programmed whereas a person improves and learns with practice. An example of the learning steps for a simple OR gate is given in the appendix. However, this type of neural network with just one neuron had a serious limitation as described by Minsky and Papert in 1969 [8]. It could learn to be a logical AND gate or an OR gate but it could not learn to be an eXclusive-OR (i.e. XOR), that is A or B but not both. More technically, this type of processing unit could only learn to solve problems that were linearly separable. The individual neuron cannot map all possible combinations of inputs to outputs. This problem was resolved by the introduction of a middle layer, or a hidden layer, of neurons in the architecture. The neural network has an input layer to receive what could be sensory input and an output layer that might feed into a control system. Without the middle layer a robotic probe could 'understand' that it had enough fuel to explore or needed to refuel but could not 'understand' that it could not do both. It would be an Artificial Idiot.

Neural Architecture

So far the behaviour of a neuron has been largely discussed as a single processing unit. By itself the neuron does not seem to offer any advantages over traditional models of computer programming and, as described above, encountered the XOR problem. In the brain each neuron is a processing unit and the architecture is massively parallel. In other words, a PC might be composed of a few processors sharing tasks such as arithmetic and video manipulation. These silicon-based processors are very fast but what the brain's biological processors lack in speed, they make up for in numbers. Clearly the XOR problem is not an issue for biological brains. Even children understand that there are times when they can have one toy or another but not both. In the 1970s AI researchers started to use combinations of software neurons in multiple layers and more sophisticated algorithms for training them, most notably a method known as 'Back Propagation'. After calculating the results of a set of inputs applied to the system, this algorithm calculates the difference between the actual output and the target output. This error amount is then fed back through the system to adjust the weighting factor of connections between neurons. At this point in the story, we are in a position where computer systems are able to learn and generalise their knowledge in order to operate with a never before encountered environmental input. However, just as AI seems to offer the miracles promised by science fiction it also comes with some of the problems found in the brains of any human astronaut.

A neural network with a hidden layer.

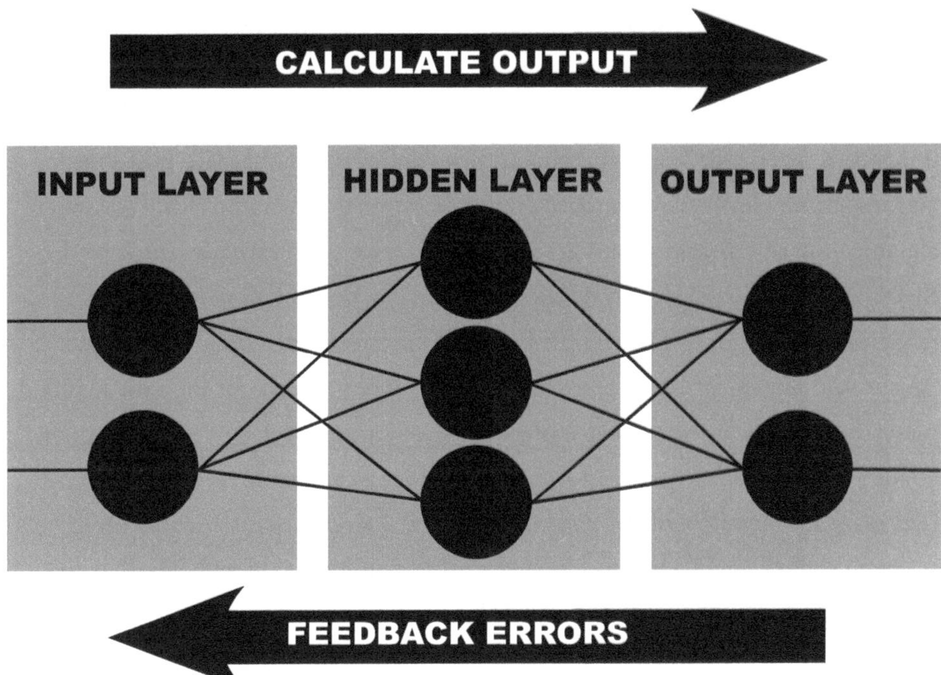

It should also be highlighted that this type of approach requires the system to be trained with the correct answers. For a spacecraft's control system there undoubtedly will be a correct answer with which it can be trained. When a human learns to drive a car the difference between a safe journey and a crash is obvious. Human behaviour allows us to learn skills, understand problems and make sense of our environment. Animals seem to make sense of their environment without explicitly being taught what makes one object a rock and another one a predator. It is too late to learn after the feedback from the experience of being eaten. In order to artificially model these concepts it is necessary to over simplify. In fact humans can learn about dangerous phenomena in the world by observing the reactions of others. If everybody else runs from the roaring animal with large teeth, it is probably best to copy their example. However, learning about different classes of animal for humans, compared to astronomical objects for probes, is a different type of problem to the supervised training described above and shall be addressed shortly.

Supervised and unsupervised learning

It could be argued that the split between art and science is one of objective versus subjective truth. We would expect that the world of space travel had to be built upon objective truth. Spacecraft should have definite and objective positions and velocities. Control by standard computer algorithm fits into this. However, classification of objects encountered would be more subjective for people and this would also be true for an AI system. Classifications of alien life or even recognising it might not be straightforward. For example, an animal that laid eggs would not normally be considered to be a mammal, except for a duck-billed platypus. Classification may also be an issue of context. People do not normally count a car seat as furniture but they do consider a car seat to be a seat and that seats are members of the class of furniture. As AI systems become more sophisticated they will start to live in a world of ambiguity already familiar to people. The cost of more resourceful and independent probes may come at the price of their being predictable.

The Cydonia region of Mars has become a well-known image. In 1976 the Viking 1 probe took a picture of this region on the red planet that appeared to show a human face. Closer to home we have probably all experienced shadows, TV static or other random phenomena that appear to look like faces, an illusion known as pareidolia. Artificial neural networks have the ability to learn and hunt for patterns but they also have the ability to find patterns where none exist. A particular example of a related problem would be a neural network that has been taught to find trends in the stock market. Anyone that gets tomorrow's market valuation today will soon be rich enough to fund their own space programme. However, here is the catch: the neural network is required to generalise its knowledge, not just only recognise the Martian rocks it has already seen. During training the network is 'learning' about the properties such as the OR gate. In the case of a logical OR there are only four possibilities for two binary inputs and the learning cycle is very short, yet the stock market has many more possibilities and far more data that needs to be analysed, just like Martian rocks. After training the network it may perform perfectly on the data with which it was trained but given new data, will fail to produce correct results. For this reason a complicated network should be both trained and then tested with separate data sets. The problem of memorising the already known rather than generalising for the yet to be encountered, is known as over-fitting. It would be highly undesirable for a spaceship's control system only to be capable of managing situations that it had previously encountered while being taught on Earth.

KOHONEN SELF-ORGANISING MAPS

So far we have discussed AI systems that can be taught and then be sent on one way trips to explore environments that would be far too hazardous for humans. However, when sent off into the depths of space the probe may encounter totally unexpected conditions. How does it classify strange new life, for example? The neural networks discussed so far are classed as systems that require supervised training. Not everything we learn requires an external teacher telling us we whether we have got it right or wrong. As an alternative approach unsupervised learning has much in common with statistical cluster analysis. It is also a common human experience to organise the world into groups of objects. The key concept behind neural networks is their ability to alter the strength of the connections between their components. A classic unsupervised network called a 'Self-Organising Map' can be thought of as taking data with a high number of independent dimensions and then reducing that to two dimensions. This is achieved with a grid of neurons. During training the neuron that reacts most strongly to some input data receives a change to the weighting of its connections to increase further that level of activation. The connections to the adjacent neurons also receive a tweak. If more than one neuron has an equally strong reaction then a random choice is made, so that only one of the neurons is adjusted. Neurons within the grid are effectively in competition with each other but also supporting their immediate neighbours, unlike the next neural network. In this way the grid of neurons is adjusted to separate out groups within the training data. What was impossible to visualise in high dimensions becomes easier to see as clusters on a two dimensional grid.

The foundations of this theory were developed in the early 1980s by Finnish academic Teuvo Kohonen. As a result, this class of neural network is often referred to as a Kohonen network. He was particularly interested in using this approach to identify phonemes in speech. A particular series of activated neurons or a track across the neural grid could then be used to identify a particular word. Maybe at some point in the future this may be the first step in understanding an alien communication. Closer to home this method has been used to discover hidden activities such as fraud. In the same way that fraud could be considered to be a financial malfunction, self organising neural networks could monitor system functions in the hope of identifying a failure before it became critical. Any advance warning of failure is useful and all the more so in the threatening conditions of space.

ADAPTIVE RESONANCE THEORY

It is perhaps not surprising to find that different classes of neural network can be combined. Having established neurons as the basic component of an AI system,

I will now look at a collection of architectures known as 'Adaptive Resonance Theory', referred to as ART from this point. The flow of information through the systems discussed so far has been information provided from the environment and then followed by a wave of adjustments to reduce the errors between a supervised target or an unsupervised cluster. More generally this is called bottom-up analysis as information is derived from data flowing up through the system. Conversely top-down analysis is about decisions made on high and then imposed upon the data. Much of our experience of the world comes from first deciding what we expect to see and then seeing what we expected i.e. top-down. Of course we also take information from the outside world and then combine both top down and bottom up information.

ART [9] makes use of two neural networks that could be considered to be in competition. The intention is to incorporate effects of neural transmitters by including feedback between the two networks. They also liken this to the functioning of short-term and long-term memory.

Before further describing the theory underpinning ART, I will briefly digress into the world of psychology. Most people will be familiar with the concepts of short-term and long-term memory. Various types of long-term memory will be discussed in the next section but for now it is enough to say that as people, we seem to have a virtually limitless ability to file away memories. It might be virtually limitless but anybody that has had to study for an exam will know it is not always easy to add or retrieve information from this store. For storage linking, information with a narrative will make a big difference.

Short-term memory on the other hand is very restricted. There are many examples of humans constructing groups of seven (days of the week, musical notes, wonders etc). According to George Miller [10] that is all the short term memory you have, plus or minus two. However, he states that these ubiquitous groups of seven may be "only a pernicious, Pythagorean coincidence". A workaround for this limited human short term memory is to group information into chunks (i.e. instead of trying to remember 3-6-3-1-2-8-4-9 try 3631 and 2849). However, the limited amount of short term memory still places constraints on how many distractions a person can handle at one time. A more complete discussion of short term memory would include memory buffered from the senses and the storage of verbal information. The point is we are easily overwhelmed by information from our environment. AI might lack our sophistication but we can certainly give it the short-term memory of a super human genius.

As stated earlier, ART systems are composed of multiple components, the first of which is a self-organising network that receives an input directly from the environment. Because of its immediate contact with the environment and the possibly rapid-changing of its organisation, Carpenter and Grossberg referred to this as short-term memory. The second net takes its input from the weights of the connections in the first and, because of its more enduring structure, is referred to as long-term memory. In effect the first network trains the second. There are additional feedback loops between the two which are intended to reflect the behaviour of neurotransmitters, plus another control loop that deflects errors between the first network and the environment. In this way the system can both monitor and improve its own performance.

Artificial neurons should be seen as building blocks. Our brain has been optimised for the conditions in which humans have traditionally lived. The size of our brains has been limited by the physical constraints of their birth. As we learn more about the demands of space travel so we will be able to evolve architectures that optimise AI performance. As for the size constraints, we can make the networks as big as we like. Although given that more connections and components requires more time for internal communications, there may be a choice between all-powerful and quick-thinking.

ADAPTIVE CONTROL OF THOUGHT

The discussion so far has largely built upon the physical properties of a biological unit in the brain. This unit has been duplicated as software. This software has been exposed to the environment and an algorithm used to optimise a mapping of 'sensory' input to an output. In a robot this output could be motion, speech etc. However, this would seem to fall very short of our own experience of thought.

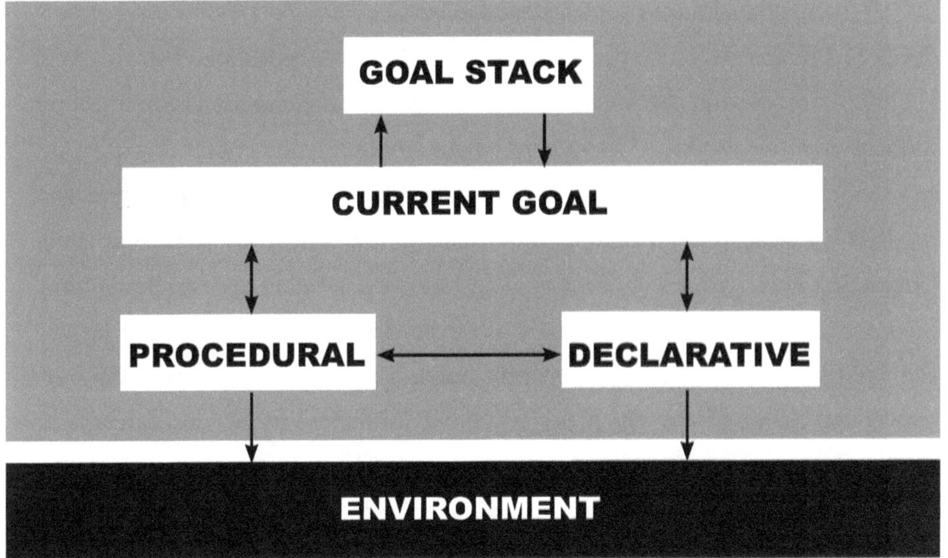

A schematic of an Adaptive Control of Thought (ACT) system.

Perhaps it is simply good enough that the AI systems so far can be flexible enough to perform a limited range of tasks with a limited amount of predefined resources. Human beings have desires and aversions, logic may tell you how to solve a problem but the inner world of people gives us a reason to want to solve them, for example a passion to overcome challenges and explore the unknown.

Computer memory stores information in exactly the same way a book does. The information is written into memory and can then be accessed by looking it up in the index. Human memory is terribly forgetful by comparison but the index is superb. Concepts are linked associatively. This is a crucial part of humankind's mental processes and hints of this can be seen in the previously described neural networks. The final AI model to be discussed was born out of this more psychological approach, combining theories of 'Human Associative Memory' with a rule-based approach to AI. At the beginning of this chapter the concept of an 'Expert System' was introduced. The application suggested was one of medical diagnoses. Expert systems are composed of conditions and actions. For example, the condition 'patient has a headache' fires the action 'prescribe headache tablet'.

The first version of the Adaptive Control of Thought (ACT) system was developed in the early 1970s. Unlike the neural networks described above, ACT [11] does not attempt to emulate any of the biological elements of the brain but models cognitive modules. There are two goal modules, one containing a list of all the system's goals and the other containing the current goal. The current goal module can either add goals to the goal stack or load one into the current state. Interacting both with the current goal and the environment are two 'long-term' memory modules. It is no doubt part of the reader's mental toolbox that he or she will remember or know how to ride a bike. This knowledge is stored in our 'procedural memory'. For the lucky few, procedural memory will also include how to pilot the space shuttle. Most readers can probably name the closest planet to the Sun without having any memory of the episode in their life when they learnt this fact. This knowledge is stored in our 'declarative memory'. These should not be seen as being separate physical systems in the brain.

Neural networks demonstrate how knowledge is dispersed across the network. This has an important impact on reliability. A word processing program on a computer might work perfectly or refuse to work at all. The human brain on the other hand degrades with age and maybe only stops completely when the individual owner is dead. Neural networks also show this ability to continue functioning when damaged, or example if some nodes are removed, the equivalent being a brain lesion. There is a debate concerning the degree to which brains are compartmentalised into special areas but this goes outside the concerns of this chapter. However, the ACT approach does take as its starting point that evolution will have optimised these logical modules for the environment.

The human experience of the world is highly integrated given that human memories are associative. How to pilot a space shuttle is linked to many facts, not least of which is what a space shuttle actually is. Clearly the facts of space flight need to be learnt before they become part of a subconscious skill. This is also true for ACT, which is capable of acquiring facts and compiling these into procedures. The procedure can then be fired as a memorised fact is triggered by a demand from the current goal in tandem with sensory information from the environment. For its part the procedural systems will operate on the environment and also manage the current goal. As stated earlier, the resolution of conflicting goals is very important for the remote probe far from home and speedy human intervention.

PROBLEMS IN SPACE AND THEIR SOLUTIONS

It is not surprising that, given our day to day experience of technology, hardware in space encounters similar problems. Who would feel confident trusting their lives to the reliability of their PC? In the case of the Hubble Space Telescope's problems, the solution boiled down to sending astronauts up as repairmen. Whereas for a recent delivery of supplies to the International Space Station by the Dragon cargo capsule, a failure of the thrusters required for docking was solved by human ingenuity. Instructions to clear a blocked valve, equivalent to a human picking up a blocked tube and blowing through it, were transmitted to the vessel. Given that the function of breathing in and out is to supply the body with oxygen, this could be construed as an example of lateral thinking, although in this case not a novel one for people to perform.

For one of the Mariner probes exploring the inner Solar System in March 1974, the solution to a thruster problem was to use a solar panel as a solar sail. Since these solutions have been used once then I feel sure that should similar problems occur, similar solutions will be attempted. However, there will always be new problems waiting to happen, problems requiring creative solutions. Will AI ever be able to help?

AI and creativity is necessarily a newer field than the AI discussed so far. In humans it is also a phenomena that is associated with conditions such as schizophrenia [12]. It may always be advisable to separate the AI systems responsible for operations from those that generate creative options. Much of the research so far has looked at music composition or generating stories. Problems in space are stories that need happy endings. Perhaps the first step will be for AI on Earth to provide human mission controls with creative options that can be used or ignored as a human mind sees fit.

CONCLUSIONS

It may seem that components at the heart of a computer are simple on–off switches and that the AI described also makes use of very simple basic components. Despite the simplicity of the basic components the result would seem to be greater than the sum of its parts, in much the same way that a home is more than a set of structural components. For AI this means that these simple algorithms start to display more complex behaviour, known as emergent behaviour. For example, psychologists have trained simple neural networks to produce correct grammatical statements. However, like a child these systems have first over-generalised (e.g. conjugating past participle of sleep as 'sleeped' instead of 'slept') before learning the exception to the rules. Just like a human astronaut, the AI system will need to be thoroughly trained and tested.

So far we have assumed that we have one spacecraft with many complex components and their various sub-systems, some of which might have varying degrees of intelligence. However, using the animal kingdom for inspiration, we have the example of insect colonies. The likes of ants and bees as individuals each have very basic tasks to perform. Yet the colony as a whole demonstrates emergent behaviour. This raises the possibility that we might create swarms of probes. Each probe would have a very limited amount of AI. However, rather like ants, these might be set tasks to go about building the equivalent of a nest. Although in this case, it may be taking raw materials found in asteroids and constructing larger machines or habitats for future human colonists.

Humankind has sent relatively few missions into deep space when compared to the daily amount of commercial air traffic. One day perhaps our experience of space travel will be comparable to that of aviation. In the meantime human computer interaction in the aviation industry provides a number of lessons. Automation cannot be completely trusted and the capability for manual override is required. In 2008 a commercial airliner landing in a strong cross-wind at Hamburg airport nearly crashed as a result of computer intervention. One of the wheels had touched the runway and the computer put the plane into ground mode, a situation where it would not be expected that the plane should need to bank and roll. However, a strong gust pushed a wing tip down into the runway while the onboard computer prevented a sufficiently large adjustment, which it was only programmed to allow while in the air. The pilot was able to increase thrust, become airborne and save all onboard. Sometimes the computer system is at fault. Another problem is that the autopilot is just too good. In 2009 a flight to Minneapolis flew happily past its destination. The autopilot worked perfectly but the pilots onboard had ceased to fly the plane, over-automation leading to pilot disengagement. Perhaps here the solution would be a Turing test but this time the AI system should monitor the pilots. When humans travels to the stars they will do so in partnership with AI.

As described above, AI systems are vulnerable and so are astronauts. Maybe the most famous AI space explorer from science fiction is HAL in *2001: A Space Odyssey*. Of course drama may require a villain whereas real life does not. However, it is likely that a real-life version of HAL would require goals such as those described in the ACT system and these goals would then evolve within the system. It might not be a question of emotional stability but it would be advisable to monitor how those goals evolved. AI systems also build up a mental landscape that includes associations between concepts. In people, problems in this area are investigated by psychoanalysis. The future might require a whole new discipline to monitor the 'mental state' of AI systems responsible for the safety and welfare of its human partners. Both space travel and AI are at the dawn of their history. There may be pitfalls to watch out for but we have every reason to believe that the potential of both is limitless.

At the start of this chapter I proposed that the two themes of AI and space exploration had been brought together in fiction. However, there is perhaps a deeper connection. Humankind is both searching the stars for other intelligent life and trying to create it inside machines, almost as if being driven from a cosmic sense of loneliness. We may find companion minds or create them and maybe these two moments in human history will occur together. Humankind is searching the skies for the former and slowly progressing towards the latter.

REFERENCES

[1] Turing, A M; 'Computing Machinery and Intelligence,' *Mind*, 1950.

[2] Turing, A M; 'On Computable Numbers, With an Application to the Entscheidungsproblem,' *Proceedings of the London Mathematical Society*, 1936.

[3] Hofstadter, D R; 'Godel, Escher, Bach: An Eternal Golden Braid' *A Metaphorical Fugue on Minds and Machines in the spirit of Lewis Carroll*, Penguin Books, 1979.

[4] Penrose, R; *Shadows of the Mind: A Search for the Missing Science of Consciousness*, Oxford University Press, 1994.

[5] McCulloch, W S and Pitts, W H; 'A Logical Calculus of the Ideas Immanent in Nervous Activity,' *Bulletin of Mathematical Biophysics*, 1943.

[6] Hebb, D; *The Organisation of Behaviour*, Wiley, 1949.

[7] Wason, P; 'Reasoning About a Rule'. *Quarterly Journal of Experimental Psychology*, Volume 20. 1968.

[8] Minsky, M and Papert, S; *Perceptrons: An introduction to Computational Geometry*, MIT Press, 1969.

[9] Carpenter, G A and Grossberg, S; *ART 3: Hierarchical Search Using Chemical Transmitters in Self-Organizing Pattern Recognition Architectures*, Pergamon Press plc, 1990.

[10] Miller, G A; 'The Magical Number Seven, Plus or Minus Two: Some Limits on Our Capacity for Processing Information,' *The Psychological Review*, 1956

[11] Anderson, J R and Lebiere, C; 'The Atomic Components of Thought,' Mahwah, NJ, Lawrence Erlbaum Associates, 1998.

[12] Kyaga S, Lichtenstein P, Boman M, Hultman C, Långström N and Landén M; 'Creativity and Mental Disorder: Family Study of 300,000 People With Severe Mental Disorder,' *The British Journal of Psychiatry*, 2011.

FURTHER READING

• Adaptive Resonance Theory at work :

www.youtube.com/watch?v=1_EepzTRIuo

• Mariner 10 timeline

www.dmuller.net/spaceflight/mission.php?mission=mariner10

• 2008 Computer system set airliner to ground mode when wheel touches down: www.youtube.com/watch?v=jfB4xyM7tMw

• 2009 Pilots completely defer to autopilot: http://news.bbc.co.uk/1/hi/world/americas/8321748.stm

• Aleksander, I and Morton, H; *An Introduction to Neural Computing Second Edition*, International Thomson Computer Press, 1995

• Penrose, R; *The Emperor's New Mind: Concerning Computers, Minds and the Laws of Physics*, Oxford University Press, 1989.

• Well, A; *Rethinking Cognitive Computation: Turing and The Science of the Mind*, Palgrave Macmiilian, 2006.

APPENDIX: ADAPTIVE LINEAR NEURON – SINGLE LAYER – 'OR' GATE

An example of the training steps for a single neuron learning to be an OR gate. Input 1 = true, 0 = false. This neuron includes a bias as well as two inputs. The four possible sets of inputs are [11 = 1, 10 = 1, 01 = 1 and 00 = –1]

Note, –1 is used as false value for output.

The inputs are repeatedly applied to the network as training data until the actual outputs match the target outputs.

Where:

Wi   = Weight for element i

α = Learning rate (set equal to 1)

The formula to calculate output by summing the product of an input and a weighting factor:

Output = W1 * Input1 + W2 * Input2 + WBias * Bias

Learning equation to calculate new weights:

Wi (New) = Wi(Old) + α(Target – Actual Output)

Table 1: Calculated inputs, weights and outputs of an adaptive linear neuron (OR gate).

	Input 1	Input 2	Bias	Target	Weight	Weight 2	Bias Weight	Output Dec	Output Bin
A	1	1	1	1	0	0	0	0	0
	1	0	1	1	1	1	1	2	1
	0	1	1	1	1	1	1	2	1
	0	0	1	−1	1	1	1	1	0
B	1	1	1	1	1	1	0	2	1
	1	0	1	1	1	1	0	1	0
	0	1	1	1	2	1	1	2	1
	0	0	1	−1	2	1	1	1	0
C	1	1	1	1	2	1	0	3	1
	1	0	1	1	2	1	0	2	1
	0	1	1	1	2	1	0	1	0
	0	0	1	−1	2	2	1	1	0
D	1	1	1	1	2	2	0	4	1
	1	0	1	1	2	2	0	2	1
	0	1	1	1	2	2	0	2	1
	0	0	1	−1	2	2	0	0	0
E	1	1	1	1	2	2	−1	3	1
	1	0	1	1	2	2	−1	1	0
	0	1	1	1	3	2	0	2	1
	0	0	1	−1	3	2	0	0	0
F	1	1	1	1	3	2	−1	4	1
	1	0	1	1	3	2	−1	2	1
	0	1	1	1	3	2	−1	1	0
	0	0	1	−1	3	3	0	0	0
G	1	1	1	1	3	3	−1	5	1
	1	0	1	1	3	3	−1	2	1
	0	1	1	1	3	3	−1	2	1
	0	0	1	−1	3	3	−1	−1	0
H	1	1	1	1	3	3	−2	4	1
	1	0	1	1	3	3	−2	1	0
	0	1	1	1	4	3	−1	2	1
	0	0	1	−1	4	3	−1	−1	0
I	1	1	1	1	4	3	−2	5	1
	1	0	1	1	4	3	−2	2	1
	0	1	1	1	4	3	−2	1	0
	0	0	1	−1	4	4	−1	−1	0
J	1	1	1	1	4	4	−2	6	1
	1	0	1	1	4	4	−2	2	1
	0	1	1	1	4	4	−2	2	1
	0	0	1	−1	4	4	−2	−2	−1

SECTION IV

INTERSTELLAR DESTINATIONS, COLONISATION AND THE SEARCH FOR LIFE

INTRODUCTION BY KELVIN F LONG

In this section we explore possible interstellar destinations and what we may do when we get there. **Keith Cooper,** a science journalist, kicks it off with an excellent review of the nearby planetary systems. Although Keith is not the first to point out, planets are not the only place that we might live as we explore the cosmos.

Researcher **Kenneth Roy**, along with his colleagues **Robert Kennedy** and **David Fields**, then examines the use of small world-ships before moving on to thinking about Plutoids, which are dwarf-like planets, and how we might someday colonise and indeed terraform them and other worlds for our own use.

Andreas Hein, an aerospace engineer, examines the ultimate aim of interstellar spacecraft design, which is to build fully-manned vessels the size of small worlds that can travel from one star system to another. He also examines how this links to the different architectures and roadmaps that we might develop, showing that there are many roads to the stars.

Finally in this section, we have a chapter by **Martin Ciupa**, a researcher, who explores the Drake Equation and the Fermi Paradox. As Martin says at the end of his chapter, the question of whether we are alone in the Universe is one with profound philosophical consequences. Discovering whether alien life, and especially intelligent life, exists elsewhere has to be seen as one of the objectives of interstellar flight. To quote Arthur C Clarke, "Two possibilities exist: either we are alone in the Universe or we are not. Both are equally terrifying," although I would like to think both are equally exciting.

CHAPTER 16

DESTINATIONS:
THE PLANETS NEXT DOOR

KEITH COOPER

The Milky Way is huge: a spiralling 100,000 light year diameter disc of two hundred billion stars and gas equivalent to 20 billion more stars [1]. The stars are spaced on average about four light years apart. It seems a cruel twist of fate that propulsion limitations would, for the foreseeable future at least, seemingly constrain us to a sphere of space a dozen or so light years in radius at best; so many deliciously wondrous destinations are beyond our reach. Yet things aren't really so bad – our immediate neighbours, and hence potential destinations, are far more interesting than they at first let on.

Out to a distance of 15 light years there are 52 stars and at least 11 brown dwarfs that we know of (brown dwarfs are 'failed stars' – objects too small to initiate the fusion of hydrogen into helium in their cores, although they do manage some deuterium fusing during the early stages of their lives, which differentiates them from gas giant planets like Jupiter) [2]. Most of these 52 fully-fledged stars are smaller than our Sun, feeble red dwarfs that emit relatively little heat but will remain lit up for trillions of years. That these stars seem so similar to our Sun is good news if we are intent on finding another planet – a future destination for a colony perhaps – that is like Earth and can support life as we know it.

Yet planets in our vicinity seem scarce. Of the 1,800-plus confirmed exoplanets at the time of writing (September 2014), only seven are found within

15 light years, and that is only if we include the red dwarf Gliese 876 at 15.3 light years, which has four orbiting worlds [3]. Another of these seven worlds is an Earth-mass planet purportedly orbiting alpha Centauri B [4], part of the nearest star system to our own, but this world exists on the border of detectability and, as we will discover, may not even be real [5]. Given that the alpha Centauri system is likely going to be one of our top future destinations, the absence of planets there would be devastating – the next nearest known exoplanet is around epsilon Eridani, 10.7 light years away. It would take a Daedalus-style starship [6], moving at ten percent of the speed of light and taking additional time to accelerate and decelerate, well over a century to reach it.

The current paucity of nearby exoplanets is, however, not the crisis that it appears to be. The results of the past two decades of exoplanet hunting are skewed with significant bias relating to the limitations of both our telescopes and the discovery techniques in use. Larger planets, closer to their stars, are easier to find than smaller planets at greater orbital radii. Furthermore, it seems somewhat paradoxical that this bias does not give preference to the nearest star systems, which you would be forgiven for thinking are the easiest stars around which to find planets. After all, things that are closer are easier to see, right?

Not necessarily. Only on rare occasions do we actually directly view an exoplanet; there have been around a dozen instances of direct imaging of exoplanets, all of which involve young, hot worlds at large distances from their stars so they are not drowned out by their sun's glare. Instead, the majority of the confirmed exoplanets have been discovered by one of two means: either radial velocity measurements or transits. (There are several other means by which planets have been found, ranging from microlensing where the passage of an unseen foreground star and orbiting planet causes the brightness of a background star to brighten for a short time, to transit timing variations whereby the regularity of the transit of an exoplanet across the face of its star is disrupted by the gravity of unseen planets in wider orbits. Additionally, the three planets discovered around pulsars gave away their presence by disturbing the regularity of their host pulsar's radio beam.)

Radial velocity measurements were the early front runner in the exoplanet detection stakes. In 1995 Michel Mayor of Geneva Observatory and Didier Queloz, at the time also at Geneva Observatory but now at the University of Cambridge, discovered the first exoplanet around a Sun-like star, namely 51 Pegasi, just under 51 light years distant [7] by radial velocity measurements. Let's look at how they did this. Think of two bodies exerting a gravitational pull on each other. Their centre of mass is somewhere between them, closer to the more massive object. In the case of a star and a planet, the difference in mass is so great that the centre of mass often lies within the star, but crucially not in its centre along its rotational axis. This leads to the peculiar scenario in which, if we're being completely accurate, the planet does not orbit the star but rather orbits the centre of mass, as does the star itself. This imparts what appears to us a wobble in the star's motion as it revolves around this centre of mass that is off its axis of rotation. As the star revolves towards us its light is Doppler shifted ever so slightly into bluer wavelengths; as it moves away, its light stretches to longer, redder wavelengths. It is a slight effect, equivalent to a motion towards or away from us of just tens of metres per second, but careful spectroscopic analysis of a star's light can show this alternating pattern of blueshifting and redshifting. In the case of 51 Pegasi and its planet the amplitude of the Doppler shift was a radial velocity of 56 metres per second – remember, the system is 480 million billion metres away, illustrating just what a tiny signal this is and yet astronomers still managed to measure not only this but also radial velocity signals of planets around other stars that are less than a metre per second in amplitude, an astounding technical feat.

51 Pegasi b (exoplanets are denoted by letters of the alphabet: the star is technically 'a' and then the first planet discovered is 'b', then 'c' and so on) is a gas giant around half the mass of Jupiter orbiting at an astonishingly close seven million kilometres to its star. Okay, to you and me seven million kilometres is a long way, but for planets it is but a small step. Put it this way: the Sun has a diameter of 1.39 million kilometres. A planet with an orbital radius of seven million kilometres would be almost on top of it. Mercury, by comparison, is the

closest planet to our Sun and has an orbital radii of 57.9 million kilometres. (Note: planetary orbits are elliptical, some more than others, with the semi-major axis being the widest dimension, so when we talk about orbital radii and distances from a star, we are really referring to the semi-major axis.)

This relatively close distance coupled with 51 Pegasi b's relatively large mass means that the radial velocity shift is quite large – a smaller planet, or a planet with a wider orbit, would produce a correspondingly smaller radial velocity signal. The most sensitive radial velocity detector currently in operation is the High Accuracy Radial velocity Planet Searcher (HARPS) attached to the 3.6-metre telescope at the European Southern Observatory's (ESO's) La Silla site in Chile and managed by Mayor's Geneva Observatory team [8]. HARPS has detected terrestrial planets with radial velocities as low as 51 centimetres (0.51 metres) per second as in the case of alpha Centauri Bb – two orders of magnitude smaller than the 51 Pegasi b detection. Upcoming instruments such as HARPS' successor, ESPRESSO (Echelle SPectrograph for Rocky Exoplanet and Stable Spectroscopic Observations), which will be installed on ESO's Very Large Telescope and will be sensitive to radial velocities as low as ten centimetres per second, will allow astronomers to search for exoplanets around nearby Sun-like stars and red dwarfs [9].

HARPS reigned as the number one exoplanet detector for mant years, having made a haul of 135 exoplanet discoveries in over ten years. It has since lost its crown though – NASA's much vaunted Kepler Space Telescope has made around a thousand confirmed discoveries, with not to mention over 4,200 candidates it has waiting in the wings. Unlike HARPS, Kepler watches for transits, en masse – it's field-of-view includes 150,000 stars in the direction of the constellations of Cygnus and Lyra. None of these stars are particularly nearby however, so Kepler will not find any potential destinations. What Kepler can do is accurately sample what is out there amongst the menagerie of exoplanets. Prior to Kepler we could count the number of planets known to be rocky on one hand, as these were much more difficult to find than giant 'hot jupiters' – large gas giants close to their stars, such as 51 Pegasi b. Now, if we factor in all those Kepler candidates then the hot jupiters are in the minority and the terrestrial worlds, from sub-Earth sized chunks of rock to the hefty 'super-earths' several times the mass of our blue planet, are clearly the majority. Seventeen percent of stars have a planet between 0.8 and 1.25 times the size of Earth in an orbit that takes 85 days or less to complete, while around a quarter of stars have a super-earth in an orbit of 150 days or less and a mini-neptune (up to four times the diameter of Earth) in a orbit of 250 days or less. On the other hand, only five percent of stars host a gas giant in an orbit of 400 days or less [10]. That said, Jupiter's orbit is over 4,300 days, so gas giants could still exist in large numbers in wide orbits, beyond the detection range of Kepler. (The space telescope needs to see three transits, to help rule out false positives, before announcing an exoplanet candidate and given that Jupiter takes around 12 years to complete one orbit, Kepler would have to be on hand for at best two dozen years to observe three transits. Alas, a failure of two of Kepler's reaction wheels has seriously affected its pointing accuracy and today it is operating in a different mode, staring at a different patch of sky.)

The sensitivity of Kepler is equally as awe-inspiring as HARPS. The dip in starlight as a planet crosses in front of its star (transits) from our point of view is tiny, a blip just one hundredth of a percent, but that's enough for Kepler's sensitive photometers to pick up. The size of the transit tells us the diameter of the planet relative to its star, so if we know the star's diameter then we can determine the planet's physical size. The regularity of the transit tells us how big the exoplanet's orbit is (planets at greater distances from their star orbit slower and transit less frequently), in turn allowing astronomers to make an educated guess regarding its surface temperature.

Measuring transits is not an entirely foolproof means of discovering exoworlds, though. We can only see a transit if an exoplanet's orbit takes it directly between its star and our Solar System, an angle so small that it means many stars with planets just have no visible transits at all (and the wider an exoplanet's orbit, the smaller that angle becomes). Secondly, stars vary in brightness all the time for purely intrinsic reasons: plagues of starspots, eruptive flares, pulsations or even eclipses by a companion star in a binary system and all of these can potentially be mistaken for an exoplanet transit. As such, transit detections often need to be confirmed by radial velocity measurements, but the upside of this is that radial velocity provides the mass of the exoplanet. With diameter and mass it is then a simple calculation to determine the planet's density, which gives us an idea of what it is made of: gas, rock or water/ice. Plotting this data on a mass-radius diagram can even provide an indication of what kind of atmosphere it has – for example, super-earths above a particular boundary condition (one that actually denotes a planet made entirely of water, likely an impossibility) are probably rocky planets with light, extended atmospheres of hydrogen and helium that can make up as much as a fifth of their radius [11], as opposed to the 1.5 percent that Earth's atmosphere contributes to our planet's radius.

Like the radial velocity technique, transits are another indirect method of determining the presence of a planet. Astronomers do not even see the silhouette of a planet since the angle it subtends is far too small to resolve – most stars are point sources, even to the Hubble Space Telescope – and the glare of the star would make it all but invisible anyway. The indirect approach does not necessarily favour proximity, but instead only requires that the star be bright enough to make radial velocity or transit measurements easier. So it is a level playing field and explains why finding exoplanetary systems close to Earth is no easier than if they were further away. Nearby stars could be teeming with planets, only for them to be small or in wide orbits where the planet-finding instruments that we wield would have little to no chance of finding them.

PLANETARY ZOO

In our Solar System we have three types of planet. There are the inner, rocky worlds – Mercury, Venus, Earth and Mars. They each exhibit great diversity – Mercury being cratered and airless, Venus sweltering underneath a thick veil of carbon dioxide, Earth blue and full of life, and Mars cold, dry and red – but essentially they are pretty much the same, spheres of silicate rock with iron cores. What happens on that thin crust we call the surface is really only a tiny part of these planets.

Next up are the gas giants, Jupiter and Saturn, in the outer Solar System beyond the handy demarkation point of the Asteroid Belt. These gas bags have no surface and could swallow Earth whole many times over inside their turbulent atmospheres. They each also have a family of moons that are like mini planetary systems in their own right. Similar, only smaller and colder, are the ice giants Uranus and Neptune, wrapped in gaseous atmospheres that may or may not hide a solid surface deep down.

Exoplanet discoveries have added to this range of planets. We've previously met the hot jupiters, such as 51 Pegasi b; these are gas giants that formed further out from their stars before migrating inwards as they swept up material from the protoplanetary disc of gas and dust that gave birth to them. Such migration is usually fatal to other planets in the system, which are run aground, either colliding with other planets or more often than not being ejected from the planetary system entirely. The hot jupiters take up position just a few million kilometres from their stars, their atmospheres heated to hundreds, sometimes thousands of degrees Celsius, which ignites ferocious storms amongst their clouds and fierce winds that blow at ten thousand kilometres per hour. These are not worlds we would be able to colonise – not even their moons if they are that close to their star – and, if they do bully the other planets out of the system, a star system with one hot jupiter planet only is perhaps not the most interesting to explore. They would likely not be amongst our first destinations from a colonising perspective.

Similarly inhospitable to you or I are the hot neptunes, which are like the ice giants of our Solar System, rich in volatiles such as water and methane but far closer to their stars, just like the hot jupiters. The next step down are super-earths, which are currently the topic of much discussion amongst planetary scientists. Ranging in mass from anywhere between two and ten times the mass of Earth, these rocky worlds are seemingly common around other stars despite their absence in our Solar System, where there is a gulf between Earth and the next planet up, Neptune at 17 times the mass of Earth. Yet their potential habitability is uncertain given that many of their characteristics remain up in the air. Would the extra heat within their interior make plate tectonics more common, facilitating a carbon-silicate cycle that draws the greenhouse gas carbon dioxide rain out of the atmosphere and into the sea where it slips through subduction zones into the mantle, allowing the atmosphere to cool before returning to the surface through volcanoes and subsequently warming the planet, causing more rain and repeating the cycle? [12] Or does all that extra heat limit the tectonic activity by increasing convection currents in the mantle that act to thicken the crust under the extra gravity? [13] Some super-earths could even be 'carbon planets', formed in systems rich in carbon where the interior pressure of the planet could crush the carbon into layers of diamond and graphite around an iron core. Others, rich in water could be water-worlds where a quarter of the planet's mass is water [14]. Take GJ 1214b, a super-earth orbiting a red dwarf 40 light years away and one of the most bizarre planets known to man [15]. Beneath a dense fog of steamy vapour, exotic phases of water exist, like crystalline 'hot ice' or superfluid water that has zero viscosity and could climb up walls, were there any on GJ 1214b to climb up. Yet there are no walls, nor any land and possibly not even a solid core. Sweltering in intense heat emitted by a star only two million kilometres distant from the planet, it nevertheless contains much more water than Earth and far less rock – three-quarters of the planet, which is 2.7 times the diameter of Earth or 34,403 kilometres across, may be composed of water judging from its density, and 20 percent of its diameter is atmosphere. Not that there would be any ocean present for intrepid interstellar explorers to go swimming in – baking at 230 degrees Celsius heat, the steamy atmosphere simply becomes ever denser the deeper one descends into it until one reaches its core. How did such a weird planet arise? It likely formed much further from its star, out beyond the 'snow line' where temperatures are cool enough for ices to form, before migrating inwards, a gigantic ball of thawing ice [16].

At the time of writing, few planets the size of Earth or smaller have been discovered. However there is no cosmic conspiracy that is preventing small terrestrial planets; it is simply a matter of our instruments not being sensitive enough to catch the indirect effects of worlds that small. Still, that doesn't mean there hasn't been some success in this area. The pulsar planets were sub-earth size, while the Kepler Space Telescope has begun to turn up a handful of miniature worlds, its tiniest so far being Kepler 37b, which is smaller than even the smallest world in our Solar System, Mercury [17]. Alas, at 210 light years away, it is likely out of our range, at least for the first few generations of any future starships.

Red Dwarfs: Diminutive but deadly?

Instead, within our potential range of 15 light years lie forty red dwarf stars. Now, red dwarfs are particularly liked by planet hunters. Being scaled down stars, their planetary systems are also scaled down, bunched together and much closer to the star than in our own Solar System, for example. That's okay; the smaller ratio of masses and diameters makes transit and radial velocity measurements somewhat easier, as does the proximity of the planet to their stars. Nor do their proximity to their star rule out their position in the habitable zone, for the cooler the star, the closer its habitable zone is. In addition, smaller stars live the longest and red dwarfs are expected to outlast all other stars, surviving for trillions of years to a time in the future when cosmic expansion has taken all the other galaxies far beyond the cosmic horizon and the Universe beyond our own Galaxy (or, rather, the product of the likely impending merger between the Milky Way and the Andromeda Galaxy in several billion years' time) is only dark. Red dwarfs seem very attractive to life on the move, searching for a new home amongst the stars. A civilisation hoping to endure into the twilight of the Universe would find few longer-lasting abodes than a planet around a red dwarf.

They're not perfect stars for life. Any planets orbiting close to them will be tidally locked, meaning that they always show the same face to their star, much like the Moon always shows the same familiar face to us here on Earth. The reason is because our Moon is in lockstep with planet Earth – as Earth's gravity acts on the smaller body of the Moon, tidal forces create a torque that causes a bulge in the Moon as it orbits our planet. The bulge tries to pull closer to Earth, in a direction that synchronises the Moon's rotation with that of Earth. Therefore the Moon always shows the same face not because it isn't rotating, but because its period of rotation is exactly locked to its orbit around Earth and the same holds true for exoplanets orbiting close to their parent stars.

For exoplanets, this can have some interesting results. If one side always points towards its star, the other side always points away into space. Absent of an atmosphere, the 'dark side' will freeze while the day side is warm (or just warm as is likely the case with a cool red dwarf). The terminator between the two hemispheres then becomes a thin sliver of temperate twilight. Add in an atmosphere and things get even more interesting, since winds can distribute heat around to the dark side of the planet. In the case of a number of hot jupiter planets where astronomers have been able to measure atmospheric properties via transit spectroscopy (light from the star reflected by the planet, or filtered through its atmosphere, is imprinted with the compositional signature of the atmosphere and, by subtracting the light of just the star when the planet is in eclipse behind it from the light of the star and planet together, the planet and its reflected/absorbed light can be isolated). The Sun will always appear to hang in the same part of the sky from the dayside – the sub-stellar point (high noon on Earth) will burn but it isn't the hottest spot on the planet, as winds moving from the warm daylight hemisphere to the cooler, dark hemisphere transport the heat. In the case of HD 189733b, a hot jupiter around 63 light years distant, the cloud tops on its dayside hemisphere reach over 900 degrees Celsius and supersonic winds whipping around the planet at ten thousand kilometres per hour redistribute the heat. Here even the permanent night side will not freeze out but will be kept balmy by the constant flow of heat brought the winds [18]. The same could occur on a tidally-locked super-earth with an atmosphere. While things may be less extreme on a planet around a white dwarf, the yin-yang climate of the day and night hemispheres will remain.

Additionally red dwarfs, being so cool (their surface temperature is less than 4,000 degrees Celsius), emit most of their light at infrared wavelengths, which is lower energy than visible light and would slow down biochemical and photosynthetic processes of any animals or plants on an orbiting planet (research indicates that plants on a world around a red dwarf could be black in order to absorb as wide a range of wavelengths as possible to increase the energy they receive [19]). Meanwhile, some red dwarfs spin very fast, a feature that lends itself to the generation of a powerful magnetic dynamo within a star's core, encouraging a strong magnetic field leading to plagues of starspots covering half the star's surface and violent X-ray flares on the star's surface that could produce radiation capable of ionising a nearby planet's atmosphere. However, these flares could be life's saviour; while they do produce X-rays, they also produce bursts of ultraviolet light that, as long as they are not too intense, could aid photosynthesis and the formation of complex molecules by breaking apart molecules in the atmosphere [20]. Of course this assumes that alien plant life on such a planet would operate under the same principles as vegetation on Earth, but if an interstellar mission arrived in a red dwarf planetary system and set up a colony on a habitable zone world, it may learn to make the most of these flares for the agriculture necessary for humans to survive. However, some models suggest that planets around red dwarfs will be bathed in radiation, their atmosphere stripped from them by a wind of radiation equivalent to our Sun's solar wind. The radiation wind may be able to attack a planetary atmosphere thanks to a red dwarf's magnetic field that could be so powerful as to push a planet's magnetic field back down to the surface, or even below the surface, leaving the planet exposed to the space elements [21].

Statistical extrapolation from HARPS observations suggests that at least 60 percent of red dwarfs have planets the size of Neptune or smaller, with 41 percent being super-earths in the habitable zone [22]. Meanwhile, Kepler data points to six percent of red dwarfs having Earth-sized worlds in the habitable zone [23]. Considering that red dwarfs number 160 billion in a galaxy of approximately 200 billion stars, even six percent amounts to a lot of stars (9.6 billion). So we can reasonably expect 16 of the 40 nearby red dwarfs to possess super-earth planets or smaller (although a little beyond that range, but in June 2013 three super-earths were found in the habitable zone of the red dwarf Gliese 667, around 22 light years away [24]) and between two and three of them to possess Earth-sized worlds in their habitable zones (note that this does not necessarily mean they are Earth-like, only Earth-sized).

How does that stack up with what we currently know for sure is out there? The red dwarf Gliese 876, located 15.3 light years away, possesses four worlds, a super-earth and three gas giants. The super-earth, Gliese 876d, has 5.88 times the mass of our planet and is the innermost world – at the time of its discovery in 2005 it was the lowest mass exoplanet known. Too close to its star, there's no chance for liquid water or life as we know it on Gliese 876d, not with a surface baking at somewhere between 157–357 degrees Celsius on its dayside hemisphere. Yet it will also be tidally locked, incurring that never-ending day/night dichotomy. The cycles of day and night, of the rising and setting Sun that influences life cycles of animals and plants on Earth, will be foreign in these lands.

As for the gas giants of Gliese 876, planet e is beyond the habitable zone, but planets b and c fall in it. This could lead to intriguing possibilities if they possess large moons that could support liquid water. In our Solar System several of the gas giants' moons are as large as small planets, such as 5,268-kilometre wide Ganymede and 5,150-kilometre Titan, both of which are actually larger than Mercury. Were they to exist around a gas giant in the habitable zone they would be equally potentially habitable as any rocky planet in the same location. David Kipping, now of Harvard University, suggested in 2009 that such 'exomoons' could be detected through transit timing variations (TTV) [25] – the notion is that the gravitational tug of the moons on the planet could marginally change the timing of when the planet should transit its stars. While unseen planets have been detected through TTV, no exomoon has yet been found and it remains uncertain whether our current capabilities are sufficient to make a detection.

Meanwhile Gliese 674, another red dwarf 14.8 light years distant, sports a Neptune-sized planet orbiting close to it. Hot and gaseous, this would be no world to settle on. Unfortunately, the star is unlikely to harbour any other worlds if the hot neptune migrated in from colder realms, barging past any other smaller worlds and ejecting them from the system.

As far as we know, that is it as far as nearby red dwarfs go. So either our statistics are wrong or there are still many more worlds around nearby red dwarfs that remain undiscovered. Can we expect expect the red dwarf Proxima Centauri, the closest star to the Sun, to feature a planet in the habitable zone? What about Barnard's Star, which is the fastest (apparent) moving star in the sky? Or Wolf 359, the location of that great fictional battle with the feared Borg in *Star Trek: The Next Generation*? Barnard's Star has been studied in great detail over recent decades but to no avail, yet the statistics certainly favour continuing to search around the 36 other red dwarfs within 15 light years.

WHAT MAKES A WORLD HABITABLE?

Two other stars within 15 light years are also known to have planets. Ten point five light years away, epsilon Eridani has a gas giant, but this time not a hot jupiter. With a semi-major axis of 3.4 astronomical units (one astronomical unit is the average distance between Earth and the Sun, 149.6 million kilometres; denoted by the letters AU, the astronomical unit is the standard unit of measurement on the scale of planetary systems) it is respectfully distant from epsilon Eridani compared to its hot-headed cousins around other stars. Still, it is beyond the habitable zone and likely too cold for life as we know it on any moons it may posses. More planets may exist closer in.

Finally there is alpha Centauri B, a member of a trio of stars including Proxima Centauri that form the closest stellar system to our own Solar System. As such they are a strong candidate for our first interstellar destination. First the lowdown on the stars: alpha Centauri A is ten percent more massive than our Sun and 1.5 times as luminous, whereas alpha Centauri B is only 90 percent as massive as our Sun and only half as luminous. Both stars orbit one another and are slightly older than the Sun at six billion light years (which explains why alpha Centauri A is so much brighter when it is only slightly larger; stars increase in luminosity as they get older and the Sun will do the same, spelling the end for animal life on Earth within a billion years, and microbial life in 2.8 billion years time [26]) but there is no reason why one or even both cannot harbour planets, especially given that we are now beginning to discover exoplanets in other binary systems; evidently there are gravitationally stable zones between companion stars where planets can reside without disruption. Then there's Proxima, an outlier that's a full tenth of a light year closer to us than the other two and probably gravitationally bound to alpha Centauri A and B, although that's not known for sure.

Given that they are effectively our 'next door neighbours', great effort has gone into searching for planets around one of the alpha Centauri stars in recent years, resulting in triumph in late 2012 when European astronomers using HARPS detected a small radial velocity signal. They attributed this to a planet with 1.13 times the mass of Earth orbiting around alpha Centauri B at a distance of 5.98 million kilometres (0.04 astronomical units), close enough that a year on the planet (i.e. the time taken to orbit the star once) lasts just 3.24 Earth days. As mentioned earlier in this chapter, the radial velocity signal was the smallest yet measured for an exoplanet, just 51 centimetres per second. Alpha Centauri Bb (we can see here how the nomenclature can grow confusing!) will also be tidally locked, so that the dayside hemisphere will swelter in excess of 870 degrees Celsius. At that temperature, its dayside surface won't be solid at all, but an ocean of magma, only hardening nearer the terminator. Its farside will, by contrast, be plunged into twilight (alpha Centauri A will be visible in the sky, but its distance ranges from around 1.5 billion kilometres and six billion kilometres and it would appear in the sky over the farside of the planet during periods of daylight lasting almost 39 hours). Temperatures will be low enough for ice on this hemisphere. Hot and cold, fire and ice. This is one world we're not going to want to colonise.

Pushing the envelope like this on the radial velocity measurement inevitably leads to questioning of the results. Does the signal and therefore the planet really exist, or is it a figment of analysing the data incorrectly? It is an important question to ask, especially for interstellar mission controllers who may target the planetary system as the first destination of a voyage to the stars. Recent work by Artie Hatzes of the Friedrich Schiller University in Jena, Germany, has suggested that by breaking down the radial velocity data and analysing segments at a time rather than lumping all the data into one big melting pot the pattern of the signal becomes no longer apparent, casting doubt on alpha Centauri Bb's existence. We should be mindful, however, that even if it doesn't exist there is still a good chance that there could be super-earth planets or smaller somewhere in the alpha Centauri system. Statistical studies point to there being a total of 100 billion planets at least in our Galaxy alone, with ten billion being smaller, rocky worlds and an estimated 1,500 expected to reside within 50 light years of the Solar System [27]. Almost all stars will have planets and where there is one planet it has been shown that there is inevitably almost more. That's why people grew excited over the possibility that alpha Centauri Bb might exist, because even though it is too hot and lifeless to ever be a planet humans would want to settle on, its existence would imply that there would likely be other planets around alpha Centauri B too, perhaps in the habitable zone where temperatures are appropriate for liquid water.

It's worth pausing at this juncture to just go into a little bit more detail about what the habitable zone can promise. In short, it actually promises nothing, which does somewhat make a mockery of the high esteem that exoplanet hunters and the media seem to hold it in. Indeed, throughout this chapter there have even been frequent references to searching specifically for terrestrial planets or exomoons within the habitable zones of their respective stars. The implication of the habitable zone, often popularly referred to as the 'Goldilocks' zone where temperatures are 'just right' is that as long as there is a planet there then, hey presto, there's bound to be water and conditions suitable for life. In an ideal scenario there would be but, if we have learnt anything from our two decade study of exoplanets, it is that planets rarely follow a typical scenario.

Part of the problem is that many people confuse 'habitable zone' with 'inhabited'. For water to exist as a liquid an insulating atmosphere is required, but not one too thick that it warms the planet too much, as with the case on Venus. This can make habitable zones decidedly uneven, with planets closer to their Sun retaining liquid water if they have thinner, more transparent atmospheres, and those on the outer edge of what is deemed to be the habitable zone retaining their water because they have a thicker atmosphere that is able to retain more heat and prevent the planet from freezing over. Plus, let us not forget that planetary orbits are elliptical and some exoplanets have been seen to exist on highly elliptical orbits, almost diving into their Sun at closest approach (perihelion) before looping back out again. Such worlds could pass through a habitable zone – would this mean they are only habitable for part of the time? Furthermore, there is much more to habitability than just the presence of liquid water, with plate tectonics resulting in a carbon–silicate cycle, a large moon and a magnetic field all being attributed to the success of life on Earth. And what of planets with sub-surface oceans, like Jupiter's cold moon Europa? These would exist beyond the traditional habitable zone. So when discussing the habitable zone, we must remember that it is not the be all and end all; at best, it's just a starting point, but an important one if we are looking for Earth mark two. However, nature could find other avenues to habitability; it is not necessarily a one-size fits all. Perhaps we shouldn't become too obsessed with finding worlds just like Earth.

PLANETARY CHAUVINISM

From a scientific point of view all planets hold great interest, just as the other planets in our Solar System are ripe for exploration. We would surely send a purely scientific probe to a planetary system that has the widest diversity within reach of Earth. This will mean a multi-planetary system and there is good news on that front. Although the closest known multi-planetary system is 15 light years away, analysis of the findings from planet-hunting surveys such as Kepler show that in most cases planets come in multiples [28] (all with the exception of hot jupiters – as we've seen these worlds ensure they are alone by ejecting the other planets as they migrate in-system). So we may only know of one planet around alpha Centauri B but, if it really exists (there is some controversy over this, given the weakness of the signals and the newness of the particular technique for teasing out the data [29]), then there were surely be more outside of our current detection range. Wouldn't it be wonderful if there were a planetary system as rich and diverse as our Solar System right on our doorstep? Particularly if it sports planets different to ours – hot neptunes and super-earths in particular are begging for more detailed study so we can better fit them into the pantheon of planets.

We are also running the risk of planetary chauvinism here. Sure, planets are interesting to us – we live on one after all, and would like to compare our world with others. We might even think we'd like to find another planet to live on. Yet planets are not the only objects to orbit stars that may be of interest. Around epsilon Eridani for instance (and I cannot mention epsilon Eridani without also pointing out that this was the system where the fictional *Babylon 5* space station was located) there exists two asteroid belts, as observed by NASA's infrared-viewing Spitzer Space Telescope [30] and initially detected with submillimetre wavelength SCUBA camera on the James Clerk Maxwell Telescope in Hawaii in 1998 [31], as well as an icy ring of comets analogous to our Solar System's own Kuiper Belt. The asteroids belts are three and twenty astronomical units from epsilon Eridani respectively, with the known planet epsilon Eridani b orbiting on the inner edge of the first belt, likely constraining the belt and ushering material into it via gravitational resonances just like Jupiter does with our Asteroid Belt. This strongly suggests the presence of at least two more planets shepherding the second asteroid belt and the icy belt of comets located between 35 and 90 astronomical units. The mass of the two asteroid belts combined has been estimated, via the Spitzer observations, to be 10^{18} kilograms – that's a billion billion kilograms. Our Asteroid Belt is more massive – 10^{21} kilograms – but even that is dwarfed by some of the other asteroid belts that we are finding, such as that around the star HD 69830, which is 41 light years away and has an asteroid belt 25 times more massive than the Solar System's belt [32]. Or the gargantuan asteroid belt around zeta Leporis, 200 times more massive [33] than our own.

Why would asteroids and comets specifically interest us? Both are full of resources, from water–ice that can be used to create rocket fuel or extracted for drinking water or oxygen, to metals such as zinc, tin, silver, gold, copper and platinum-group metals. Robotic probes could use these resources to refuel or build replicas of themselves – self-replicating machines styled after the physicist John von Neumann's idea of a machine that can build identical copies of itself from spare parts. Alternatively, perhaps after decades or centuries travelling through deep space, human voyagers will have grown accustomed to living in a space environment and will no longer see a need to land on a planet and make a home there. These interstellar nomads would find asteroids and comets to be easily accessible sources of resources, much preferred over dealing with the energy-consuming gravitational wells of planets. (As an aside, science journalist Michael Chorost has postulated that the escape velocities of super-earths may be so high as to make a journey into orbit extremely expensive and difficult, requiring lots or resources for the rocket power necessary [34] – that said, Chorost also points out that a civilisation on a super-earth would likely have access to far more resources than we have on earth. Either way, for our exploratory mission, landing on a super-earth could potentially trap the lander there.)

Asteroids and comets are not the only non-planetary targets that may tempt us. There has been some speculation that there could be a brown dwarf, or perhaps even more than one, closer to us than the alpha Centauri system and indeed astronomers have been searching using infrared surveys, in particular NASA's now defunct Wide-field Infrared Survey Explorer (WISE) spacecraft. Brown dwarfs are even cooler than red dwarfs, with the coolest, discovered by WISE, being no more than room temperature [35]. This renders them extremely faint and difficult to spot and it is not beyond reason that one could exist within one or two light years of the Sun. If such a brown dwarf were discovered, it would become the next logical step after our Solar System's Oort Cloud of comets, which is thought to stretch almost a light year.

Then there are rogue planets – all those worlds ejected by rampaging hot jupiters have to go somewhere, right? The idea is that they are sent into interstellar space to wander alone. Some will find themselves captured on the edge of other planetary systems they come across, while others will float amongst the stars for eternity. Indeed, recent evaluations of these free-floating planets predicts that they could number 100,000 times more than there are stars in the Galaxy [36] – and considering there are 200 billion stars in the Galaxy, that's a lot of rogue planets. If that number is correct, it almost inevitably means there will be one, or possibly more than one, close to our own Solar System, but like brown dwarfs they will be difficult to find ahead of an interstellar mission. Unlike brown dwarfs, however, rogue planets could be habitable, even without a star to breath warm light on them.

VOYAGING BEYOND THE BOUNDARY

Let's not forget the stars themselves. Both from a scientific and a resource perspective they garner a great deal of interest. Most of our knowledge about the fine details of how stars function comes from our studies of our nearest star, the Sun. The other stars in the Galaxy are so distant that in most cases they cannot even be resolved as anything but point sources of light. Astronomers have observed magnetic fields in other stars, detectable by the way the magnetic field polarises the star's light, but have only observed a complete magnetic cycle of activity like the eleven year sunspot cycle on the Sun in one star, tau Boötis [37]. To be able to get up close to a star and observe its own activity cycle, replete with flares, prominences, coronal mass ejections, starspots (as we have seen, red dwarfs can sport high degrees of magnetic activity and cover fifty percent of their surfaces in darker, cooler blemishes) and high and lulls in magnetic activity would provide a much-need comparison to understanding the behaviour of our own Sun.

Stellar magnetic fields are able to create magnetic bubbles around their planetary systems. Our Sun does exactly this, its magnetic field pulled and stretched away from it by the solar wind to create an environment known as the heliosphere that extends beyond 18 billion kilometres from the Sun. The two Voyager probes, launched by NASA in 1977, are currently exploring this distant realm and studying the interaction between the Sun's magnetic environment, filled with charged and neutral particles from the solar wind, and the interstellar medium. In 2004 Voyager 1 crossed the 'termination shock' – the entrance to the outer 'heliosheath' – where the solar wind drops below supersonic speed, a zone that can ebb and flow with the tide of the solar wind pressure against the opposing pressure of the exterior interstellar magnetic field. On 25 August 2012 Voyager 1 departed the heliosphere, passing beyond the boundary where solar wind particles are leaking out and escaping the Sun's magnetic environment and crossing into interstellar space. Voyager and NASA's Interstellar Boundary Explorer (IBEX) have also observed cosmic rays – high energy particles accelerated by supernovae remnants – coming the other way, penetrating the heliosphere and reach the inner planets. Fortunately for life on Earth there are several obstacles to these potentially hazardous cosmic rays, not least the magnetic field of the heliosphere itself, which is able to deflect them, and Earth's own atmosphere. Voyager 1's sister probe, Voyager 2, is expected to also clear the heliosphere and enter interstellar space in the next few years.

The Voyagers will not encounter any stars on their journeys; the closest they will come to one is around 1.6 light years, which is the distance that Voyager 1 will pass the red dwarf star Gliese 445 in the year 40,272 AD, and Voyager 2 will pass within 1.7 light years of another red dwarf star, Ross 248 around the year 42,200 AD. [38] (Note: both these stars are currently quite far from the Sun – Gliese 445 is 17.6 light years and Ross 248 is 10.3 light years distant. However, the stars are also moving and will come relatively close to the Sun – closer than the current closest star – within forty millennia.)

From the point of view of a dedicated, high-velocity, interstellar starship the Voyager findings on the boundary of the Solar System indicate what we may expect to find as we approach the magnetic bubble of another star, called is 'astrosphere'. According to NASA's Arik Posner, the Voyager Programme Scientist for the agency's Heliospheric Division, a starship speeding towards an astrosphere will first detect the interstellar magnetic field beginning to buffet up against the star's, while at the same time detecting the onrushing flux of neutral particles from the star that can pass through the magnetic fields unhindered (whereas charged particles end up following the magnetic field lines) [39]. The starship will therefore be able to study the star's stellar wind before it even enters its magnetic domain, telling us about conditions on the star and how it might affect planets close to it.

Stars, brown dwarfs, planets – rogue or in situ – asteroids, comets; is there anything else within 15 light years that could garner our attention and call to us to fly to it? Not to our knowledge. Yet what about further afield? Vast nebulae dozens, hundreds, even thousand of light years across shelter stellar embryos, new stars and planets in the process of forming. There are dead white dwarfs, the core remnants of Sun-like stars that have gone extinct and around which some astronomers are even searching for planets [40], and neutron stars and black holes that are the exotic products of the violent deaths of massive stars. Some researchers have speculated that rather than colonising planets or spending the centuries travelling in the darkness between stars, technologically advanced extraterrestrial civilisations may gather around such objects, using their steep gravitational wells to siphon off energy or fabricate giant quantum computers [41]. Right now, to human beings, planets seem attractive, but after centuries in the microgravity of space or to any intelligent, thinking space probes that we may launch in the future, things might seem different.

Estimates suggest that we could colonise the entire Galaxy – assuming there were no other intelligent life-forms competing for the same stars, planets and resources – in anywhere between two and fifty million years [42]. We could have moved on to the nearest galaxies, the Large and Small Magellanic Clouds, in similar timescales and have set course for the other large galaxies in the Local Group of Galaxies, the Andromeda and Triangulum galaxies, 2.5 and 2.9 million light years away respectively, sometime after that. Of course, when I say we, our descendants will be vastly changed compared to humanity now. But hopefully they will carry that same seed, that same spark of curiosity, to yearn to want to learn what is out there, just over the stellar horizon and beyond the boundary. We won't find out unless we go.

CONCLUSIONS

Planets around other stars are proving just as wonderfully different as the greatest science fiction stories have depicted. Worlds burning on the doorstep of stars, planets where day and night are permanent, spheres made of water, ice or fire, swathed in deep atmospheres where storms rage or exposed naked to the environment of space, with moons and rings and maybe even sea and land and possibly, somewhere that we could call home. Our first interstellar wanderers, NASA's Voyager probes, are now beginning to leave the Sun's heliosphere; when we one day follow them will we find a planet close to us that is sufficiently a twin of Earth?

RECOMMENDED READING

- Stephen Baxter, *Proxima*, Gollancz, 2013.
- Stephen Baxter, *The Science of Avatar*, Gollancz, 2013.
- Ray Jayawardhana, *Strange New Worlds: The Search for Alien Planets and Life Beyond Our Solar System*, Princeton University Press, 2011.
- James Kasting, *How To Find a Habitable Planet*, Princeton University Press, 2010.
- Charles H Langmuir and Wally Broecker, *How To Build a Habitable Planet*, Princeton University Press, 2012
- Kelvin F Long, *Deep Space Propulsion*, Springer, 2012.
- Stephen J Pyne, *Voyager: Seeking Newer Worlds in the Third Great Age of Discovery*, Viking Books, 2010.
- Dimitar Sasselov, *The Life of Super-Earths*, Basic Books, 2012.
- Peter Ward and Donald Brownlee, *Rare Earth*, Copernicus, 2000.

REFERENCES

[1] William Waller and Paul Hodge; *Galaxies and the Cosmic Frontier*, Harvard University Press (2003).

[2] The One Hundred Nearest Star Systems (by RECONS – Research Consortium On Nearby Stars) (online) www.recons.org/TOP100.posted.htm (last accessed 15 October 2014).

[3] E Rivera et al; 'The Lick-Carnegie Exoplanet Survey: A Uranus-mass Fourth Planet for GJ 876 in an Extrasolar Laplace Configuration', *The Astrophysical Journal*, 719, (2010), pp890–899.

[4] X Dumusque et al; 'An Earth Mass Planet Orbiting Alpha Centauri B', *Nature*, 491, 7423 (2012) pp207–211.

[5] A Hatzes; 'Radial Velocity Detection of Earth-mass Planets in the Presence of Activity Noise: The Case of Alpha Centauri Bb', *The Astrophysical Journal*, 770, 133 (2013).

[6] A Bond et al; 'Project Daedalus – The Final Report on the BIS Starship Study', *Journal of the British Interplanetary Society Interstellar Studies*, Supplement (1978).

[7] M Mayor et al; 'A Jupiter-Mass Companion To a Solar-Type Star', *Nature*, 378, (1995) pp 355–359.

[8] M Mayor et al; 'Setting New Standards With HARPS', *ESO Messenger*, 114, (2003) pp20–24.

[9] F Pepe et al; 'ESPRESSO: The Echelle Spectrograph for Rocky Exoplanets and Stable Spectroscopic Observations', *Proceedings of the SPIE*, 7735, (2010).

[10] F Fressin et al; 'The False Positive Rate of Kepler and the Occurrence of Planets', *The Astrophysical Journal*, 766, 2013.

[11] D Kipping et al, 'A Simple Quantitative Method to Infer the Minimum Atmospheric Height of Small Exoplanets', *Monthly Notices of the Royal Astronomical Society*, in press, 2014.

[12] H J van Heck et al; 'Plate Tectonics on Super-Earths: Equally or More Likely Than On Earth', *Earth and Planetary Science Letters*, 310, (2011) pp252–261.

[13] V Stamenković et al; 'The Influence of Pressure–Dependent Viscosity on the Thermal Evolution of Super-Earths', *The Astrophysical Journal*, 748 (2012).

[14] S Seager et al; 'Mass–Radius Relationships for Solid Exoplanets', *The Astrophysical Journal*, 669 (2007) pp1279–1297.

[15] D Charbonneau et al; 'A Super-Earth Transiting a Nearby Low Mass Star', *Nature*, 462 (2009) pp891–894.

[16] Z Berta et al; 'The GJ 1214 Super-earth System: Stellar Variability, New Transits and a Search for Additional Planets', *The Astrophysical Journal*, 736 (2011).

[17] T Barclay et al; 'A Sub-Mercury Sized Exoplanet', *Nature*, 494 (2013)

pp452–454.

[18] C J Grillmair et al; 'A Spitzer Spectrum of the Exoplanet HD 189733b', *The Astrophysical Journal*, 658 (2007) pp767–769.

[19] J T O'Malley-James et al; 'Life and Light: Exotic Photosynthesis in Binary and Multiple-Star Systems', *Astrobiology*, 12 (2012) pp115–124.

[20] A Buccino et al; 'UV Habitable Zones Around M Stars', *Icarus*, 192 (2007) pp582–587.

[21] A Vidotto et al; 'The Effects of M Dwarf Magnetic Fields On Potentially Habitable Planets', *Astronomy and Astrophysics*, 557 (2013).

[22] X Bonfils et al; 'The HARPS Search for Southern Extra-Solar Planets XXXI: The M-Dwarf Sample', *Astronomy and Astrophysics*, 549 (2013).

[23] C D Dressing et al; 'The Occurrence Rate of Small Planets Around Small Stars', *The Astrophysical Journal*, 767 (2013) pp225–250.

[24] G Anglada-Escudé et al; 'A Dynamically-Packed Planetary System Around GJ 667C With Three Super-Earths In Its Habitable Zone', *Astronomy and Astrophysics*, 556 (2013).

[25] D Kipping; 'Transit Timing Effects Due to an Exomoon', *Monthly Notices of the Royal Astronomical Society*, 396 (2009) pp1797–1804.

[26] J T O'Malley–James et al; 'Swansong Biospheres: Refuges for Life and Novel Microbial Biospheres on Terrestrial Planets Near the End of their Habitable Lifetimes', *International Journal of Astrobiology*, 12 (2013) pp99–112.

[27] A Cassan et al; 'One Or More Bound Planets Per Milky Way Star From Microlensing Observations', *Nature*, 481 (2012) pp 167–169.

[28] J Lissauer et al, 'Almost All of Kepler's Multiple-Planet Candidates are Planets', *The Astrophysical Journal*, 750 (2012).

[29] A Hatzes et al, 'Radial Velocity Detection of Earth-Mass Planets in the Presence of Activity Noise: The Case of Alpha Centauri Bb', *The Astrophysical Journal*, 770 (2013).

[30] D Backman et al; 'Epsilon Eridani's Planetary Debris Disc: Structure and Dynamics Based on Spitzer and Caltech Submillimeter Observatory Observations', *The Astrophysical Journal*, 690 (2009).

[31] J Greaves et al; 'A Dust Ring Around Epsilon Eridani: Analogue to the Young Solar System', *The Astrophysical Journal*, 506 (1998) pp133–137.

[32] C A Beichman et al; 'An Excess Due to Small Grains Around the Nearby K0 V Star HD 69830: Asteroid or Cometary Debris?', *The Astrophysical Journal*, 626 (2005).

[33] M Moerchen et al; 'Mid-Infrared Resolution of a 3 AU-Radius Debris Disc Around Zeta Leporis', *The Astrophysical Journal*, 655 (2007) pp109–112.

[34] M Chorost; 'Do Super-Earths Trap the Civilisations on Them?', *Psychology Today*, (online) www.psychologytoday.com/blog/world-wide-mind/201211/do-super-earths-trap-the-civilizations-them (last accessed 16 October 2014).

[35] J D Kirkpatrick et al; 'The First Hundred Brown Dwarfs Discovered by the Wide-Field Infrared Survey Explorer (WISE)', *The Astrophysical Journal Supplement Series*, 197 (2011).

[36] L Strigari et al; 'Nomads of the Galaxy', *Monthly Notices of the Royal Astronomical Society*, 423 (2012).

[37] A Vidotto et al; 'The Stellar Wind Cycles and Planetary Radio Emission of the tau Boo System', *Monthly Notices of the Royal Astronomical Society*, 423, 2012.

[38] NASA; *Voyager: The Interstellar Mission*, (online) http://voyager.jpl.nasa.gov/mission/interstellar.html (last accessed 16 October 2014)

[39] K Cooper; 'Retrospective: Voyager', *Principium*, issue 2 (2013).

[40] M Burleigh et al; 'Imaging Planets Around Nearby White Dwarfs', *Monthly Notices of the Royal Astronomical Society*, 331 (2002).

[41] S Lloyd; 'Almost Certain Escape from Black Holes in Final State Projection Models', *Physical Review Letters*, 96 (2006).

[42] M Hart; 'Interstellar Migration, the Biological Revolution and the Future of the Galaxy', *Interstellar Migration and the Human Experience*, University of California Press (1986).

CHAPTER 17

WHAT DO WE DO ONCE WE GET THERE?

KENNETH ROY, DAVID FIELDS AND ROBERT G KENNEDY

There are two basic reasons to journey to a distant star. The first is to explore, to learn, to understand something of our Universe that we cannot gain by staying at home. The second is to colonise, to carve out a new home for humanity. In the sixteenth century the explorers of the new world had a third goal: to get rich. A few of the lucky ones were able to acquire (loot) large quantities of gold and silver and, in a few months, be back in Europe to enjoy their wealth and fame. Unless faster-than-light travel (FTL) becomes possible, this third option is not a consideration because of the time and distance limitations. Any scientists or colonists that venture to another star must accept the reality that they aren't coming back. However, unlike the explorers of the new world, they will be in constant communication with the home world. True, it could take a decade to get an answer to a question but they'll be plugged into the pop culture, literature, movies, music, scientific advances, software updates and news from Earth, just a few years later than Earth dwellers. There is also the possibility of faster-than-light communication. If feasible, instant communication with the experts on Earth could greatly aid a colony in its struggle to survive and prosper.

If FTL travel becomes possible then exploration and colonisation of many distant stars becomes pretty much inevitable. Humanity will explode into the Galaxy like a tsunami. That may not be a good thing, for it will indicate that there is something fundamental about the Fermi Paradox that we don't understand. The Universe teaches us daily that what we don't understand can indeed hurt us. Perhaps we should hope that FTL travel is very difficult, if not outright impossible, for that at least offers a weak, but benign, explanation for the Fermi Paradox.

So we assume that starships travel at a sub-light velocity. Current technologies hold out the possibility that we can power these starships up to 10 percent of the speed of light, coast for a while, then decelerate to zero at their destination. The first ships will be robot scouts such as those envisioned by Project Icarus [1]. The ones that follow them will be crewed. Humans impose constraints on the ship design and a number of designs for crewed starships have been proposed. One early concept, the Enzmann starship, has a habitation module ~91 metres in diameter and 273 metres long and contains 575,000 cubic metres of living space [2]. The habitation module is spun-up to provide artificial gravity. The initial crew is estimated to be around 200 with the capacity to grow to 2,000 during the voyage. It uses a fusion drive system that allows a cruising speed of around 10 percent of the speed of light. At this velocity the travel time to alpha Centauri is 43.7 years. Allowing for acceleration and deceleration phases, the total trip time is probably closer to 50 years. Of course more distant stars will take longer to reach. If we limit the trip duration to 100 years, these starships have a range of ten light years. There are eight star systems within this radius, three of which are binary, plus one brown dwarf system named Lohman 16. Proxima Centauri, the nearest star at a distance of 4.24 light years, seems to be either loosely coupled gravitationally or, perhaps, even independent of the alpha Centauri system and is thus considered to be its own star system. It is the closest star to our Sun.

If the mission planners are depending on finding an Earth-type planet, ready for colonisation by humans, they will most likely be disappointed. Earth-type worlds are possibly rare and in all likelihood tens or even hundreds of star systems will need to be searched before we find the first one. However, most stars seem to have planets and sterile planets like Mercury and possibly Mars are probably fairly common. There are terraforming concepts being discussed today that could allow for the colonisation of almost any star system with such planets.

Such terraforming projects will take many centuries and possibly millennia before the newly terraformed worlds are ready for human habitation. During the terraforming project the new colony will need a secure base in which to live and grow, a base from which to explore their new star system, a base that can provide resources to support the terraforming project and from which to construct and send out additional robot probes and crewed starships to star systems further away. Building such a base will probably be the first order of business for the new colony. It is in this effort that the colony will survive or perish, but several issues need to be considered.

MATTERS OF GRAVITY

Humans need gravity to thrive and, perhaps, even to survive. One researcher has noted that many of the effects of exposure to a long term zero-g environment appear to mimic the symptoms of accelerated aging [3].

Exposure to zero-g for even a few months triggers major changes in the musculoskeletal system; it causes significant loss of muscle mass (even with strenuous, regular in-flight exercise), loss of bone density and produces a marked reduction of the functional capacity of the cardiovascular system. The immune system becomes less effective while some bacteria seem to become more dangerous. The function of the inner ears and the eyes is affected. Metabolic changes affect how drugs are absorbed into the body and even how they function, making the effects of even common drugs something of an unknown [4].

Once Earth-normal gravity is restored the damage to the musculoskeletal system seems to gradually recover, but the damage to sensory organs is problematic. The zero-g environment affects the inner ear systems in a way that in some cases leads to tinnitus and hearing loss. A number of male astronauts have developed permanent eyesight problems after long duration, zero-g missions [5].

The International Space Station and earlier space flight activities in Earth orbit have provided some opportunity to study human biology in a zero-g environment. However, there is effectively no experience with humans in a low-g (as opposed to a zero-g) environment. The time that astronauts have spent on Earth's Moon is too limited to draw any conclusions. Low-g environments cannot be created on Earth for more than a few minutes at a time. There are many open questions relating to how much gravity a human being needs to survive, grow and reproduce. Developing children seem to need gravity similar to that of Earth to develop bone, muscle and balancing abilities. They would probably do just fine in a gravity that is 80 or 90 percent that of Earth, but at some unknown level children will not develop properly.

For a colony to be successful it must allow for human reproduction. However, we know very little about human reproduction in zero-g environments. Is a successful human pregnancy even possible in a zero-g environment? Nothing is known about human embryogenesis under such conditions. Studies in vertebrate animal models seem to indicate that there are critical periods of development where gravity is important for proper embryogenesis [6, 7]. Radical changes occur in the structure and connectivity of nerve cells during the early development of the nervous system. Those changes are regulated by mechanical and biochemical factors, perhaps dependent to some extent on gravity. It has been observed that animals deprived of the opportunity to walk during their first weeks after birth never learn to walk properly [4]. The same questions also apply to low-g environments.

A viable base for the colony should be able to provide Earth-normal gravity, at least for pregnant women and growing children [4, 8]. Unless the colonists are lucky enough to find a planet or a moon that can provide natural gravity approaching Earth-normal, then some type of artificial gravity will be required. For a settlement or coasting spaceship the only form of artificial gravity available (barring a major scientific advance) is centrifugal acceleration resulting from circular motion. Long-term human habitation in a rotating environment poses challenges to the inner ear as a result of the Coriolis effect. Because of this and for general comfort a maximum rotation rate of one rpm is recommended [8, 9]. A specific rotation rate coupled with the required acceleration physics that are a result of that rotation then requires a specific radius, as we see in the table on the opposite page.

Table One: Radius of rotating body (in metres) given the rotation rate and acceleration

Gravity (in gees) (Earth = 1g)	0.5rpm	1rpm	1.5rpm	2rpm	2.5rpm
1	3,578	895	397	223	143
0.9	3,220	805	357	201	128
0.75	2,683	670	298	167	107
0.5	1,789	447	198	111	71
0.25	894	223	99	55	35

The challenge for starship designers is obvious and the proposed radius of the Enzmann starship (45 metres) is too low. The colony base, if it uses a rotating body to create an artificial gravity of one g, will need to have a radius of 800 to 900 metres. A crewed starship intended for a journey that could last for a hundred years probably needs something similar.

RADIATION ISSUES

Radiation from solar flares and cosmic rays poses a major problem when designing space ships and space settlements. The US Federal Government has established a total effective radiation dose limit of 0.5 rem/year (0.005 Sv/year) for the general population [10]. A rem (Sv, or sievert) can be thought of as a unit of damage done to living tissue by radiation. This limit is in addition to natural background radiation to which we are all exposed. On average, each human receives 0.3 rem/year (0.003 Sv/year) from natural sources and if they are lucky they get an additional 0.05 rem/year (0.0005 Sv/year) from medical x-rays. If a safe and effective cure for cancer is ever found, or if protective drugs are developed, these limits could be raised somewhat. Also, there is evidence that adult exposure to chronic radiation levels up to 50 rem/year (0.5 Sv/year) results in no detectable damage. However, when dealing with pregnant women and growing children it is better to be careful. For a proposed space colony it is reasonable to require that the colony base (intended for general populations including children and pregnant women) limit the total effective radiation dose to 0.5 rem/year (0.005 Sv/year) or less [8].

Our Sun occasionally produces flares consisting of gamma-rays, electrons, protons and some alpha particles. A NASA design study concluded that 550 grams per square centimetre of shielding is required to keep radiation levels below 0.5 rem/year when exposed to this type of radiation [8].

Cosmic rays are isotropic and bombard all objects in space constantly. They are relativistic particles that consist of about 90 percent protons and 10 percent heavier nuclei. These particles are stopped by modest shielding but they do create a shower of secondary particles (electrons, positrons, neutrons, muons, pions, gamma rays, as well as additional secondary ions) that continue on to create tertiary radiation and additional ions and so on.

Generally, high-atomic-mass materials are not a good choice for radiation shielding because when they are struck by the relativistic ions, the nuclei of these materials can fragment and produce a shower of secondary particles. Low-atomic-mass materials are less likely to fragment. Good choices would be hydrogen, boron, and lithium. Water or ice, because of its hydrogen content, is also a good choice, particularly when shielding against gamma-ray radiation. Studies have determined that 42.3cm (17 inches) of water reduces the intensity of gamma-ray radiation from fission reactors by a factor of ten [11]. Thus ten metres of water would reduce the intensity of gamma-ray radiation by a factor of at least 10^{23}. This is sufficient to protect against even a fairly intense gamma-ray burst. The minimum amount of water estimated by Eugene Parker and M Kim to shield from the general radiation hazard is a kilogram per square centimetre of water [12, 13]. As water has a mass of about one gram per cubic centimetre, a barrier of water 1,000cm (10 metres) thick is required. If we use ice, the thickness increases slightly to 10.6 metres. It would be prudent for the preliminary design objective for any proposed colony base to have at least a kilogram per square centimetre of water as shielding, with more being even better.

The Enzmann starship design used large quantities of deuterium (heavy hydrogen) as a fuel [2]. In the original design this fuel was stored in a frozen or slush state within a sphere in front of the habitation modules. It could serve as shielding by inserting the habitation modules in this sphere during the cruise phase.

Recent studies are beginning to suggest that relativistic heavy ions tearing through brain tissue can lead to significant cognitive impairment similar to Alzheimer's disease [14]. Thus, additional shielding beyond that described above may be needed to protect against the few relativistic heavy (iron) ions that make up the tail end of the cosmic-ray spectrum. The need for adequate shielding becomes a primary driver certainly for space settlements but also for starship designers.

PLUTOIDS

In August 2006 the International Astronomical Union (IAU) reclassified Pluto from 'planet' to 'dwarf planet'. The politics for this demotion are interesting irrelevant, although it did raise the question of what exactly to call this type of celestial body. Currently Pluto and other dwarf planets in the outer Solar System have been straddled with the IAU's somewhat controversial term 'plutoid'. Plutoids are defined as celestial bodies in orbit around the Sun with a semi-major axis greater than that of Neptune's and that have enough mass to allow gravity to overcome the rigid body forces and achieve hydrostatic equilibrium (a spherical shape), but that have not cleared their neighborhood of other objects. By stretching the current IAU definition somewhat, it is reasonable to call Pluto-type worlds around other stars plutoids.

Plutoids are important not because they are in any way Earth-like, but because they seem to have a predictable structure and to be fairly ubiquitous. At the time of writing only four objects are officially described as plutoids, namely Pluto, Eris, Makemake and Haumea. In addition a number of other candidate plutoids in the Kuiper Belt and beyond have also been discovered, including 2007 OR10, Sedna, Quaoar, Orcus and 2012 VP113 with the expectation that many more will be added to the list in coming years. Mike Brown of the California Institute of Technology, whose co-discovery of Eris prompted the reclassification of Pluto, maintains a list of possible dwarf planets in the outer Solar System. The above named bodies are listed as certain, but Brown also lists 30 more as highly likely, 60 that are likely, 103 that are probable and 394 as possible dwarf planets [15]. Brown also estimates that there are between 40 to 120 Sedna-sized objects (at least 1,000 kilometres in diameter) in the Oort cloud and five to eight such objects in the Kuiper Belt.

To understand why plutoids seem to be so common it is necessary to understand the idea of the snow-line, which is the distance from a star where volatiles (materials with a relatively low boiling point, such as water) freeze. Stars originate from a large cloud of gas and dust that is compressed in a small enough region of space. Gravity then causes this cloud to contract further into a protostar and an accretion disc. Once the protostar becomes massive enough and dense enough it initiates fusion reactions in its interior and a star is born. The radiation from this new star then begins to warm the particles that form the accretion disc. These particles consist of ices and dust. Ices inside the snow-line melt vaporise and are pushed out of the inner system by the solar wind and radiation pressure. Ices beyond the snow-line warm slightly but remain frozen. Meanwhile, particles throughout the disc continue to accrete into larger and larger bodies, eventually resulting in planets. Those inside the snow-line are formed mainly of dust and become rocky worlds. Planets beyond the snow-line are formed of both dust and ices. If they become large enough, they can also begin to collect hydrogen and grow into gas giants. Bodies that form in the outer reaches are hampered by the lower density of the accretion disc at this distance and are thus limited in how big they become. However, the volume of the outer reaches of the accretion disc is vast and so the number of plutoids formed is large relative to the number of rocky worlds in the inner disc. It is reasonable to assume that most other stars, including binary stars, have large numbers of plutoids orbiting out beyond their snow-line.

By definition, plutoids are large enough to have sufficient mass that self-gravity pulls the body into a nearly spherical shape. It is probable, although not

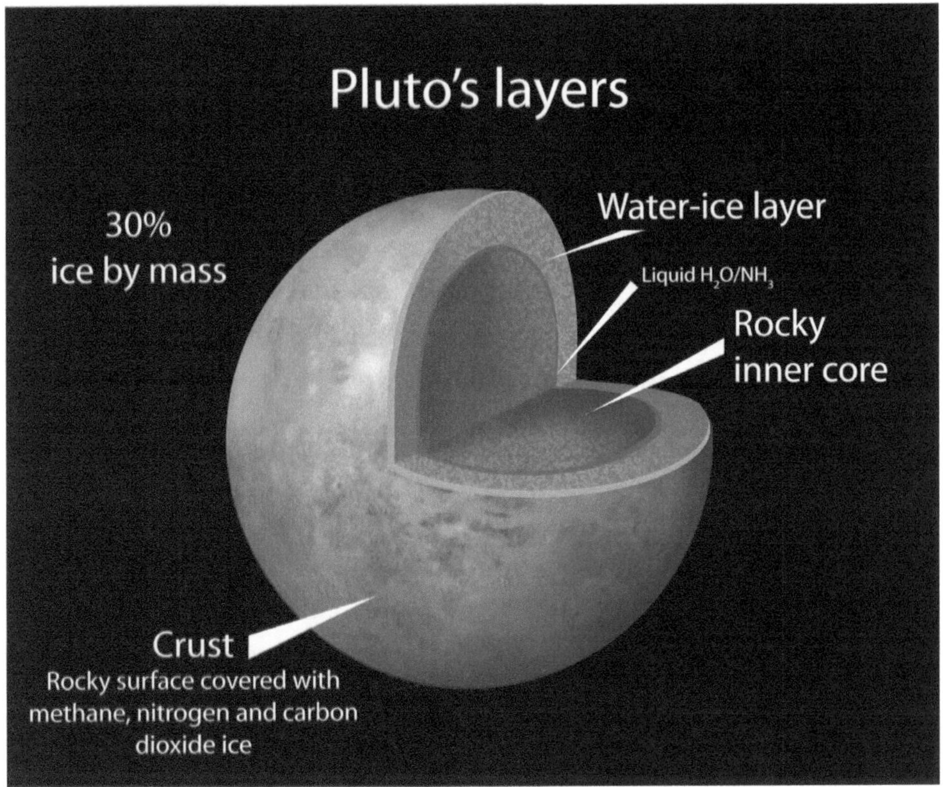

A cutaway view of the interior of Pluto. Image: Debbie Hughes.

proven, that this process of gravitational contraction releases enough heat to melt the ices outright or at least soften the plutoids enough to allow the heavier rocky materials to sink through toward the centre. This results in a layer of liquid, or mushy, ice surrounding a rocky core. Slowly the liquid, or mushy, ice refreezes.

Not all plutoids are identical. Eris and Pluto are both plutoids and are roughly the same diameter (~2,300 kilometres) but Eris seems to be about 25 percent more massive than Pluto. It probably has more rock and less ice. So while no two plutoids are exactly the same, they are probably similar in basic structure. In the billions of years since their formation they have probably collected additional dust and ice and maybe a few comets and asteroids as well. This material forms an outer crust over the ice mantle. At this point it is unknown if this outer crust is metres or kilometres thick. In general, it is probably safe to assume that all plutoids probably resemble the structure shown in the accompanying diagram, with some variation in overall size, mass and composition.

Plutoids are cold. Estimated temperatures at Pluto's surface range between 40 to 60 kelvin (−233 to −213 degrees Celsius). Pluto has a varied surface being black, dark orange and white in color. It seems to be covered with nitrogen-ice, although methane, carbon monoxide and ethane have been detected. This is mixed with dust and tholin (a hydrocarbon sludge resulting from the irradiation of methane). The tholin gives the surface the dark orange color. It is unknown how deep this surface crust goes, but plutoids are probably geologically stable.

Under the crust is an ice mantle many kilometres thick. This ice mantle is probably salty and contains gases in the hydrate state. In the case of Pluto, this ice mantle is estimated to be 325 kilometres deep (assuming that the core has a density similar to the planet Mars). Under the ice mantle is the rocky core that contains minerals, silicates, and metals. It also contains some radioactive elements that produce heat. If the core produces enough heat and if the ice mantle provides enough insulation, then it is possible that the interface between the rocky core and the ice mantle is an ocean of liquid water. This layer will be salty water with dissolved gases such as ammonia under high pressure. If life exists on Pluto (or on any other plutoid), this dark, benthic zone is where it is likely to live.

It is easy to think of plutoids as small, insignificant celestial bodies. They are indeed small when compared to Earth or Mars but each is a world unto itself.

Charon, Pluto's largest moon, has a land area equal to half that of the continental United States. It is estimated that Charon's ice mantle has a depth of over 200 kilometres and a volume of water equal to half that of Earth's oceans. It has an overall diameter of 1,207 kilometres and a surface gravity of about three percent that of Earth. The diameter of the rocky core (assuming that it has a density similar to Mars) is estimated to be about 790 kilometres. There are dozens and perhaps hundreds of such worlds in our Kuiper Belt and Oort Cloud.

Knowledge of exactly what type of resources will be available to colonists residing on a plutoid is limited. They would have water-ice (H_2O) in some

PLUTO, CHARON and CERES

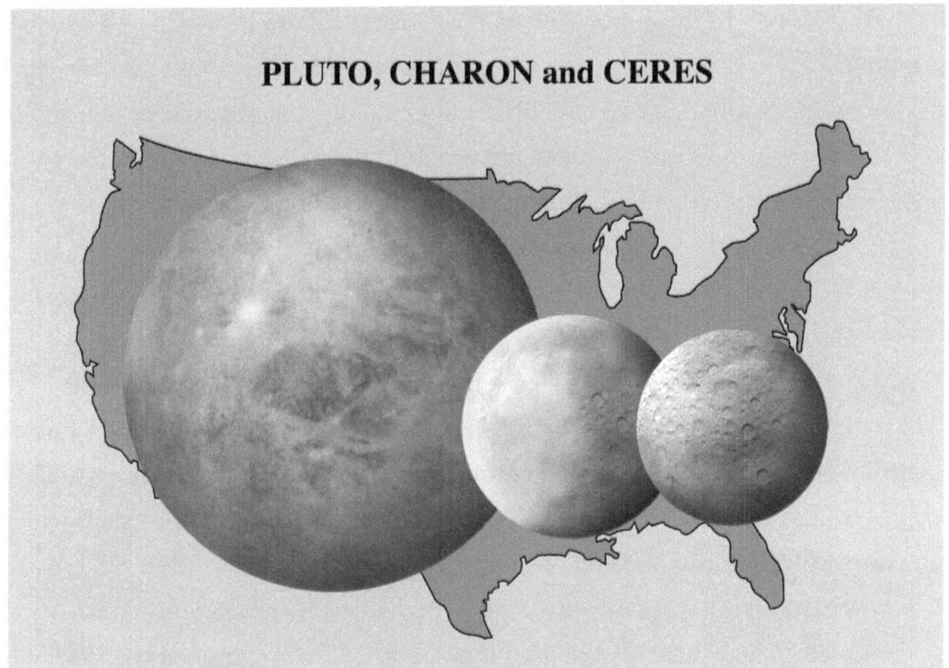

A size comparison between (left to right) Pluto, its moon Charon and Ceres in the Asteroid Belt. Image: Debbie Hughes.

quantity. Other volatiles that are likely to be found in the water-ice, perhaps as clathrate hydrates, include carbon dioxide (CO_2), carbon monoxide (CO), methane (CH_4), nitrogen (N_2), ammonia (NH_3), argon, xenon and krypton [16, 17]. Comets have been studied and found to contain these volatiles as well as dust and complex hydrocarbons. The surfaces of plutoids are likely dusty ices and other materials accumulated after the plutoids formed. This crust probably contains some minerals, silicates and metals. It might be possible for the colonists to obtain all the minerals and metals they need by mining this outer crust and tunnelling down as deep as the rocky core. At the interface layer between the crust and core there is the possibility of a liquid ocean, or a former liquid ocean that is of scientific interest and may itself contain useful resources [17]. Furthermore, many plutoids have moons that also may contain resources that could be available to the colonists. Pluto has five moons with more probably waiting to be discovered. The escape velocity of Pluto is only ten percent that of Earth so access to these moons from the surface of Pluto is possible. With enough energy and technology to process these resources into useful products, the colony could probably survive with little additional support once it is established.

Water and especially nitrogen are two limiting resources necessary for any terraforming projects in the inner Solar System. These are available in large quantities on the plutoids, amongst other locations.

THE BASE

As other planetary systems are explored it is expected that Earth-type worlds will be found to be very rare, that Mars-type worlds may be more common, but that plutoids are fairly common. A crewed starship arriving at a new star system should be able to find a suitable plutoid fairly easily. Their fuel will be exhausted and they will be nearing the end of their supplies. If the fuel was their shielding, it is now gone and they are now exposed to higher levels of radiation.

Establishing a base on a plutoid is going to be challenging but we have many plutoids in our system on which to practice. The colonists setting out to build a base would/should have a good experience base on which to draw. The obvious solution to the radiation protection problem is for the colonists to drill down into the plutoid's icy mantle and then hollow out a cavern. If a kilogram per square centimetre of ice is required for overhead shielding then the colonists will need to drill down 10.6 metres and hollow out a cavern, keeping the roof of the cavern at least 10.6 metres under the surface. The presence of the dusty outer crust could complicate this somewhat. If this layer has a density and hydrogen content similar to ice, the numbers are the same. If it is a thick layer the colonists will be required to work with it instead of ice. It should be possible to drill down deep enough that a cavern carved out of ice or other material will provide the necessary radiation shielding.

This cavern will be cold and airless. To make it useful it must be lined with an air-tight insulator and pressurised with oxygen and nitrogen to Earth-normal levels. Other pressures are possible but it is desirable to maintain as Earth-like an environment as possible. With enough heat and light, this cavern becomes a shirt-sleeve environment in which humans can work with radiation levels well within acceptable limits.

However, with one atmosphere of pressure inside the cavern the air is pressing up on the ceiling with 101 kilo-pascals (kPa) while the 10.6 metres of ice (or dust of equal density) is pressing down with only 12.7 kPa. This downward pressure is calculated using equation one, which is $p = \rho gh$, where 'p' is the pressure in pascals, 'ρ' is the density of material overhead in kilograms per cubic metre (assumed to be 950kg/m^3 for dirty ice or dust), 'g' is gravity in metres per second squared (the surface gravity of Pluto is estimated to be 0.67 ms^{-2}) and 'h' is the thickness of material from the plutoid's surface to the ceiling of the cavern in metres.

This equation is fairly accurate for fluids and even solids where there is little sheer stress. Ice and dust are fairly weak in sheer, so this approximation is probably reasonable. This example assumes that the surface gravity is 0.67ms^{-2} (the expected gravity at Pluto is used here as an example).

The forces are not equal. Something will probably give and the ceiling will rise, releasing the contained atmosphere. The solution is to go deeper. By setting the pressure equal to 101 kPa (Earth-normal) it turns out that, in the case of Pluto, if the ceiling of the cavern is 160 metres under the surface then the pressure of the material above the ceiling presses down with a pressure equal to one atmosphere. In other words, the ceiling is entirely supported by the atmosphere contained in the cavern. In theory, the cavern can be made to be any size without worrying about supporting the ceiling, provided that pressure in the cavern can be maintained at the specified level.

The example given above is for Pluto. Smaller plutoids will have a smaller gravity and thus the cavern's ceiling must be correspondingly deeper. In the case of Charon, the required depth is 365 metres. Other variables that have not been considered in this example include the actual density of crust material, the depth of the crust material, the actual density of ice encountered and perhaps the spin of the plutoid and location of the colony with respect to the spin axis. Pluto is one of the larger plutoids that we know of and even for Pluto the required depth of 160 metres of ice is much more than the 10.6 metres required for radiation protection. It is very possible that the background radiation level in such a cavern could be less than the background level of radiation on Earth.

As we have mentioned, it is necessary that the cavern be lined with an air tight insulator in order to contain the atmosphere and heat. The air in the cavern is assumed to be Earth-normal in both pressure and composition, although other values are indeed possible. Insulation is an engineering problem, one that we can solve today. Some heat will be lost and it will have to be made up but the important design requirement is that the ice that forms the walls, ceiling and floor of the cavern does not soften or melt. It will be necessary to provide artificial light and to recycle the atmosphere. However, in theory, the cavern can offer a large, shirt-sleeve environment in which to work. If the floor is covered with the proper minerals and then seeded with Earth organisms living soil can be created. With the correct lighting (colour and intensity) it should be possible to grow crops, or trees, or gardens. Specially-built caverns can be dedicated to agriculture, forestry, manufacturing, storage, energy production, chemical processing and so on. It may be possible to construct large ponds, small seas, or unusually deep pools within such a cavern. All will be shirt-sleeve environments; spacesuits would not be needed. The temperature could be varied as needed. The ceiling level with respect to the surface is fixed, but the size and interior height of the caverns are essentially unlimited.

The gravity within these caverns will be that of the surface gravity of the plutoid. This may be acceptable for plants but, as discussed above, is inadequate for a long-term human colony. Assuming the absence of any scientific breakthrough that provides artificial gravity, the only option is acceleration resulting from circular motion. Space settlement designs to achieve one g have been developed using this approach [8, 9]. Generally, space settlement designs specify one rpm as the upper limit of permissible rotation rates for the living quarters of the colonists. If the rotational speed is constrained to one rpm then we can obtain the necessary radius using equation two, which is $R = 3600a/4\pi^2$ where 'R' is the radius in metres and 'a' is the centrifugal acceleration in metres per second squared.

Thus, to experience one g in a settlement rotating at one rpm then the radius must be 895 metres. There is no reason why such a rotating settlement cannot be located in one of the pressurised caverns described above. A number of settlement designs have been worked out in some detail [8, 9]. One of their major constraints centered on the fact that the habitat had to contain an atmosphere and thus the outer membrane had to act like a pressure vessel in tension. If the rotating habitat is located in a pressurised cavern, this constraint goes away. However, two complexities arise to replace it: 1) the habitat is now rotating within a weak gravity field that to some extent complicates equation two and 2) the habitat now requires some support to counter this weak gravity field.

The gravity experienced in the rotating structure is then the sum of two perpendicular acceleration vectors. For Pluto the gravity acceleration vector is minor compared to the acceleration vector from rotation. It does allow us to reduce the radius of the rotating structure from 895 metres to 891 metres and the 'floor' needs to be tilted four degrees from the vertical. These effects are minor but important enough that they cannot be ignored. With smaller plutoids, the effects would be even smaller.

Even in the weak gravitational field of a plutoid it will be necessary to support this large rotating structure. The outer edge of a habitat, given the above parameters, is moving at 112 metres per second (251 miles per hour). This is within the speed envelope of maglev (magnetic levitation) trains currently in service. It should be possible to modify this technology to levitate a large rotating structure against the slight gravity of a plutoid. Redundant systems will be required so one or two of them can be taken offline for maintenance and repair without having to stop the habitat.

A maglev system is efficient, with most of the energy that is lost going to air resistance. With a large rotating torus or disc the only air resistance will be as a result of skin friction, which can be minimised with drag reduction technologies. For the most part, friction turns to heat which on a plutoid is a welcome byproduct.

The habitation modules proposed for the Enzmann starship will probably have to be scaled up. If the radius of these modules is increased to 890 metres they could be reused as rotating habitats for the colony base, allowing for rapid deployment upon arrival. Future habitats would have to be manufactured using local resources.

The colonists would live on the rotating habitat(s) and commute to work on a daily basis to other caverns as needed. Most of the actual labor done in the caverns might be conducted with robots controlled and monitored by the colonists from their habitat(s), although there is no reason humans could not do this work as needed or wanted.

The size of the colony raises some interesting issues. From strictly a genetic standpoint, anthropologist John H Moore has estimated that a population of 150–180 healthy individuals would allow normal reproduction for 60 to 80 generations. If frozen sperm and embryos are included, fewer individuals could allow normal reproduction for more generations. If the numbers of necessary professions, such as doctors, dentists, machinists, chemists, teachers, robot repair technicians, etc., are considered it is clear that this number needs to be much higher. Each profession must have enough members to do the work, remain proficient in their area of expertise and train the next generation. Considering the number of skills and professions necessary for a modern society to function, it is likely that the initial minimal colony size would number in the tens of thousands. From there it should be able to grow into the millions, without crowding or running out of resources.

EXPLORATION

One of the main reasons to journey to the stars is to explore and find out what is there. Each target star system will have been imaged using the best optical and radio telescopes available. Using gravitational lensing it is possible to obtain good images of all the major planets in the target system [18]. Robotic probes will be launched and they will provide more detailed information. By the time the crewed starship leaves our Solar System, they will have a good idea of what awaits them. They will no doubt peer through powerful telescopes as they near their destination and launch additional robot probes during their deceleration phase. While the colonists and crew are busy constructing the base, the planetary scientists will be busy analysing vast amounts of data that the probes are providing. This data will be shared with Earth, a small down payment for the resources invested in the project.

After the base is established and its industry is up and running, more robotic probes will be constructed and launched to explore all the planets and moons in the new star system. Once interesting, or curious, sites are identified then crewed vehicles can be constructed and sent out. Until we get there we have no idea what these sites may be. Perhaps a planet has signs of alien life. Perhaps there are signs of a long dead civilisation. Perhaps there are planets or moons that shouldn't exist but do. All of these and more would be worth exploring. The objection may be made that the manufacturing capability of a remote base would be orders of magnitude too limited to survive, much less support sending probes, engaging in base replication and massive terraforming projects. This is a valid concern if using current technology. However, the colonists will have access to technology somewhat advanced over what we have today. They will probably depend on 3-D printers using data sent from Earth. Nanotechnology is another possibility. They will also need energy sources more advanced than what we have today. Robots and advanced computers will provide them capabilities that we are hard pressed to envision today. Even then, there is no guarantee of success, but if they do survive they will have an entire planaetary system to explore and what they learn will be shared with Earth. They will make contributions to the human experience and culture, for they are a part of it. Eventually they will choose a planet to make their own, a planet to terraform into a new Earth or perhaps into something they conceive of as even better.

TERRAFORMING

Terraforming literally means 'Earth-shaping'. It is the process of deliberately modifying a planet or moon's temperature, atmosphere, ecology and topography to make it habitable by humans (and other Earth lifeforms) without the benefit of a spacesuit. There have been books written on terraforming [19, 20, 21]. By the time starships are travelling to nearby stars, humanity will have some experience with terraforming Mars and maybe Venus. Terraforming will involve the importation of vast quantities of materials lacking on the target world, such as water, nitrogen and oxygen. Or, in the case of a Venus-type world, removal of vast amounts of carbon dioxide. If a planet's spin is too fast or too slow it will have to be adjusted. Giant mirrors in space, positioned at the correct spot could heat or cool the planet. Oceans will need to be created, sometimes from nothing. Eventually life from Earth will be introduced.

Seven of the thirteen stars within ten light years of Earth are red dwarf stars (M class). Red dwarfs are far more variable and violent than their more stable, larger cousins. The have been observed to put out violent flares very destructive to life. These intense flares will also strip off a planet's atmosphere. The dim red light put out by red dwarf stars and occasional flares make traditional terraforming methods problematic.

One approach allows for the terraforming of worlds around such stars [22, 23]. That approach is to construct a shell of matter around the planet supported by atmospheric pressure. The shell will be thick enough to protect the atmosphere and life under it from radiation, both cosmic and flare-related. Shell worlds must provide their own heating and lighting and even their own climate but they are effectively independent, to a large extent, from the star they orbit. Shell worlds, in theory, don't actually need a star.

Mars-type worlds are expected to be far more common than Earth-type worlds. They make an ideal candidate for terraforming. Let's assume that one such world is formed within the snow-line (thus it is a rocky planet with limited volatiles) of a red dwarf star that often produces violent solar flares that irradiates its surface and has long ago stripped off any atmosphere. Let's also assume it has the same mass and diameter as Mars. This gives it about the same surface area as Earth's land area. If we set aside 25 percent of this area for oceans then the remaining land area is equal to about 73 percent of Earth's land area. Mars has a gravity that is about 38 percent of Earth. It is unknown whether this is adequate for humans and, if not, the colonists may have to resort to the rotating habitat trick for their children and pregnant women.

The shell is not just a thin membrane. It must be massive enough that the weight of the shell due to gravity pulling down equals the pressure of atmosphere pressing up. In the case of a shell that is fifteen kilometres above the surface, the shell must have a mass of 14,100 kilograms per metre squared in order to contain an atmosphere that is Earth-normal at the surface [20]. This is equivalent to almost 15 metres of ice. This plus 15 kilometres of air provides a high degree of radiation shielding for surface dwellers. Ozone is not necessary to protect the surface from harmful ultraviolet radiation and a shell world could withstand a fairly intense gamma-ray bust. Shell worlds are vulnerable to asteroid strikes but so is an Earth-type world. Just ask the dinosaurs.

To provide the oxygen and nitrogen necessary for this world, gases equal to 38 percent of Earth's atmosphere will have to be imported. Water to fill the oceans and perhaps supply the shell will also have to be introduced. All of this is available from the plutoids. Light and heat must be provided and machinery to maintain the atmosphere and simulate the water cycle will be required. It will be an engineered environment but one that is very Earth-like, except for gravity. Because the shell is in a low state of stress, it should last for a long time before needing replacement.

Construction of the shell and transportation of the necessary materials will require technology unavailable today. Establishing a viable ecology, using Earth life on a formerly airless and lifeless body is a complex and challenging problem, but one that has a solution.

Terraforming a world, whatever the approach, will take vast amounts of energy and very advanced technology. It will also take many centuries or even millennia, but eventually the descendants of the original colonists can move out of their 'temporary' base to their new home planet.

THE NEXT STEP

Travelling to a distant star, establishing a base on a plutoid, exploring a new Solar System and terraforming their final home will occupy the colonists and crew and their descendants for a long time. They will be in constant contact with Earth this whole time, so they will be part of the larger human culture. Their numbers will grow from millions to billions. Perhaps they will start a second terraforming project – they have the time. At some point during this process they will construct probes to send to more distant stars and will then follow them with crewed starships full of colonists. The colony will establish and support its own colonies. And so on.

In this way, given time, a galaxy is settled.

REFERENCES

[1] Long, K F, M Fogg, R Obousy et al., 'Project Icarus: Son of Daedalus – Flying Closer to Another Star', *Journal of the British Interplanetary Society*, 62, 403–414, 2009.

[2] Crowl, A, K F Long, R Obousy; 'The Enzmann Starship: History and Engineering Appraisal', *Journal of the British Interplanetary Society*, 65, 185–199, June 2012.

[3] Vernikos, J, and V S Schneider; 'Space, Gravity and the Physiology of Aging: Parallel or Convergent Disciplines? A Mini-Review,' *Gerontology*, 56 (2), 157–166, March 2010.

[4] Guitton, M J; 'Challenges in Space Medicine', *Journal of Public Health Frontier (PHF)*, 1 (3), 73–77, September 2012.

[5] Mader, T H et al; 'Optic Disc Edema, Globe Flattening, Choroidal Folds and Hyperopic Shifts Observed in Astronauts After Long-Duration Spaceflight', *Ophthalmology*, 118 (10), 2,058–2,069, October 2011.

[6] Wakayama, S et al; 'Detrimental Effects of Microgravity on Mouse Preimplantation Development In-Vitro,' *PLoS One*, 4 (8), 6,753, August 2009.

[7] Horn, E R and M Gabriel; 'Gravity-Related Critical Periods in Vestibular and Tail Development of Xenopuslaevis,' *Journal of Experimental Zoology Part A: Ecological Genetics and Physiology*, 315, issue 9, 505–511, November 2011.

[8] NASA; 'Space Settlements: A Design Study', NASA SP-413, Washington, DC, 1977.

[9] NASA; 'Space Resources and Space Settlements', NASA SP-428, Washington, DC, 1979

[10] United States Nuclear Regulatory Commission, Code of Federal Regulations, 10 CFR, Part 20.

[11] Goldstein, H; 'The Attenuation of Gamma Rays and Neutrons in Reactor Shielding', U S Atomic Energy Commission, May 1957.

[12] Parker, E; 'Shielding Space Explorers From Cosmic Rays', *Space Weather*, 3, S08004, 2005.

[13] Kim, M et al; 'Performance Study of Galactic Cosmic Ray Shield Materials', NASA-TP-3473, November 1994.

[14] Cherry, J et al; 'Galactic Cosmic Radiation Leads to Cognitive Impairment and Increased Aβ Plaque Accumulation in a Mouse Model of Alzheimer's Disease', *PLoS One*, 7 (12), e53275, 2012.

[15] Brown, M; 'The Largest Kuiper Belt Objects', online, http://web.gps.caltech.edu/~mbrown/papers/ps/kbochap.pdf, California Institute of Technology (last accessed 12 June 2014).

[16] Mousis, O and Y Alibert; 'On the Composition of Ices Incorporated in Ceres', *Monthly Notices of the Royal Astronomical Society*, 358, 188–192, 2005.

[17] Cole, G H A; 'Interior Structures of the Icy Satellites and of Pluto', *Quarterly Journal of the Royal Astronomical Society*, 25, 19–27, 1984.

[18] Eshelman, V; 'Gravitational Lens of the Sun: Its Potential for Observations and Communications over Interstellar Distances', *Science*, 205, 1,133–1,135, 1979.

[19] Fogg Martyn J; *Terraforming: Engineering Planetary Environments*, Society of Automotive Engineers, Warrendale, PA, 1995.

[20] Beech, Martin; *Terraforming: The Creating of Habitable Worlds*, Springer Science + Business Media, LLC, New York, 2009.

[21] Oberg, James Edward; *New Earths: Restructuring Earth and Other Planets*, Stackpole Books, Harrisburg, Pennsylvania, 1981.

[22] Roy, K I, Robert G Kennedy III and D Fields; 'Shell Worlds: An Approach to Terraforming Moons, Small Planets and Plutoids', *Journal of the British Interplanetary Society*, 62, 32–38, 2009.

[23] Roy, K I, Robert G Kennedy III and D Fields; 'Shell Worlds', *Acta Astronautica*, 82 (2), 238–245, February 2013.

Chapter 18

THE GREATEST CHALLENGE:
MANNED INTERSTELLAR TRAVEL

ANDREAS M HEIN

Besides transporting humans, a manned interstellar mission is distinct from an unmanned mission in two aspects: its objectives and the required technologies. The main objective of an unmanned interstellar mission is the collection of scientific data and sending this data back to the Solar System. A manned mission usually has the objective of colonising a star system. 'Colonisation' is defined here as the process of establishing a civilisation by creating colonies. 'Colonies' are defined here as populated habitats, similar to the way Gerard O'Neill used the notion of 'space colonies' [1]. The logical steps that need to be fulfilled can be seen in figure one. The rectangular boxes represent the objects whose state is changed by processes. Processes are depicted as rectangles with rounded corners.

A civilisation can only be established once the two main objectives of unmanned and manned missions have been achieved. How are these functions allocated to different spacecraft and technologies? Spacecraft would not only have to transport humans but also human culture. Furthermore, all the necessary means for constructing colonies must also be transported. Future interstellar mission designers need not only be experts in designing spacecraft but they also have to be experts in 'civilisation engineering'. Purposefully engineering the prerequisites for a new civilisation might be a much bigger challenge than the transportation of humans themselves. However, before this issue is addressed it must first to be determined where colonies can be established within a star system.

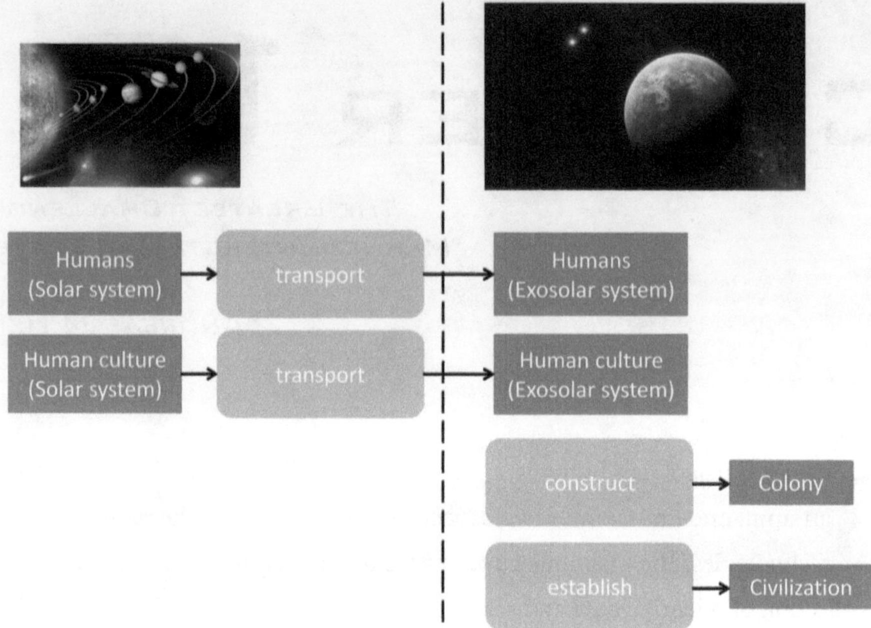

Functions that have to be fulfilled by a manned interstellar mission. Images: NASA/ JPL–Caltech/ESO and L Calçada.

COLONISATION TARGETS

There are two extreme cases of colonising a star system. One extreme is a star system with Earth 2.0, a planet that resembles the Earth and readily provides the resources humans need to survive and prosper. Furthermore, humans are compatible to the ecosystem that already exists on the planet. Paul Davies has recently pointed out that it is extremely unlikely that such a biocompatible planet exists [2]. Humans have evolved in parallel to an ecosystem over thousands of years. This ecosystem does not only consist of animals and plants but of an incredible variety of microorganisms. We currently have no idea what the conditions for compatibility are and what the minimum ecosystem that would have to be transported is.

The other extreme is a star system devoid of any large planetary bodies but with an arbitrary number of small bodies, for example asteroids and comets. These bodies could be mined and the resources processed in order to construct colonies.

Between these extremes one can imagine star systems with planets that have an atmosphere and an ecosystem but are incompatible with humans. Other planets might be in the habitable zone (see Chapter 16) but no ecosystem may exist on their surfaces. Depending on what resources in the form of planets and other bodies can be found in a star system, there are various options for colonising them, as shown in the figure on the opposite page [3].

The engineering effort depends on the type of colony. The effort ranges from the benign adaptation of humans to the planetary environment to full-blown terraforming [4]. Off-surface colonies might be established by hollowing out asteroids or comets, as proposed by Dandridge Cole [5]. Or it might consist of creating artificial colonies out of space resources, as elaborated by O'Neill and Matloff, [6, 7].

TRANSPORTING HUMANS TO THE STARS

The difference between an unmanned and a manned interstellar spacecraft is that, by definition, the latter transports humans. In principle, all manned interstellar spacecraft concepts can be distinguished by their mode of transporting humans. The specific mode determines the required technologies, not only for transportation itself, but also for establishing the initial colony. Table One, overleaf, gives an overview of transportation modes and the transportation concepts that are introduced in this section.

There are four modes of transportation that seem to be feasible with technologies that could become available in the future: humans living in a large, self-sustained habitat; hibernation; sending human as zygotes or embryos; and digital 'uploading' and storage. Each of these modes has profound implications on the required technologies and mission architectures. Furthermore, the required resources differ greatly.

WORLD-SHIPS

Spacecraft transporting humans in a large, self-sustained habitat are often termed world-ships, generation ships, interstellar arks or colony ships. There is a

Table One: overview of different states and their attributes relevant for manned interstellar travel

Mode categories		World-ship	Hibernation	Embryo	Digital
Developmental state	Zygote	✓	✓		
	Embryo	✓	✓		
	Infant	✓			
	Child	✓			
	Adult	✓			
	Elderly	✓			
Metabolic state	Reduced		✓		
	Stopped		✓		
Substrate	Biological	✓	✓	✓	
	Artificial				✓

significant overlap between each of these concepts and clear distinctions are difficult to make. All concepts propose a habitat, similar to a space colony with a propulsion system attached. The first concepts date back to the early twentieth century, proposed by Konstantin Tsiolkovsky and John Desmond Bernal [8]. O'Neill suggested the conversion of free-floating space colonies into interstellar arks in his seminal 1977 book, *The High Frontier*. He imagined incremental space colonisation, gradually expanding from the inner to the outer Solar System. At one point, one of the colonies might decide to leave the Solar System. Although O'Neill leaves many engineering questions unanswered, his vision of the gradual transition of humanity into an interstellar species had a significant impact on the way we think about interstellar colonisation today. He even hypothesises that the inhabitants of such a colony might not be inclined towards an existence on a planetary surface at all and instead will choose to live within their artificial habitats indefinitely.

How does such a habitat work? Popular habitat shapes are cylinders, spheres and tori, all creating gravity by rotating the habitat and creating centrifugal force [9]. The inhabitants live in an environment that resembles Earth. Light is provided by an artificial light source, powered by fusion reactors, for example. Life would mostly take place in small settlements with stretches of green in between. A highly sophisticated bio-regenerative life support system takes care of water supply, food provision, air recycling, and waste disposal.

Habitat sizes and the number of inhabitants greatly differ from concept to concept. Cameron Smith recently determined a lower limit to population size from a biological perspective [10]. He concludes that a minimum population for a trip lasting about two centuries needs to be of the order of tens of thousands. This is about the population size proposed for existing space colony designs such as O'Neill's Bernal Sphere or the Stanford Torus. The size of these habitats ranges from several hundreds of metres to a few kilometres in diameter. Much larger versions have habitable areas comparable to small states, with a length of several dozens of kilometres and hundreds of thousands of inhabitants [11].

It is reasonable to think of a rather evolutionary transition from a Solar System-wide civilisation to an interstellar species, although this transition is going to happen less smoothly as O'Neill believed. Transforming a colony into an interstellar ark is not a trivial issue and requires considerable redesign. For example, power has to be entirely provided internally and radiation shielding has to be adjusted to the interstellar radiation environment. Thus, it is a rather futile attempt to retrofit an existing colony for this purpose. Nevertheless, what can be reused is the experience gained with designing, manufacturing and operating large habitats in space for extended timeframes. Without this experience, sending a generation ship on its way to the stars is suicidal. It is thus a necessary condition that humanity is a Solar System-wide civilisation in order to conduct such a mission successfully.

Another technology that has to mature is autonomous in-space manufacturing of complex systems. We cannot expect that a population which has travelled for 200 years is still capable of constructing space or surface colonies at the target star. Such colonies are immensely complex technical systems and they require highly diversified skills and knowledge of construction. The only viable option is the autonomous construction of such colonies with the required skills and knowledge codified in machines. Such manufacturing facilities could be built up by macroscopic replicators, as described by Robert Freitas for replicating an interstellar probe and a lunar infrastructure [12], [13]. However, Freitas does not talk about a single type of self-replicating machine. He imagines a whole zoo of machines. Each has a different function and is part of a highly complex supply chain. Thus, it is better to talk about a self-replicating supply chain. Such long and complex supply chains are prone to malfunctions and difficult to control. However, recent advances in additive manufacturing, alternatively called '3D-printing', could substantially shorten these supply chains. This could be accomplished by the capability of 3D-printers to manufacture a large variety of components and being able to process a large number of different materials, both without reconfiguring the production line. Advances in additive manufacturing already enables the manufacture of highly complex geometries, for example rocket engine injectors. Further progress might enable the manufacturing of highly integrated electronic and mechanical components. For example, solar cells can already be manufactured with desktop-sized printers. One can imagine that several decades from now, space systems will be manufactured in space with only a small percentage of components supplied from Earth. Self-replicating infrastructures are not only required for world ship mission architectures but for all other transportation concepts as well. Thus, they are a necessary technology for manned interstellar travel.

There are two basic decisions to make for world-ship mission architectures:
• Whether or not to split the mission into an unmanned and a manned spacecraft;
• How to partition the various payloads among the spacecraft.

Splitting a manned mission into an unmanned and manned part is a strategy used in Robert Zubrin's Mars Direct mission architecture [14]. Splitting the mission is interesting if something is prepared at the target destination that requires time in the order of the mission duration. In the case of Mars Direct, it is propellant for the return trip that takes about a year to produce. In the case of an interstellar mission, it is the build-up of a colony. As described earlier in this

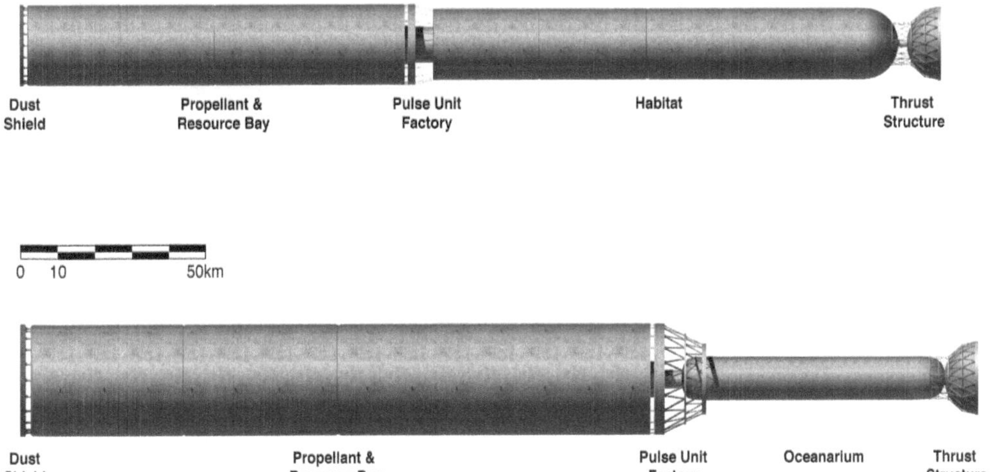

Bond and Martin world-ships. Images: Adrian Mann (www.bisbos.com).

chapter, the difficulty of building up a colony depends on the star and planetary system. Some colonies take a lot of time to prepare – for example, terraforming a world might take hundreds to thousands of years. Even constructing 'modestly-sized' O'Neill-type colonies would probably take decades. Thus, sending a precursor mission for developing a habitat makes sense as the world ship population does not have to wait in their vessel for the colony to be completed. Furthermore, such an approach reduces the risk of mission failure if the colony construction process failed. When people are sent to the target star, ideally the colony has already been proven to function. Note that a precursor in this context is an unmanned spacecraft that is preparing the star system for the later arrival of the crewed ship. This type of precursor is distinct from precursors exploring the star system. Precursors for exploration are necessary in any case, as sending a manned spacecraft without sufficient data about the star system is too risky.

In this split mission architecture, the precursor probe arrives at the target star system in advance of the world-ship. Prior to the world-ship's arrival, it constructs a number of space colonies that are ideally ready for use. One disadvantage of this architecture is the need for a highly-automated habitat

Table Two: colonisation targets mapped to payload and mission architectures.

		Terraforming Payload	Seeding payload	Self-replicating payload	Factory
For easily terraformable planet		✓	✓	✓	✓
For biocompatible planet			✓	✓	✓
For habitable planet				✓	✓
Architecture 1	World-Ship precursor	✓	✓	✓	✓
Architecture 2	World-Ship precursor	✓	✓		✓
Architecuture 3	World-ship	✓	✓	✓	✓
Architecture 4	World-ship	✓	✓		✓

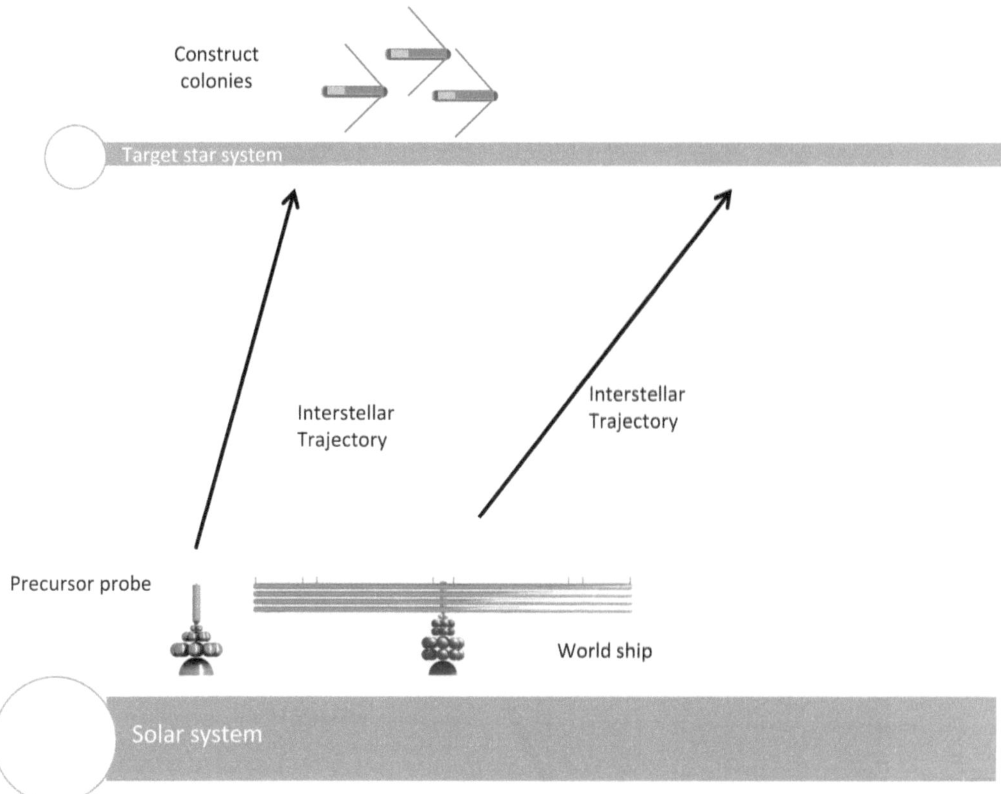

The architecture of a world-ship split-mission with a precursor component, which finds the infrastucture all ready for when the colonists arrive.

construction process without human intervention. Extensive amounts of local resources have to be mined, processed and turned into habitat components. Nevertheless, high degrees of automation are desirable in any case, as there is no guarantee that the world-ship population is capable of performing such sophisticated tasks after their journey.

The second decision is concerned with how to partition the habitat construction payloads among the spacecraft. It is obvious that the precursor is required to transport all the payload necessary for constructing a habitat. However, additional construction payloads can be delivered with the world-ship. This redundancy would reduce the mission risk, in case the precursor's payloads fail at some point. Table Two, on the opposite page, shows payload options for

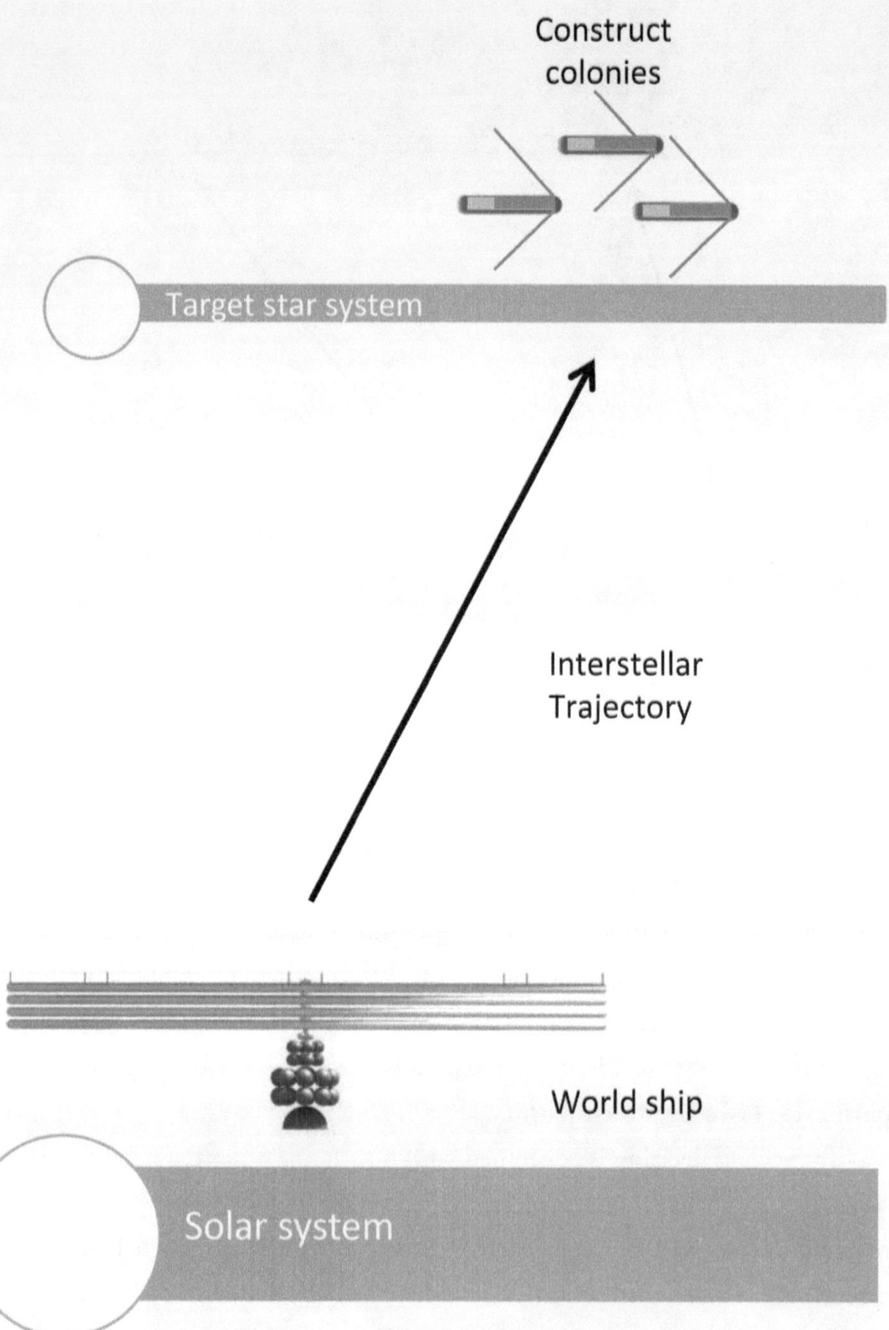

What a world-ship mission would look like without the precursor components; upon arrival at its destination, the colonists would have to build a civilisation from scratch.

various planet types. For a star system without suitable planets, a self-replicating payload and a factory would be required. In the best case, the self-replicating payload replicates itself up to a point where sufficient industrial capability has been built prior to the start of construction of the habitat.

Icarus Interstellar's Project Icarus and Project Hyperion recently delivered more detailed designs for interstellar vehicles that are required for a world-ship mission architecture. One such example was designed by students at the Technical University of Munich for the Project Icarus Concept Design Competition in 2013. The spacecraft is capable of delivering a 150 tonne payload to alpha Centauri within a century and is propelled by a deuterium–deuterium fusion engine. A payload of this size might be sufficient to start mining and process local resources to build up the manufacturing capabilities required for constructing a colony.

From a technological perspective, world-ships are the most mature option for manned interstellar travel. Nevertheless, they also require the largest amount of resources as a result of their large payload size.

HIBERNATION

Hibernation is a common form of saving metabolic energy among mammals during winter. Hibernating mammals basically reduce their metabolic rate and therefore energy consumption by falling into a sleep-like state. The key idea is to induce humans into artificial hibernation. They are then 'woken up' at the target destination. Popular accounts to hibernation have featured in the movies *Alien* and *Avatar*.

The potential advantages of using hibernation consist of a hypothesised longer life-span and a reduction in required resources onboard the spacecraft. Resource reduction can be achieved by a smaller-sized habitat and a reduced amount of food consumption because of the lower metabolic rate. A habitat is nevertheless still necessary, as humans have to be woken regularly in order to avoid the accumulation of harmful substances in their bodies. Furthermore, degradation of muscles and other parts of the body have to be dealt with. If periodical awakening is done successively, the habitat only needs to be the size of the awakened population. Thus, a hibernation ship might be significantly smaller than an interstellar ark. Nevertheless, it still requires all the systems that are required for an ark and an even more sophisticated life support system for the hibernating humans.

We can estimate how much mass could be saved simply by calculating the average number of humans that are awake at any given point in time. As we need on the order of thousands of people onboard the ship and, if we assume that they have to be kept awake for a duration of a few weeks per year, we can assume that about two percent of the population is awake on average, which translates to tens of people. Assuming a linear scaling of the habitat mass, the savings from using hibernation are about two orders of magnitude, which is quite significant.

However, it is difficult to judge the feasibility of hibernation today. There are several challenges. For instance, it is currently unknown how harmful the effects of hibernation on the human body are. Hibernation over extended periods of years might result in irreversible damage to the human body. Experiments with animals in some cases resulted in substantial damage to the brain. Furthermore, the accumulation of harmful substances within the human body might lead to its continuous degradation, and therefore limiting the duration of hibernation. [15], [16].

Another issue is aging. Hibernation does not stop aging and has only the potential to slow down aging in humans to a certain degree. This might somehow lead to the disturbing consequence that the hibernating crew actually 'oversleeps' almost its entire life.

An alternative to hibernation might be cryogenics. However, we even have less knowledge about the feasibility of using cryogenics to preserve the human body over extended timeframes. It can only be speculated whether or not it is possible to 'reanimate' frozen humans.

In summary, hibernation has the potential to reduce the payload mass of an interstellar vehicle by one to two orders of magnitude. However, there are currently significant hurdles in applying it to an interstellar context and much is unknown. The accumulation of harmful substances and damage to the human body might render hibernation infeasible for this purpose.

Mission architectures for hibernation ships are similar to world-ship mission architectures.

ZYGOTES AND EMBRYOS

Payload mass can be further reduced by not transporting a mix of humans at all developmental stages but rather humans at their early stages of development, namely as zygotes or embryos [17]. The zygotes or embryos are transported in a cryogenic state and nurtured at the destination. It is obvious that from a pure transportation perspective, this concept might lead to a radical payload mass reduction down to a few dozens of kilograms or a few tonnes at most. This is a million-fold mass reduction, compared to an interstellar ark. The cryogenic storage and reactivation of cells is feasible, at least for durations of several years. Whether it is possible for decades or centuries is an open question.

However, this reduction in mass comes at a price: the actual colonisation is a lot more challenging. Colonists have to be 'bred' at the target destination, possibly using artificial wombs. Then, a human society has to be built up from scratch. Everything we consider 'culture' has to be transferred to the newborn. Otherwise, it will be impossible to establish a sustainable colony. This is an extremely daunting task. Humans by their very nature depend on culture for their survival. More concretely, this means the manufacture and proper use of tools. For our past ancestors, this meant, for example, finding and creating habitats, hunting prey and collecting fruits and vegetables. Even for establishing a Stone Age civilisation, a large amount of knowledge has to be transferred. Ironically, the transfer of explicit knowledge is much easier than the transfer of tacit knowledge. Explicit knowledge is already available in a transferrable form, e.g. scripture, digital documents and videos. Tacit knowledge is usually transmitted by mimicking another human and ideally being corrected by the 'teacher'. However, this so-called psychomotor education is currently one of the hardest problems in developing artificial intelligence. Chess computers already have outperformed human chess players. However, soccer players do not need to watch out for similar artificial competitors. Thus, artificial intelligence and robotics that is far

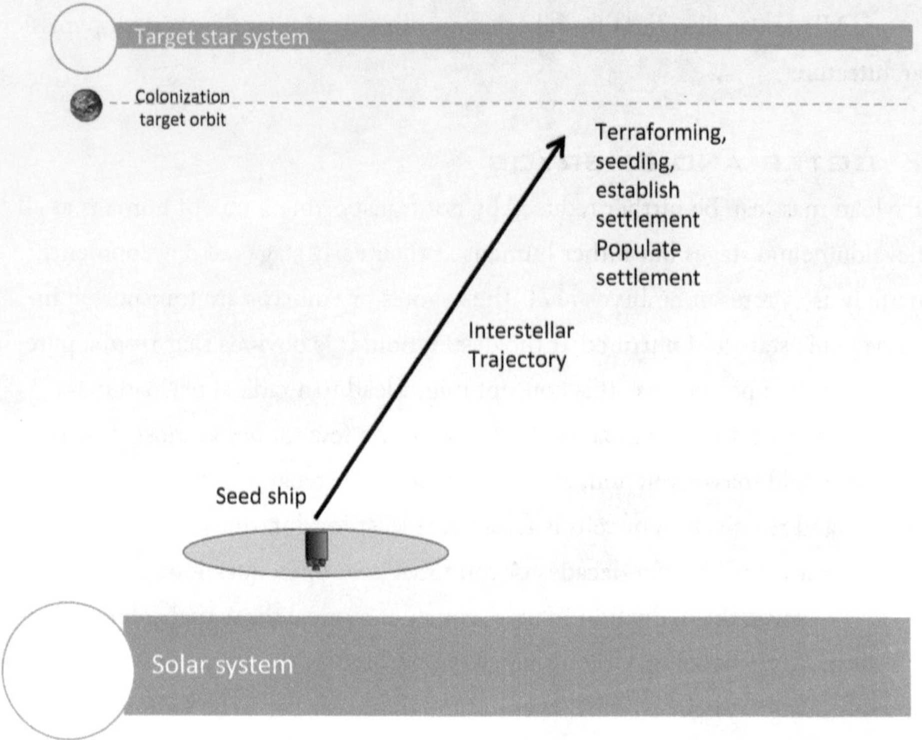

A single-stage architecture for an interstellar mission that would carry embryos and/or zygotes, plus all the equipment and knowledge base to artificially generate a civilisation.

more advanced than currently available is required for this type of interstellar mission. One might argue that such an initial colony might entirely rely on machines and humans would not need to learn to survive. Indeed, initial survival has to be assured autonomously in any case. Nevertheless, the transfer of tacit knowledge is also a problem that has to be addressed in any case, as some form of cultural continuity has to be established.

The mass required for constructing such an infrastructure is probably much larger than the human payload itself. However, even then, the savings that come along with this concept are enormous. From today's perspective, it seems that the risk is shifted from the 'front-end' as in the case of world-ships and hibernation 'sleeper' ships, to the 'back-end': it appears relatively easy to send out a spacecraft

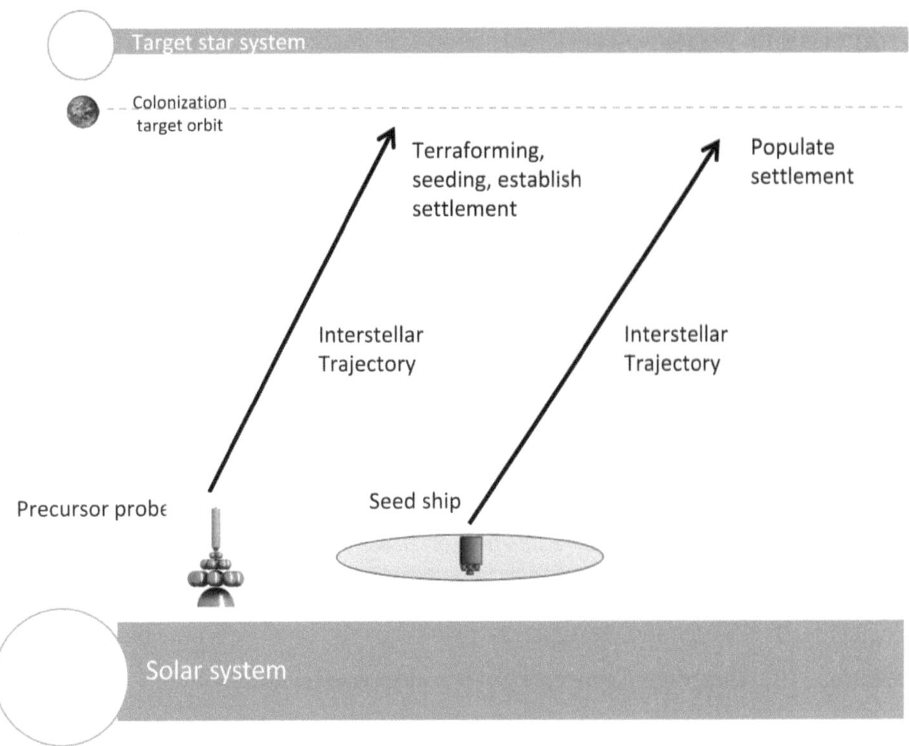

An interstellar embryo mission in two stages, with the precursor mission establishing a settlement, followed by a 'seed ship' containing the embryos which can then grow and populate the settlement.

with a zygotic/embryonic payload, but the challenge of successfully establishing a new civilisation at the target destination gets even bigger.

A technology that can considerably reduce the mass of the initial infrastructure is the use of macroscopic or nano-replicators. They would be either deployed in advance or with the genetic payload and would use local resources for creating the initial infrastructure.

The possible mission architectures are again quite similar to the world-ship architectures. Nevertheless, sending out one spacecraft with a replicator payload and the zygotes or embryos is less problematic if these can be stored for the duration of colony construction. However, for long-duration colony construction, for example terraforming, a split mission architecture is still more attractive.

DIGITISED HUMANS

The basic idea of digital transportation of humans is to digitise the human mind and/or body and to recreate it at the target destination. Seen from a perspective of technological evolution, the digital transportation of humans is a logical consequence of trying to reduce the resources needed for transporting humans to the stars.

Transporting humans in digital form has huge benefits for interstellar travel: it leads to extreme mass savings, comparable or even larger than for transporting zygotes. At the same time it offers the capability to 'resurrect' living humans at the target destination, including their knowledge and therefore culture. It therefore bypasses the disadvantages of transporting zygotes. A hybrid concept is the combination of embryos and zygotes with knowledge and technology transferred via emulated brains at the target destination, either by education or 'hard-wiring' emulations into biological brains. A more advanced concept is the complete replacement of biological life by artificial life. In this scenario, the spacecraft would be the 'seed' for a non-biological civilisation [18]. Thus, digital concepts can be categorized as follows:

Transport humans
- Hardware static storage
- Brain emulations
- Hybrid: genetic material + emulations
- Electromagnetic waves
- Transmit electromagnetic waves/nano-spacecraft via wormholes

Construct colony
- Macroscopic Replicators
- Micro/nano-replicators

Establish civilization
- Emulations + biological humans
- Cyborgs
- Emulations + biological humans
- Emulation cities
- Matrioshka brain

In order to transport humans in digital form, they need to be digitised. Digitisation might be accomplished by some advanced from of scanning. Hans Moravec was one of the first to envision a form of brain scanning procedure, by which the human brain would be incrementally transformed into a digital copy in a destructive way [19]. Ray Kurzweil and others envisage non-destructive ways of creating a digital copy, for example by using nanoscopic robots that scan the brain from within.

Creating a digital copy of the brain is a daunting task. It is far more than copying just the structure of the brain, but also the neurons and their linkages to other neurons. What is further needed is to copy the behavior of individual cells and larger structures in the brain. This is similar to a technical system. The understanding of how the parts of a car are related to each other does not give any idea of how they operate together to perform the desired function of transporting passengers. It can only be inferred by painstakingly assessing how individual components and larger groups of components perform sub-functions. These sub-functions together perform the top-level function. This reverse engineering method is called a bottom-up approach. As an alternative, one can analyze functions top-down, by first assessing the top-level functions and then to decompose these functions into sub-functions. Similar approaches exist for creating digital copies of the brain, which are called 'brain emulations' [20]. Such emulations could be created, copied, run and also switched off as desired [21, 22]. For an interstellar mission, emulations could be stored and activated at the target destination. This would save energy for running emulations during flight. Once arriving at the target destination, one can imagine activated emulations that first assess the environment within the target star system and then determine the best strategy for beginning colonisation. Maybe a whole population of emulations is activated, which debates possible strategies and analyses their potential outcomes. Robin Hanson imagines various types of emulations that also form hierarchies, depending on the speed of their simulation speed. Such emulation cities on Earth would consume a huge amount of power to sustain the emulations and their virtual environment in which they exist. Manipulations of the physical world are performed by various types of manipulators and robots.

A strategy for an interstellar mission would be the reactivation of an initial small population of emulations that make the initial decisions regarding how to proceed with colonization. Then, resources would be mined and processed, in order to increase computational capability and to create a larger number of emulations, which are also able to create biological humans or a mixture of biological and non-biological parts, a so-called 'android' or 'cyborg', along with their habitats.

Another option is the simultaneous transportation of zygotes and brain data. After having arrived at the destination, humans would be created in a way similar to the embryo colonisation concept. However, in contrast to that concept, the brain would be 'downloaded' into these humans by 'rewiring' the structure in a similar way as one of the emulations. Because of possible physiological constraints, such humans would probably be clones of the person whose brain was emulated. Although this seems to be far-off, today stem cells can already be used to create human organs in a laboratory and first attempts have been made to 'print' organs.

A more advanced version of such a mission is the initial creation of an infrastructure within the target star system by using replicators coupled with the construction of a receiver for electromagnetic signals, for example a laser beam. Once established, data for objects could be transmitted at the speed of light. Of course, this is the concept of teleportation. Teleportation was often deemed infeasible, as the amount of information to be transmitted for assembling a human body molecule by molecule would be prohibitive. However, this argument is not very convincing. For example, Roberts et al. argue that a total of 2.6×10^{42} bits are necessary for recreating the human brain [23]. The data for recreating the rest of the human body is insignificant compared to that number, just (1.2×10^{10} bits). With a data rate of about 3.0×10^{19} bits per second, it would take 4.85 trillion years to transmit a human. However, a close look into the assumptions made in the paper reveals that the so-called 'Bekenstein bound' was used for calculating the data required to recreate the brain [24, 25]. The Bekenstein bound describes the maximum information that is required to recreate a physical

Digital mission architectures

Architecture A
- Send replicator plus emulator/storage spacecraft
- Create colony and resurrection infrastructure
- Create population

Architecture B
- Send replicator
- Create colony and resurrection infrastructure
- Send emulator/storage on follow-up spacecraft
- Create population

Architecture C
- Send replicator spacecraft
- Create receiver dishes in target star system
- Receive data for creating technical systems and humans

Architecture D
- Send replicator spacecraft
- Build receiver
- Use wormholes to transmit information to receiver
- Create technological systems and humans

system down to the quantum level. However, it is doubtful that such an extremely detailed description is necessary. Current estimates for describing the brain down to a molecular level are rather in the range between 10^{22}–10^{27} bits [26]. This amount of data could be transmitted within an hour to ten years, assuming the same data rate of 3.0×10^{19} bits per second. Thus, teleportation might not be as far-off as suggested by the current literature.

One of the more speculative approaches to enable manned interstellar travel with almost no travel time is to use some form of faster-than-light

Table Three: mission architectures and their enabling technologies

Technologies	World-ship	Hibernation	Embryos	Digital
Replicator technical systems	required	required	required	required
Replicator/Grow biological systems	required	required	required	required
Brain emulation	required	required	required	required
Teleportation			required	required
Worm holes				required

approach. There is a whole plethora of conjectured faster-than light approaches [27]. Nevertheless, sending pure data or nano probes through shortcuts in space-time might be far easier than doing so with large manned spacecraft. Ray Kurzweil describes how microscopic wormholes might enable the transmission of sufficient data or nano probes to another place in the Universe.

However, all these concepts require the occurrence of a so-called technological singularity, which is often associated with the emergence of general artificial intelligence and its exponentially increasing capabilities. Whether or not it is reasonable to expect such a singularity to happen is the matter of intense debate among scholars [28, 29]. Personal conversations with a range of brain researchers reveal a skeptical outlook on progress in creating brain emulations in the near future. Nevertheless, there is no doubt that progress is being made. Brain emulation and general artificial intelligence should not be discarded on the grounds of current or near-future unfeasibility, as we are dealing with timeframes of several centuries here.

Roger Launius and Howard McCurdy point out that a posthuman civilisation does not necessarily possess the motivation to conduct an interstellar mission [30]. Thus, the technological progress that might lead to interstellar travel could turn against itself by making the human motivation to go to the stars itself obsolete.

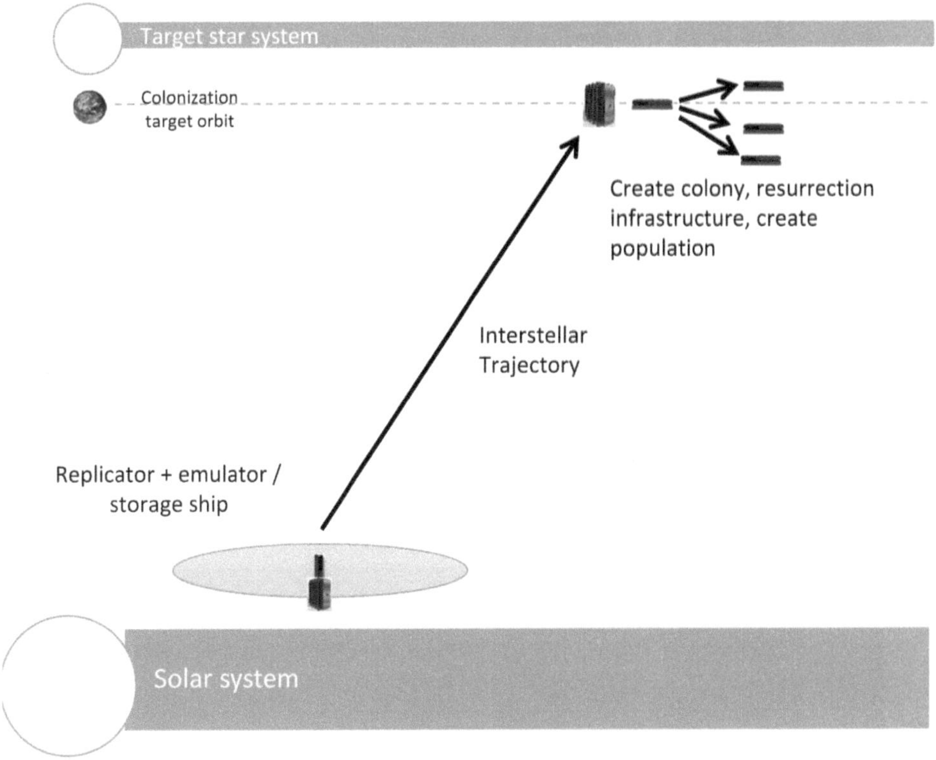

The Greatest Challenge

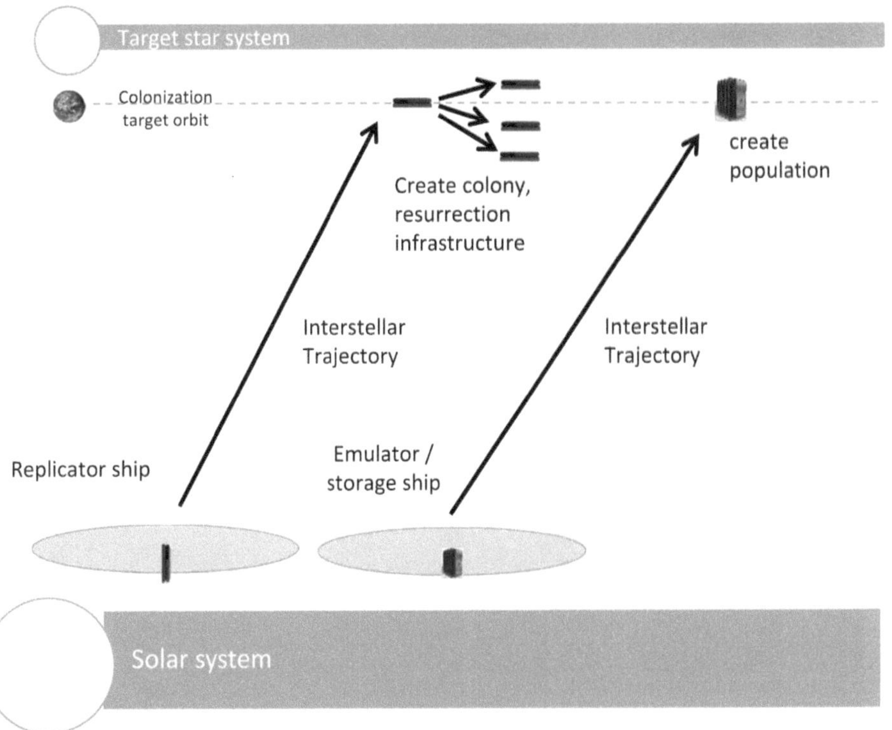

Above and below: digital architectures A and B (see page 367).

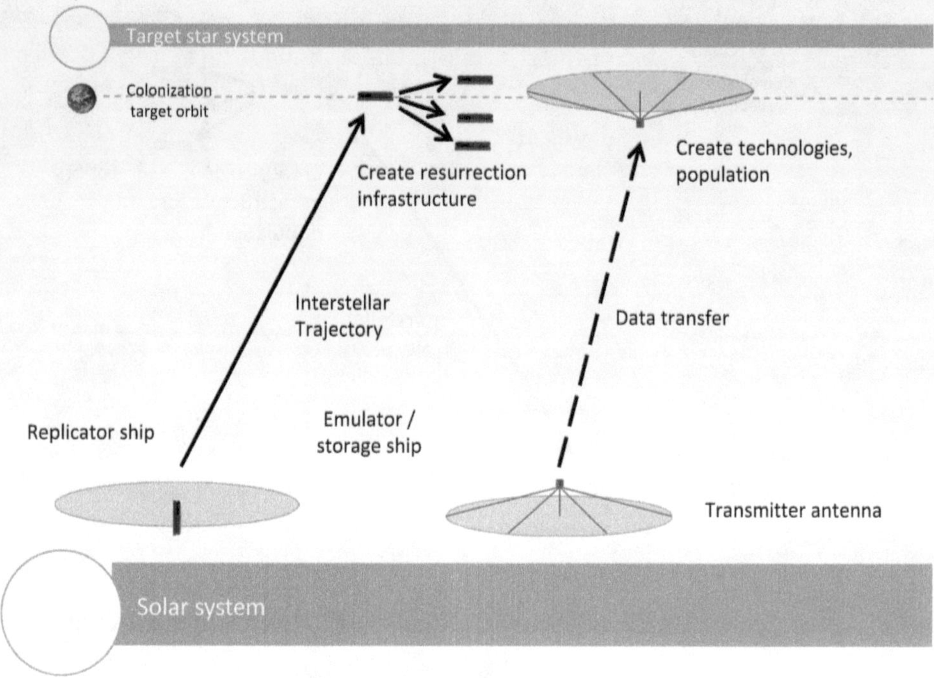

Above and below: digital architectures C and D (see page 367).

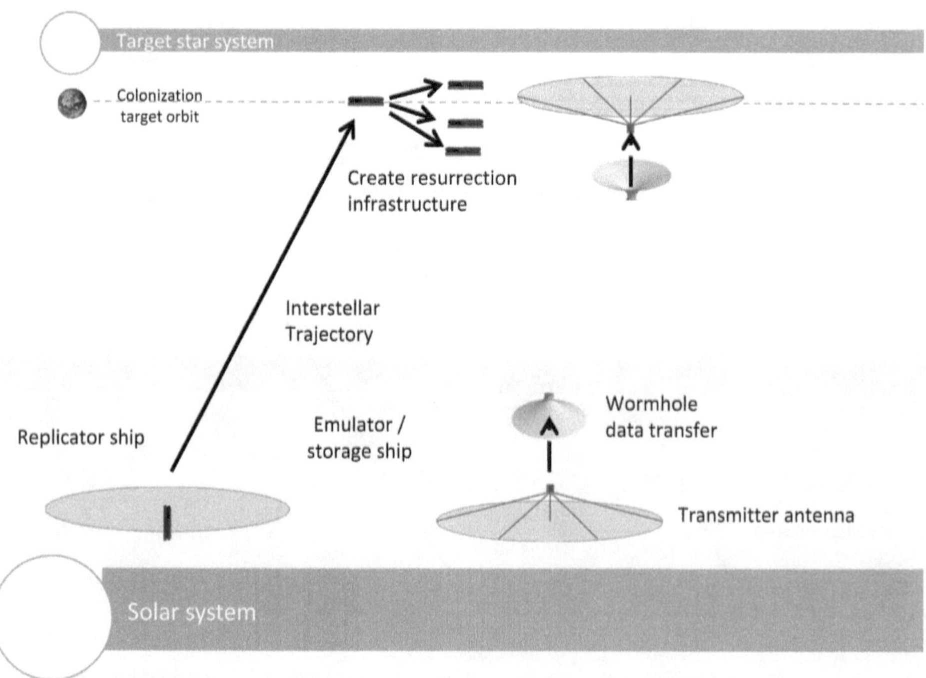

Digital interstellar missions open up a space of interesting mission architectures. Depending on the technological level of human civilisation, various architectures are feasible, as shown in Table Three on page 368.

PROSPECTS OF OPTIONS

One approach to assessing the prospects of technological options is to assess how ideal a technology is with respect to a set of criteria. The degree of ideality can be represented as how good a technology satisfies a set of desired functions at what cost and occurrence of undesired functions. Gasoline cars fulfill the useful function of transporting people and goods successfully. However, they incur cost for gasoline and maintenance. They also contribute to pollution, which is an undesired function. By doing this exercise, it is possible to identify the parameters one wants to optimise and which to minimise or take out.

USEFUL FUNCTIONS

The primary function of the spacecraft is to transport humans. Another is to transport human culture. The more humans and the more culture can be transported, the better. It is better, as more humans and more culture enable a faster build-up of a human civilisation, which is the final objective. It is also desirable to transport them safely. Thus, safety is also a factor that needs to be maximised.

On the other hand, it is desirable to complete the trip as fast as possible, ideally instantaneously. Furthermore, the cheaper this can be accomplished, the better. It would be best if no cost at all is associated with transportation. Cost can be translated into mass and energy. Heavy spacecraft require more energy to accelerate than lighter ones. Heavier spacecraft are also more complex, in case they consist of components of comparable size. More complex spacecraft are more expensive to develop. As we are looking into the far future here, we have to be aware of cases where complexity might be decoupled from spacecraft size, for example if smaller spacecraft consist of smaller components and thus have similar complexities as larger spacecraft.

Therefore, the ideal manned interstellar spacecraft would have the following characteristics:

• No mass

• Instantaneous transportation of humans and human culture

• No cost for development

• Needs no energy

Using these criteria, we can rank the concepts as seen in Table Four, below.

Table Four: Crude ranking of concepts with respect to closeness to ideality

	WS	Hib	Emb	SBS	DB	EMW	EMW+FTL	Nano
Mass	5	4	3	3	3	1	1	2
Trip time	5	4	3	3	3	2	1	3
Energy	5	4	3	3	3	1	2	3
Safety	4	5	3	2	1	1	1	1

Where WS = world-ship; Hib = hibernation; Emb = embryo; SBS = static brain storage; DB = distributed brain; EMW = electromagnetic wave; EMW+FTL = electromagnetic wave plus faster-than-light travel; Nano = nano brain storage.

This crude analysis reveals that digital or non-biological transportation approaches have the highest degree of ideality. They are also the most speculative options regarding the presupposed technological basis. The highest degree of ideality is achieved by no longer transporting matter but pure energy in the form of electromagnetic waves. As the velocity of information transmission is limited by the speed of light, approaches to circumvent this limit have the potential to further increase ideality. Nevertheless, creating artificial wormholes, for example, is probably much more costly than sending electromagnetic waves out to a receiver. Thus, a future interstellar engineer might have to trade between fast trip time and required energy.

TECHNOLOGICAL ROADMAP

The technologies that are used in the various architectures have implications for a society capable of developing and launching such spacecraft. For example, the widespread use of emulations will have a profound economic impact [31]. Furthermore, if a specific technology is already in widespread use within a society, it is much cheaper to adopt for an interstellar mission than it is to develop it solely for this purpose.

CONCLUSIONS

In this chapter, it was demonstrated that the challenge of manned interstellar flight does not only consist of transporting humans to the stars but also consists of how to transmit human culture. Furthermore, how to build up colonies and establish a new civilisation are questions that also have to be answered. Potential colonisation targets range from star systems with a planet similar to Earth to star systems with no planet suitable for colonisation. Depending on the colonisation target, different technologies have to be developed, ranging from terraforming to space colony construction. Next, the currently proposed modes

Table Five: Mapping between key technologies of a civilization with transportation technologies

	World ship	Hibernation	Embryo	Digital
Solar System-wide civilisation				
• Space colonies	Precondition	Precondition	Precondition	Precondition
• Planetary surface colonies	Precondition	Precondition	Precondition	Precondition
• Space manufacturing	Precondition	Precondition	Precondition	Precondition
• Macroscopic replicators	Precondition	Precondition	Precondition	Precondition
Post-singularity civilisation				
• General artificial intelligence	Nice to have	Nice to have	Nice to have	Nice to have
• Brain emulations	Nice to have	Nice to have	Precondition	Precondition
• Nano replicators	Nice to have	Nice to have	Nice to have	Precondition for advanced concepts

in which humans are transported between the stars are introduced: living in habitats over generations, hibernating, in a zygotic or embryonic state, and as digitalised entities. World-ships with huge habitats can be built with near-term technologies. However, they also require enormous resources for construction. Hibernation ships require smaller habitats but it is unclear whether hibernation is feasible for timeframes of decades and longer. Transporting humans as zygotes and embryos offers tremendous mass savings. However, transmitting human culture gets much more difficult. Advanced robotics and artificial intelligence are required for establishing some cultural continuity. Digital starships unify enormous mass savings with cultural continuity. They are also based on the most advanced technology. Digital starships also offer fascinating prospects if they are combined with very advanced technologies such as teleportation or speculative faster-than light technologies. With regards to potential mission architectures and their associated pros and cons, in most cases the basic architectural decision is between launching a single spacecraft with all payloads and launching two spacecraft, one with a replicator payload and the other with humans.

As often mentioned by futurologists, technological advances are often overestimated in the short run and underestimated in the long run. If we assume technological progress to continue, we might indeed be much closer to the advanced options presented in this chapter than we expect. The prospects for us and our successors would be mind-boggling and exceed our wildest dreams.

REFERENCES

[1] G K O'Neill; 'The Colonisation of Space,' *Physics Today*, 27, 32, (1974).

[2] P Davies; 'Afterword' in *Starship Century: Toward the Grandest Horizon*, Microwave Sciences/Lucky Bat Books (2013).

[3] A M Hein et al; 'World-Ships — Architectures and Feasibility Revisited,' *Journal of the British Interplanetary Society*, 65 (4), 119 (2012).

[4] M J Fogg; 'Terraforming As Part of a Strategy for Interstellar Colonisation,' *Journal of the British Interplanetary Society*, 44, pp183–192 (1991).

[5] D M Cole; *Extraterrestrial Colonies*, Martin Company, Denver Division (1960).

[6] G K O'Neill; *The High Frontier: Human Colonies in Space*, New York (1977).

[7] G L Matloff; 'Utilisation of O'Neill Model I Lagrange Point Colony as an Interstellar Ark,' *Journal of the British Interplanetary Society*, 29, pp775–785, (1976).

[8] A R Martin; 'World-Ships – Concept, Cause, Cost, Construction and Colonisation,' *Journal of the British Interplanetary Society*, 37, pp243–253, (1984).

[9] R D Johnson and C Holbrow (eds); *Space Settlements: A Design Study*, NASA SP-413, NASA (1977).

[10] C Smith; 'Estimation of a Genetically Viable Population for Multigenerational Interstellar Voyaging: Review and Data for Project Hyperion,' *Acta Astronautica*, in press, (2014).

[11] A Bond and A R Martin; 'World-Ships – An Assessment of the Engineering Feasibility,' *Journal of the British Interplanetary Society*, 37, pp254–266, (1984).

[12] R A Freitas Jr; 'A Self-Reproducing Interstellar Probe,' *Journal of the British Interplanetary Society*, 33, pp251-264, (1980).

[13] R A Freitas Jr and W Zachary; 'A Self-Replicating, Growing Lunar Factory,' in Princeton/AIAA/SSI Conference on Space Manufacturing, Volume 35, pp18–21, (1981).

[14] R Zubrin, D Baker and O Gwynne; 'Mars Direct: A Simple, Robust and Cost Effective Architecture for the Space Exploration Initiative,' in 29th Aerospace Sciences Meeting, pp91–326, AIAA, (1991).

[15] A M Strijkstra; (2006); 'Good and Bad in the Hibernating Brain,' *Journal of the British Interplanetary Society*, 59, pp119–123, (2006).

[16] M Ayre, C Zancanaro and M Malatesta; 'Morpheus–hypometabolic Stasis in Humans for Long-Term Space Flight,' *Journal of the British Interplanetary Society*, 57, pp325–339, (2004).

[17] A Crowl, J Hunt and A M Hein; 'Embryo Space Colonisation to Overcome the Interstellar Time Distance Bottleneck,' *Journal of the British Interplanetary Society*, 65, pp283–285, (2012).

[18] R Kurzweil; *The Singularity Is Near: When Humans Transcend Biology*, Penguin Books, (2005)

[19] H Moravec; *Mind Children*, Harvard University Press, (1988).

[20] A Sandberg and N Bostrom; 'Machine Intelligence Survey,' *Technical Report*, #2011-1, Future of Humanity Institute, University of Oxford, pp1–12 (2011).

[21] R Hanson; 'Economics of Brain Emulations,' in *Tomorrow's People: Proceedings of the James Martin Institute's First World Forum*, EarthScan (2008).

[22] R Hanson; 'Economics of the Singularity,' *Spectrum*, IEEE, 45(6), pp45–50, (2008).

[23] D Roberts, J Nelms, D Starkey and S Thomas; 'P4_4 Teleportation,' *Physics Special Topics*, North America, (2012).

[24] J D Bekenstein; 'Black Holes and Entropy,' *Physical Review D*, 7 (8), 2333, (1973).

[25] G J Lokhorst; 'Why I Am Not a Super-Turing Machine,' InHypercomputation Workshop, University College London, Volume 24, (2000).

[26] A Sandberg and N Bostrom; 'Whole Brain Emulation: A Roadmap,' Future of Humanity Institute, Oxford University (2008). Available at: http://www.fhi. ox.ac.uk/Reports/2008-3.pdf (Last accessed 22 February 2014).

[27] E W Davis, F K Lu and M G Millis; *Frontiers of Propulsion Science*, American Institute of Aeronautics and Astronautics, (2009).

[28] A Sandberg; 'An Overview of Models of Technological Singularity,' in Roadmaps to AGI and the future of AGI Workshop, Lugano, Switzerland, (2010).

[29] B Goertzel; 'Human-Level Artificial General Intelligence and the Possibility of a Technological Singularity: A Reaction to Ray Kurzweil's *The Singularity Is Near* and McDermott's critique of Kurzweil,' *Artificial Intelligence*, 171(18), pp1161–1173, (2007).

[30] R D Launius and H E McMurdy; *Robots in Space: Technology, Evolution and Interplanetary Travel*, Johns Hopkins University Press, (2008).

[31] R Hanson; 'Economic Growth Given Machine Intelligence,' *Journal of Artificial Intelligence Research*, (2001).

Chapter 19

THE DRAKE EQUATION,
FERMI'S PARADOX AND ACTIVE SETI

MARTIN CIUPA

One of the biggest questions in science and philosophy asks, "Are we alone in this Universe?" We have reasonable hopes of being able to answer this question within the next 50 years or so. Until then, we can speculate what the answer could be based on the limited data we have already uncovered. Even the little knowledge we currently have can influence how we go about certain SETI-related programmes. We can build scenarios that can be considered likely given this data and are able to further explore their implications.

One line of reasoning is the Drake Equation, which helps us speculate on the possible number of communicating cvilisations in the Galaxy. Depending upon what numbers we input into the equation we can derive a variety of results and subsequently consider them with the constraining effect of the Fermi Paradox (that asks the question, "If communicating intelligences are likely, why are we in fact not in clear and open communication with such intelligence?"). Finally, within the possible scenarios that an informed imagination can speculate over, we can ask one more question: "Should we listen and search in a non-invasive discrete manner, or actively announce our existence and desire to be in communication?"

There is always the risk of over-exercising our imagination in lieu of hard results and it is tempting to dismiss and shelve these questions until we have more data. However, since the consequences of Active SETI (the name given to the act of transmitting our own signals from Earth into space, also known as 'messaging extraterrestrial intelligence, or METI) are important in their potential impact, both positive and negative, upon humankind then the fact that we are embarking on such programmes (e.g., Arecibo signal of 1974 and various messages transmitted from the 70-metre Evpatoria radio dish in the Crimea by Alexander Zaitsev of the Russian Academy of Sciences) means that we are obliged to speculate and consider scenario implications now. Decisions are needed and we must work with the information in hand.

So, in that context, this chapter will examine these issues by discussing the Drake Equation, the Fermi Paradox and Active SETI and drawing-up some of the more likely scenarios and assessing their impacts.

THE DRAKE EQUATION AND FERMI PARADOX

In 1961, radio astronomer Frank Drake, then of the National Radio Astronomy Observatory in Green Bank, West Virginia, USA conceived of a method for estimating the number of technological, communicating civilisations in the Galaxy (denoted as 'N'). The Drake Equation, as his method has since become known, considers seven different factors that Drake considered to be vital for the development of intelligent life. The beauty and the frustration of the Drake Equation is that for many of the factors there are no confirmed values, meaning the equation has no unique solution – it depends entirely on our estimates for each factor.

Let us explore the equation in more detail. Written in full, the Drake Equation is:

$$N = (R^* \times f_p \times n_e \times f_l \times f_i) \cdot (f_c \times L)$$

I have separated the terms into two parts. The first bracketed set are terms that are astrophysical or astrobiological in nature and which we can hope to analytically and empirically establish to a reasonable degree of precision. Indeed, some are quite close to being considered fairly well established at this time. The second bracketed set of terms are more sociological and harder to establish given we have only one example of a long-lived communicating intelligence, i.e., ourselves. We can speculate, but there is nothing to say that the values that the human experience provides for these factors is typical of other civilisations.

To fully understand the strengths and weaknesses of the Drake Equation, let us explore each term.

R*

This is the average rate of star formation per year in our Galaxy. In the Milky Way this currently is speculated to be between 7-10 solar masses per year [1].

F$_p$

The next term describes the fraction of stars that have planets. Recent analyses of microlensing surveys [2] and data from the Kepler Space Telescope [3] has found that f_p may approach one; that is, stars are orbited by planets as a rule, rather than the exception; and that there are one or more bound planets per Milky Way star.

N$_E$

This is the average number of planets that can potentially support life per star. We should also consider that the moons of planets may sometimes be just as good candidates. Initially it seemed that most exoplanets discovered had very eccentric orbits, or orbits close to their star where the temperature is too high for life as we know it [4]. However, in recent years NASA's Kepler Space Telescope has revealed hundreds if not thousands of candidate planetary systems that are beginning to look more like our Solar System, while non-Kepler planetary systems that also fall into this category include HD 70642, HD 154345, Gliese 849 or Gliese 581.

There may well be other, as yet unseen, Earth-sized planets in the habitable zones of these stars. Also, the variety of star systems that might have habitable zones is not just limited to solar-type stars and Earth-sized planets; it is now estimated that even tidally-locked planets close to red dwarfs might have habitable zones. It has been speculated that about ten percent of star systems in the Milky Way galaxy are hospitable to life, by having heavy elements, being far from supernovae and being stable for a sufficient time [5]. Statistical estimates from partial data suggest that 22 percent of all G-type and K-type stars (i.e. stars like the Sun or slightly smaller and cooler) have a terrestrial planet in their habitable zone [6] and that at least 5.4 percent of all stars may host a terrestrial planet [7].

Even if planets are in the habitable zone, the number of planets with the right proportion of elements is difficult to estimate. Also, the Rare Earth hypothesis [8], which posits that conditions for intelligent life are quite rare, has advanced a set of arguments based on the Drake Equation that suggest that the number of planets or moons that could support life is small and quite possibly limited to Earth alone; in this case, the value of n_e would be tiny. The discovery of numerous gas giants in close orbit around their stars had introduced doubt that life-supporting planets commonly survive the formation of their stellar systems, given that these planets migrate in towards their stars, barging smaller terrestrial worlds out of the way. However, results from Kepler show that these systems are not the majority and only initially seemed to be because of observational bias that meant these systems are easiest to discover [9]. In addition, most stars in our galaxy are red dwarfs, which often flare violently, mostly in X-rays, a property not conducive to life as we know it. Simulations also suggest that these bursts erode planetary atmosphere. The possibility of life on moons of gas giants (such as Jupiter's moon Europa, or Saturn's moon Titan) adds further uncertainty to this figure.

F_L

Next is the fraction of planets that could support life that actually develop life at some point. Geological evidence from the Earth suggests that f_l may be high; life on Earth appears to have begun around the same time as favourable conditions arose, suggesting that abiogenesis may be relatively common once conditions are right. However, this evidence only looks at the Earth (a single model planet) and contains anthropic bias, as the planet of study was not chosen randomly, but by the living organisms that already inhabit it (ourselves).

From a classical hypothesis–testing standpoint, there are zero degrees of freedom, permitting no valid estimates to be made. If life were to be found on Mars that had developed independently from life on Earth it would imply a value for f_l close to one. Consider, if independent life (e.g., not due to a panspermia effect) was found on other planets in our Solar system then we might speculate that given generally less favourable conditions in: scarcity of liquid water; temperatures either higher or lower than that of the Earth's orbit; and, perhaps without protecting magnetic radiation belts, then the chances of life being virtually omnipresent in many systems might be supported. . While this would improve the degrees of freedom from zero to nearly one perhaps, there would remain a great deal of uncertainty on any estimate because of the small sample size and the chance they are not really independent. Countering this argument is that there is no evidence for abiogenesis occurring more than once on the Earth – that is, all terrestrial life stems from a common origin.

If abiogenesis were more common it would be speculated to have occurred more than once on the Earth. Scientists have searched for this by looking for bacteria that are unrelated to other life on Earth – a so-called 'shadow biosphere' – but none have been found yet. It is also possible that life arose more than once but that other branches were out-competed, or died in mass extinctions, or were lost in other ways. Biochemists Francis Crick and Leslie Orgel laid special emphasis on this uncertainty: "At the moment we have no means at all of knowing" whether we are "likely to be alone in the galaxy (Universe)" or whether "the Galaxy may be pulsating with life of many different forms" [10]. As an alternative to abiogenesis on Earth, they proposed the hypothesis of directed panspermia, which states that Earth life began with "micro-organisms sent here deliberately by a technological society on another planet, by means of a special long-range unmanned spaceship" [11].

In 2002, using a statistical argument based on the length of time life took to evolve on Earth, Charles H Lineweaver and Tamara M Davis (at the University of New South Wales and the Australian Centre for Astrobiology) estimated f_l as greater than 0.13 on planets that have existed for at least one billion years [12].

F$_i$

The fraction of planets with life that actually go on to develop intelligent life – in other words, develop civilisations – remains particularly controversial. Those who favour a low value, such as the biologist Ernst Mayr [13], point out that of the billions of species that have existed on Earth, only one has become intelligent and from this infer a tiny value for f_i. Those who favour higher values note the generally increasing complexity of life and conclude that the eventual appearance of intelligence might be imperative, implying an f_i approaching one. Sceptics point out that the large spread of values in this factor and others make all estimates unreliable. In addition, while it appears that life developed soon after the formation of Earth, the Cambrian explosion, in which a large variety of multi-cellular life forms came into being, occurred a considerable amount of time after the formation of Earth (around 550 million years ago), which suggests the possibility that special conditions were necessary.

Some scenarios such as the 'Snowball Earth' model or research into extinction events have raised the possibility that life on Earth is relatively fragile. Research on any past life on Mars is relevant since a discovery that life did form on Mars but ceased to exist would affect estimates of these factors. This model also has a large anthropic bias and there are still zero degrees of freedom. Note that the capacity and willingness to participate in extraterrestrial communication has come relatively recently, with the Earth having only an estimated 100,000 year history of intelligent human life and less than a century of technological ability. Estimates of f_i have been affected by consideration that the Solar System's orbit around the galactic centre is probably circular, such that it likely remains out of the spiral arms for tens of millions of years (evading radiation from novae). Also, Earth's large Moon may aid the evolution of life by stabilizing the planet's axis of rotation. This may or may not be an important issue, given lack of data it is a speculative concept.

Now, for the second part of the Drake Equation that deals with the more sociological aspects.

F_c

The fraction of civilisations that develop a technology that releases detectable signs of their existence into space is denoted by f_c. With regard to deliberate communication, the one example we have (humanity) does not do much explicit communication, although there are some efforts covering only a tiny fraction of the stars that might potentially harbour habitable planets. There is considerable speculation why an extraterrestrial civilisation might exist but choose not to communicate. However, deliberate communication is not required and calculations indicate that current or near-future Earth-level technology might well be detectable to civilisations not too much more advanced than our own. By this standard, the Earth is a communicating civilisation. It is worth noting that whilst it is true that our current technology is limited in its sensitivity in detecting 'signals', though this limitation would be reasonably expected to be mitigated with technological progress.

L

Michael Shermer has estimated L, the length of time for which civilisations release detectable signals into space, to be 420 years based on the duration of sixty historical Earthly civilisations [14]. Using 28 civilisations more recent than the Roman Empire, he calculates a figure of 304 years for 'modern' civilisations. It could also be argued from Shermer's results that the fall of most of these civilisations was followed by later civilisations that carried on the technologies, so it is doubtful that they are separate civilisations in the context of the Drake Equation.

In the alternative versions of the Drake Equation, , this lack of specificity in defining single civilisations does not matter for the end result since such a civilisation turnover could be described as an increase in the reappearance number rather than an increase in L, stating that a civilisation reappears in the form of the succeeding cultures. Furthermore, since none could communicate over interstellar space, the method of comparing with historical civilisations could be regarded as invalid. David Grinspoon has argued that once a civilisation has developed enough, it might overcome all threats to its survival [15]. It will then last for an indefinite period of time, making the value for L potentially billions of years. If this is the case, then he proposes that the Milky Way Galaxy may have been steadily accumulating advanced civilisations since it formed. He proposes that the last factor L be replaced with $f_{IC} \times T$, where f_{IC} is the fraction of communicating civilisations that become 'immortal' (in the sense that they simply do not die out) and T represents the length of time during which this process has been going on. This has the advantage that T would be a relatively easy to discover number, as it would simply be some fraction of the age of the Universe. We should note that such speculations on the time for such civilizations to develop is far from being certain.It has also been hypothesised that once a civilisation has learned of a more advanced one, its longevity could increase because it can learn from the experiences of the other [16].

The astronomer Carl Sagan speculated that all of the terms except for the lifetime of a civilisation are relatively high and the determining factor in whether there are large or small numbers of civilisations in the Universe is the civilisation lifetime or, in other words, the ability of technological civilisations to avoid self-destruction [17]. In Sagan's case, the Drake Equation was a strong motivating factor for his interest in environmental issues and his efforts to warn against the dangers of nuclear warfare. Personally, I am pessimistic on this, not so much because of self-destruction, but that civilisations do not extend their consumption exponentially, as per the Kardashev scale. (The Kardashev scale, developed by the Soviet astronomer Nikolai Kardashev in 1964, measures a civilisation's level of technology based on the amount of energy it can utilise. A type I civilisation is capable of harnessing all the resources of its own planet – humanity is considered to be at level 0.73, so still not quit a type I civilisation yet. A type II civilisation can appropriate all the energy of its star, using for example a Dyson Sphere, whereas a type III civilisation can do the same thing for all the stars within its galaxy.

Being advanced and self-sustaining is more likely in my opinion, as societies may turn their interests inwards and still be techno-progressive, seeking out life fulfilled in terms of 'Matrix like' simulations. If such a speculation is viable I think a much lower level for L may then be justified – say around 100 years (in humanity's terms this corresponds from the time of powerful television transmissions – on 2 November 1936 the BBC began transmitting the world's first public television in London) to a point when we 'go quiet', e.g. perhaps 2136, a date also coincident with some predictions of a 'technological singularity' [18]. This argument is subject to technology progressing to high sensitivity for detection of low energy signals.

SCENARIO ANALYSIS

Let us consider some scenarios. What ranges can be considered for the Drake Equation given what we have discussed above?

$R^* = 7$;
$f_p = 1$;
$n_e = (0 \rightarrow 1)$;
$f_l = (0.13 \rightarrow 1)$;
$f_i = (0 \rightarrow 1)$;
$f_c = (0 \rightarrow 1)$;
$L = (100 \rightarrow 1 \times 10^9)$.

Thus at the lower end we have: $N = (7 \times 1 \times 0 \times 0.13 \times 0) \bullet (0 \times 100) = 0$

And the upper end we have: $N = (7 \times 1 \times 1 \times 1 \times 1) \bullet (1 \times 10^9) = 7 \times 10^9$, or seven billion civilisations.

Obviously this is a massive range of scenarios. However, we can be reasonable in reducing this speculative range to a few assumption 'classes', of which here we present three. (We can extend this, but I trust this analysis can generate useful speculation, since we do have one data point for N, i.e., the Fermi Paradox observation).

SCENARIO #1: RARE INTELLIGENT LIFE

We cannot reasonably dismiss the factors of n_e, f_l and f_i being very small. Life may be exceedingly rare. We may find life on Mars, Europa and elsewhere within our Solar System. If this life is independent of life on Earth, then we can raise n_e and f_l to be non-zero with reasonable expectation. Even if we do, the possibility exists that f_i could remain very low indeed.

This scenario does not mean that intelligent life is impossible, but it does posit that n_e, f_l and f_i together may limit the Galaxy, irrespective of the values of f_c and L to be a number near 1–10, i.e., ourselves, alone or very nearly so in the Galaxy. This fits the Fermi Paradox finding.

SCENARIO #2: ABUNDANT INTELLIGENT LIFE THAT IS SHORT LIVED

We can consider the alternative astrophysical/astrobiological scenario with values of n_e, f_l and f_i being high, even say at one (unlikely but if, for example, it is more than 0.5, we are within an order of magnitude to consider this a scenario 'class', i.e., $0.5 \times 0.5 \times 0.5 = 0.125$).

Here we bring the second, sociological side of the Drake Equation into play and this particular scenario is a pessimistic one that, thanks to a variety of possible sub-scenarios, the life time of an albeit communicating civilisation is low. Consider the following:

$$N = (7 \times 1 \times 1 \times 1 \times 1) \cdot (1 \times 100) = 700$$

Within the scale of our Milky Way Galaxy, which has approximately 250 billion stars, this means one communicating civilisation per 360 million stars. In our outer region of the Galaxy, where stars are younger and are distributed less densely than in the core of the Galaxy, this means a scenario of significant scarcity of long lived civilisations. Again this is also consistent with the Fermi Paradox.

Scenario #3: Abundant intelligent life that is not communicating

Again we can consider the alternative astrophysical/astro-biological scenario with values of n_e, f_l and f_i being high, set at one, as in the 'Abundant Intelligent Life' of scenario #2. Likewise, again we can bring the second, sociological side of the Drake Equation into play, but this scenario, whilst optimistic on the lifetime of civilisations, is a pessimistic one in terms of the factor of communicating civilisations being low because of a variety of possible sub-scenarios. Consider the following:

$$N = (7 \times 1 \times 1 \times 1 \times 1) \cdot (1 \times 10^{-6} \cdot 1 \times 10^{9}) = 7{,}000$$

Following the previous argument of scenario #2, within the scale of our Milky Way Galaxy with 250 billion stars this means one communicating civilization per 35 million stars. Again, this means a scenario of significant scarcity of communicating civilisations. Whilst modelling of the distribution of civiliations in such a scenario should be done, this is consistent with the simplistic observations of the Fermi Paradox.

Wider Fermi Paradox solution scenarios

An in-depth analysis of 50 possible solutions to the Fermi Paradox has been made by Stephen Webb [19]. He considered three broad categories:

Category 1: they are already here. Included in this section are the ideas that extraterrestrials have visited us in the past (as in Erich von Däniken's 'ancient astronauts') or in the present (as in flying saucers). This is by far the shortest of the three categories because Webb is interested in science and not pseudoscience. Webb includes ideas for the sake of completeness, but they are quickly dismissed as not having any well accepted supporting evidence. However, although we do not give this category credibility, there are scenarios within it that morph to the 'quiet' non-invasive category 3 below.

Category 2: they do not exist. This broadly encompasses scenario #1 in my earlier discussion and takes a look at what is known about each step in abiogenesis and subsequent evolutionary processes: the evolution of habitable planets, of life, of complex multi-cellular life, of intelligence and of technology. If several steps in this process are exceedingly rare, we could be the only technological civilisation in the Universe.

Category 3: they are out there but we can't detect them. Again this broadly encompasses scenarios #2 and #3 of my earlier analysis. Among the many possibilities discussed by Webb are the possibility that we have not looked (or listened) long enough, or that they are deliberately and without exception hiding from us. He considers the possibilities that technological civilisations always destroy themselves before becoming space-faring/communicating.

Personally, of these three categories the third is the most interesting to speculate over, certainly from point of view of the sociological impacts. My personal bet (for what it is worth) is that there is abundant life in the Universe, but it is not communicative. Why? Two possible sub-scenarios present themselves. One is intrinsic (a self-developed rationale for withholding communication), the other is not.

THE INTRINSIC NON-COMMUNICATING SUB-SCENARIO

Intelligent 'techno-progressive' civilisations with the capability to communicate may choose not to, perhaps because of a convergent cultural evolution in 'ethical-progressive' terms. Consider the following possibilities:

a) Arising from the 'ethics of categorical imperative' [20], civilisations do not intervene with other cultures invasively, as they would not wish to have been intervened with during their adolescent civilisation stage. The argument here is one of reciprocal concern, i.e., a civilization having struggled to survive independently might be expected to value its independence and uniqueness. If this is the case (and one can consider alternatives) then reciprocation of the ethic to other civilisations might be expected to be extended to them, so that their uniqueness not be 'skewed'.

b) An ecological sustainability argument that posits advanced civilisations do not seek to expand beyond a Kardashev scale of 1 in terms of their energy consumption.

c) An argument based on the likelihood that a civilisation develops a simulation-based reality for their culture to recede into and replace the need for physical expansion [21].

It should be noted that this scenario, is not necessarily an abandonment of possible intercommunication between galactic civilisations, rather it is a cautious one. Civilisations go quiet so as not to interfere with other developing adolescent civilisations. Civilisations might seek out others and non-invasively study them, either by long distance observation or probes that are discrete. Once a civilisation meets a certain criteria of maturity it might make cautious diplomatic attempts to openly communicate and socialise in a manner that is not broadcast on an 'open' channel.

EXTRINSIC NON-COMMUNICATING SUB-SCENARIO

There may be obligations, or actions, dictated from technologically advanced interstellar civilisations to prevent adolescent civilisations from communicating before they are ready.

a) Advanced civilisations may not have ethics that are tolerant of independent action in our terms and they may actively stop adolescent civilisations from being able to communicate, by imposing their terms, or aggressively annihilating those adolescent civilisations.

b) Advanced civilisations, aware of intolerant civilisations mentioned above, may hide their presence.

Personally I expect that aggressively intolerant civilisations to be less likely than tolerant ones, I expect them to self-annihilate or be opposed, though it might be argued that stable aggressive civilizations can be sustainable. However our understanding of the evolution of morality is in its early stages, assumptions from the history of mankind is not necessarily mappable to other non-human civilizations. But if morality is a consequence of game theoretic driven behavior of group species then it may be seen to be supportable.. Our speculations in this area are not well founded, but we can consider the cases nevertheless in an existential risk analysis.

SETI and Active SETI

The Search for Extraterrestrial Intelligence (SETI) is the collective name for a number of activities undertaken to search for intelligent extraterrestrial life. SETI projects use scientific methods in this search. For example, electromagnetic radiation is monitored for signs of transmissions from civilisations on other worlds. Active SETI (Active Search for Extra-Terrestrial Intelligence) is the attempt to send messages to intelligent aliens. Active SETI messages are usually in the form of radio signals (such as the Arecibo Message [22]). Physical messages like that of the Pioneer plaque may also be considered an Active SETI message. Active SETI is also known as METI (Messaging to Extraterrestrial Intelligence) or positive SETI. Active SETI is contrasted to passive SETI, which only searches for signals, without any attempt to send them.

Sporadic Active SETI projects have beamed brief 'messages' at individual sky targets. For example, one high-powered digital radio signal was sent by consortium led by Alexander Zaitsev on 9 October 2008 towards Gliese 581c, a large terrestrial extra-solar planet that was discovered orbiting a red dwarf star 20 light years away by the HARPS (High Accuracy Radial Velocity Planet Searcher) at the European Southern Observatory. This was the fourth message that Zaitsev had sent into space using the 70-metre Evpatoria dish.

Currently Jacob Haqq-Misra of Penn State University is leading a team of scientists and entrepreneurs who are launching a new initiative called 'Lone Signal' that aims to send the first continuous mass 'hailing message' out into space. They will be specifically targeting one star system, Gliese 526, which has been identified as having a potentially habitable planetary system. Lone Signal will be sending two signals: one is a continuous wave signal, a hailing message that sends a slow binary broadcast to provide basic information about Earth and our Solar System using an encoding system created by astrophysicist and planetary scientist Michael W Busch. The binary code is based on mathematical 'first principles' that reflect established laws that, theoretically, are relatively constant throughout the Universe. [13]

It can be argued that the current Active SETI projects lack serious intent, though the principle and ethics are tested by it. Given the earlier analysis of the sub-scenarios of the Drake Equation and the Fermi Paradox, the following question naturally arises: is transmitting into space a safe project or does it hold existential risk?

Indeed, this project has caused considerable dissent in the scientific community. Various members of the International Academy of Astronautics Post Detection Protocol Committee, including former diplomat Michael Michaud and former head of NASA's SETI programme in the 1970s, Dr John Billingham, have resigned from the committee over what they see as an unwillingness to discuss the rights and wrongs of Active SETI [24].

Personally, I think the chances are that we will either not get any response to Active SETI or that if we do it will be benign and positive. However the 'extrinsic non-communicating sub-scenario' suggests there are risks, even if they are remote. If this scenario is correct and there is the possibility for existential risk then it would be wise not to dismiss it without further consideration. In that respect it is recommended that a consultative approach be engaged in which sociological and diplomatic input be sought to explore these scenarios further and, if considered necessary, revise, regulate or abandon the Active SETI programme as it is currently conceived.

In 1965 the American Political Scientist Hans J Morgenthau contributed a seminal paper to *Foreign Policy*, titled 'To Intervene Or Not To Intervene' [25]. We are no longer in the realm where science alone can guide our policy safely.

On reflection it is perhaps wiser to take a non-invasive discrete position until we can get more empirical data to fill out our knowledge of what lies behind the Drake Equation/Fermi Paradox and the scenarios that they give rise to.

CONCLUSIONS

The question of whether we are alone in the Universe is one with profound philosophical consequences. We have some answers to the question fairly well established, while others remain very speculative.

The Drake Equation helps us focus this speculation on the possible number of communicating intelligences in the Galaxy and is constrained by the existence of the Fermi Paradox. We have discussed examining the Drake Equation in two aspects: astrophysical/astrobiological and sociological. The sociological scenarios are difficult to resolve but can be subject to scenario analysis. Three scenarios have been considered most useful in analysis, namely:

a) Rare intelligent life

b) Abundant intelligent life that is not long lived

c) Abundant intelligent life that is not communicative.

Whilst other scenarios are possible, they are consistent with these broad classes. Within class c, two sub-scenarios received further consideration, namely the intrinsic and extrinsic non-communicating sub-scenarios. This highlights the remote but possible scenario of existential risk for humanity.

In that context the Active SETI/METI programme needs to consider these sub-scenarios as an urgent priority, via a consultative approach that engages sociologists and diplomats to explore these scenarios further, with a view towards regulation and mitigation of existential risk until we know more.

REFERENCES:

[1] C Wanjek; 'Radioactive 26Al From Massive Stars in the Galaxy,' *Nature*, 439 (2006), pp45–47.

[2] A Cassan et al; 'One or More Bound Planets Per Milky Way Star from Microlensing Observations,' *Nature*, 481 (2012), pp167–169.

[3] C D Dressing and D Charbonneau; 'The Occurrence Rate of Small Planets around Small Stars,' *The Astrophysical Journal*, 767 (95), 2013.

[4] Petigura, Eric, et al; 'Prevalence of Earth-Size Planets Orbiting Sun-like Stars,' *Proceedings of the National Academy of Sciences of the United States of America*,

31 October 2013.

[5] C H Lineweaver et al; 'The Galactic Habitable Zone and the Age Distribution of Complex Life in the Milky Way,' Science, 303 (2004), pp59–62.

[6] W J Borucki, et al; 'Characteristics of planetary candidates observed by Kepler, II: Analysis of the First Four Months of Data', *The Astrophysical Journal* 736, 19 (2011).

[7] D Brownlee and P Ward; *Rare Earth*, Copernicus, 2000.

[8] F Crick and L Orgel; 'Directed Panspermia,' *Icarus*, 19, 3 (1973), pp341–348.

[9] C H Lineweaver et al; 'Does the Rapid Appearance of Life on Earth Suggest That Life is Common in the Universe?' *Astrobiology*, 2, 3 (2002), pp293–304.

[10] E Mayr; 'Can SETI Succeed? Not Likely,' (online) www.astro.umass.edu/~mhanner/Lecture_Notes/Sagan-Mayr.pdf. Last accessed 9 January 2014.

[11] J Lissauer et al; 'Obliquity Variations of a Moonless Earth,' *Icarus*, 217 (2011), pp77–87.

[12] M Shermer; 'Why ET Hasn't Called,' *Scientific American*, August 2002.

[13] D Grinspoon; 'Lonely Planets: The Natural Philosophy of Alien Life', *Ecco* (2003).

[14] M Ćirković; 'Against the Empire', *Journal of the British Interplanetary Society*, 61, 7 (2008).

[15] C Sagan; 'The Abundance of Life-Bearing Planets,' (online) www.astro.umass.edu/~mhanner/Lecture_Notes/Sagan-Mayr.pdf. Last accessed 9 January 2014.

[16] R Kurzweil; *The Singularity is Near: When Humans Transcend Biology*, Viking, 2005.

[17] S Webb; *If the Universe Is Teeming with Aliens, Where is Everdbody? Fifty Solutions to the Fermi Paradox and the Problem of Extraterrestrial Life*, Copernicus, (2002).

[18] Kant's Moral Philosophy, Stanford Encyclopedia of Philosophy (online), http://plato.stanford.edu/entries/kant-moral/. Last Accessed 9 January 2014.

[19] N Bostrom; 'Are You Living In a Computer Simulation?' *Philosophical Quarterly*, 53, 211 (2003), pp243–255.

[20] 'Early SETI: Project Ozma, Arecibo Message,' (online) www.seti.org/seti-institute/project/details/early-seti-project-ozma-arecibo-message. Last Accessed 9 January 2014.

[21] M Busch and T Reddick; 'Testing SETI Message Designs,' *Communication With Extraterrestrial Intelligence*, State University of New York Press (2011), pp419–423.

[22] D Brin; 'Shouting At the Cosmos: How SETI has Taken a Worrisome Turn Into Dangerous Territory' (online) www.davidbrin.com/shouldsetitransmit.html. Last accessed 9 January 2014.

[23] H J Morgenthau; 'To Intervene or Not to Intervene,' *Foreign Affairs*, April 1967.

Section V

Communication of the Vision

Introduction by Kelvin F Long

In life, our decisions are not always guided by logic and reason, but simply by emotive force. Artistic visions, expressed through poetry, music, art or otherwise, have the power to move the human heart in a way that well written scientific papers or mathematical equations do not, although those too can have an artistic dimension. In this chapter we explore the communication of the vision through two mediums by two of our leading artistic lights.

David A Hardy has been producing magnificent artwork for interstellar spaceflight for a very long time, since the 1950s in fact, and he has had the honour of working with some great names including Sir Arthur C Clarke and Sir Patrick Moore. Here David gives us a tour of his many wonderful creations, from asteroid travelling habitats to interstellar ramscoops and energetic fusion-like vessels. These images help us to believe in a world that has yet to come to pass, one where starships are travelling between the planets and stars beyond.

Compared to David, **Alex Storer** is a new kid on the block, but is equally compelling in what he produces. He is also a space artist, but in this chapter we asked him to focus on his musical compositions. Here he takes us on a tour of his own work, but also what musical pieces influenced him, from Ziggy Stardust to *Tubular Bells*, and it is clear that music of all sorts has the potential to influence us, once it has our ear.

When we are talking about a future that is yet to come to pass, it is more important than ever that we find ways to engage our senses so as to transport us to that future, even if just for a moment. This is the role that art has to play, as a form of 'time machine' for our spirits – let the paint brushes and melodies begin.

CHAPTER 20

DAVID A HARDY

The earliest books on space travel concentrated, not surprisingly, on getting into space and to the Moon and the planets of our Solar System, which meant that artists usually painted only those subjects. The books by Arthur C Clarke, illustrated by R A Smith (but also with colour versions by Leslie Carr) and those by Willy Ley or Wernher von Braun and his *Colliers* team, illustrated by Chesley Bonestell, Rolf Klep and Fred Freeman, are cases in point. However, Clarke's 1951 *The Exploration of Space* [1] does contain one painting by Carr of the planet of a multiple star system similar to beta Lyrae, while the 1950 book *The Conquest of Space* [2] has one by Bonestell of the red supergiant Mira, with its small white companion. So we were considering the possibility of other stars possessing planets, even though it was usually pointed out that these were probably unlikely in multiple star systems (all the mass being assumed to be concentrated in the stars themselves) and that it was even less likely that any nearby planets would be Earth-like. It was not until 1964 that Bonestell and Ley produced *Beyond the Solar System*, [3] in which they described and depicted a number of exotic star systems and even a 'deep-space craft', launched from Earth and placed in a parking orbit 725 kilometres above it. This was last shown five days after crossing the orbit of Neptune, but "The type of propulsion mechanism required for such a mission is still to be developed." Its triangular, faintly glowing 'fins' were in fact radiators.

I think my own first serious attempt at depicting a starship was for the 1972 book *Challenge of the Stars* with Patrick Moore [4]. Near the end of

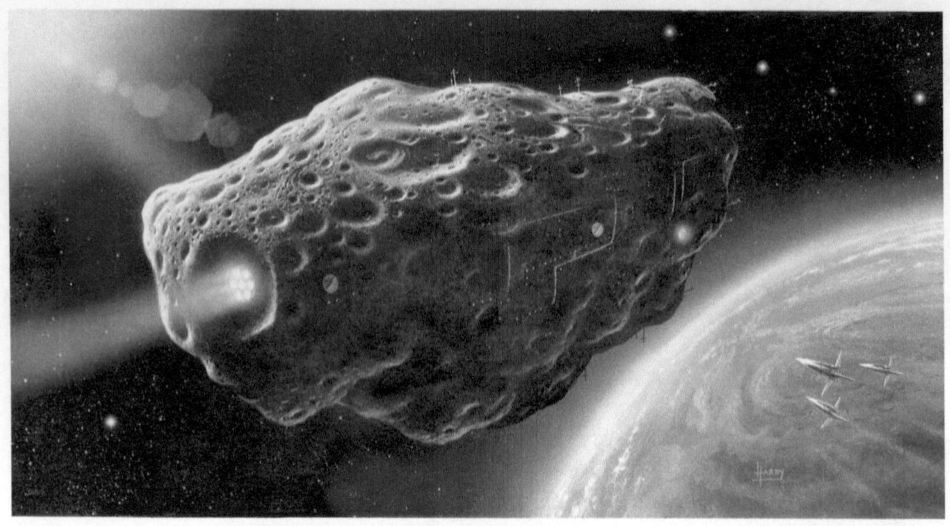

Asteroid Ark. © David A Hardy.

this we showed an alien radio telescope conducting its own version of SETI (Carl Sagan bought the original) and a number of what we would now call exoplanets, including one of Proxima Centauri with our Sun as an extra star in the constellation of Cassiopeia; the red supergiant zeta Aurigae; zeta Cancri; and even alien life in a globular cluster and the planet of a 'stray star' outside the Milky Way galaxy. Also featured was a 'Photon Starship' arriving at an Earth-like planet with a large moon and the panorama of the Trifid Nebula behind it. The ship has "a giant parabolic reflector that emits a beam of light particles," fuel and crew spheres, and at its tip shuttle-craft to make a landing.

The second, revised edition, published in 1978, imaginatively entitled *The New Challenge of the Stars* [5], was much more adventurous. It showed a spaceship that has made a crash-landing on an asteroid orbiting a black hole that draws material from a giant star; a hollowed-out 'Asteroid Ark' arriving at a terrestrial planet and being met by a reception committee of rockets, only the descendants of the original crew and colonists of course surviving the long journey; and a variation of the Death Star from *Star Wars* – a film which, not coincidentally, hit the cinemas just around the time that the book was published (but I had not seen it before painting the illustration!). That movie of course had as its core premise the proliferation throughout the Universe of inhabited planets, some with bizarre inhabitants, others who appear totally human; and also the assumption that some form of faster-than-light travel has been developed so that they can all visit each other at will, conduct wars, and so forth.

In the following years I was called upon to illustrate various concepts for interstellar vehicles. Notable among these are the Enzmann starship, a concept for a manned vehicle proposed in 1964 by Dr Robert Enzmann. A three million tonne ball of frozen deuterium would fuel nuclear fusion pulse engines (similar to those of Orion, qv.), which were contained in a cylindrical section with the crew quarters. It would be built in orbit and would be of modular construction in three units, each about 90 metres long and with a crew of 200 but room for more. One suggestion, illustrated here, is that the vehicles could be completed in orbit around Jupiter, obtaining its fuel from the gas giant's atmosphere.

Another illustration, commissioned for an encyclopedia, was for Project Daedalus, an unmanned starship study conducted by the British Interplanetary Society (BIS) between 1973 and 1978 by a team led by Alan Bond [6]. The main criteria were that it had to use current or at least near-future technology and must be able to reach its destination – Barnard's Star, about six light-years away – within a human lifetime. A two-stage vehicle, it again used fusion propulsion, and it was expected to take fifty years to reach its destination. Originally I showed it passing Neptune, but I later edited it digitally (all of the above art was produced 'traditionally' – i.e. with paint, brushes and airbrush!) to show Earth in the background. I also changed the colour of the exhaust from reddish to blue–violet, following expert advice.

As a sort of sequel to *The New Challenge of the Stars*, in 1979 I started work on *Galactic Tours* [7], a sort of interstellar travel brochure with my paintings in place of the photographs of resorts etc., that one would normally expect. In other words, it assumed that one could hop on a starship to the planets and moons of our Solar System, or others, as easily as we now jet off to Majorca or Ibiza. Because this was definitely science-fictional rather than pure scientific extrapolation, I chose SF author Bob Shaw to co-write it with me, rather than

The Sagan ramscoop. Note the damage from meteroid impacts. © David A Hardy.

Patrick. I produced small colour 'roughs' that I sent to Bob with my notes on how I imagined each world and the sort of delights to be found there, and he expanded this in his own inimitable manner into witty and readable text. The example shown above is of the Sagan ramscoop, limping in meteor-scarred to the planet of Avalon – having left Èarth in 2173 and taken two centuries to arrive. Meanwhile, of course, technology on Earth has developed a faster-than-light drive, so instead of finding a virgin world ripe for colonisation, the Sagan is met by a human reception committee – in effect, a coastguard (I was particularly pleased with the vast nebula background to this).

Thomas Cook took up this title and opened a list for potential customers who wanted to sign up to be the first tourists on the Moon. Delayed by publication issues, the book finally appeared in 1981 but unfortunately by then other books with a similar theme – treating it as if we are already living in the future (for example *Spacecraft 2000–2100 AD* and even *Tour of the Universe!*) had already started to appear.

An ion-drive starship as seen in *Futures: 50 Years in Space*. © David A Hardy.

In 1983 I was asked to provide a cover for the US *Analog* magazine of science fiction and fact [8] (my second in fact; the first, in 1981, being to illustrate *Mars in 1995!*, an article by Dr Bob Parkinson of the BIS). This was also to illustrate a factual article, *To the Stars!*, by Gordon R Woodcock. In this the author reviewed the current thinking on methods of propulsion for interstellar travel, including Orion, fusion, Bussard ramjets and matter–antimatter power. He did not propose any specific design for a starship, so I wrote to him (there was of course no e-mail in those days) and he suggested that it would be a triangular wedge-shape in order to pose the least resistance to the interstellar medium. I showed it arriving at the Earth-like moon of a large Jovian gas-giant (see opposite page); the cover, in June 1983, was voted the best of the year.

It seems amazing now that for all of this time we had no tangible proof that planets actually exist around other stars. It was almost an act of faith that because our Sun has rocky or gaseous companions, planets could also be found in orbit around other stars in our Galaxy, or indeed in any other galaxy. It was in 1999 that I was asked to produce an impression of a planet belonging to the star tau Boötis (dubbed by the press and media 'The Millennium Planet' because its announcement came around the time of the [false] millennium). This was a very large Jupiter-type gas giant and, for several years, this was the only type of planet that could be detected by the methods then available. Since then, largely thanks to the orbiting Kepler telescope, we have discovered literally thousands of extra-solar planets of all sizes but mainly large gas giants and have started to detect Earth-sized, possibly even Earth-type planets. Whereas we used to assume that only yellow, G-type Sun-like stars were likely to possess these, a recent theory is that dimmer, red dwarf stars such as Gilese 581 may be even more likely candidates. Since over 80 percent of the stars in our Galaxy are of this type it could be good news, if true!

When in 2004 Patrick Moore and I decided to produce a new edition of *Challenge of the Stars*, very much updated and now entitled *Futures: 50 years in Space – The Challenge of the Stars* [9, 10], it was naturally a very different book. Celebrating the 50 years in which we had worked together since 1954, it explains

To the Stars! © David A Hardy.

how, rather than people building a space station and going on to the Moon, building bases there, going on to Mars and building bases there and so on, space travel had evolved very differently, thanks not only to political and financial reasons but to the advance of micro-miniaturisation and the development of automated probes and robotic rovers. It contains my work from the 1950s all the way through to the twenty-first century – the latter mainly being produced for the book digitally, using *Adobe Photoshop* and *Terragen* (a terrain generator). It includes a number of exoplanet discoveries including the 2002 'warped disc' around Fomalhaut, shown as containing a Neptune-sized, ringed planet; tau Gruis; a brown dwarf; eta Carinae; and the strange, red-hot world that had recently been discovered orbiting a pulsar. The final illustration is of a starship, probably with some sort of ion drive, its radiator fins glowing, but described only as 'yet to be designed', again shown arriving at an Earth-like world with a large moon.

REFERENCES

[1] A C Clarke; *The Exploration of Space*, Harper (1951).

[2] W Ley and C Bonestell; *The Conquest of Space*, Viking (1950).

[3] W Ley; *Beyond the Solar System*, Viking Adult (1964).

[4] P Moore and D A Hardy; *Challenge of the Stars*, Mitchell Beazley/Sidgwick and Jackson (1972).

[5] P Moore and D A Hardy; *The New Challenge of the Stars*; Mitchell Beazley (1978).

[6] A Bond and A Martin; *Project Daedalus, Journal of the British Interplanetary Society* (1978).

[7] B Shaw and D A Hardy; *Galactic Tours: Thomas Cook Out of This World Holidays (Vacations in US)*, Proteus (1981).

[8] G R Woodcock; *To the Stars!, Analog Science Fiction/Fact*, June 1983.

[9] D A Hardy and P Moore; *Futures: 50 Years in Space*, AAPPL, 2004 (Hb).

[10] D A Hardy and P Moore; *50 Years in Space*, AAPPL, 2006 (Pb).

Chapter 21

Musica Universalis:
The interstellar relationship
between space, music and popular culture

Alex Storer

The glorious, awe-inspiring images of the planets in our Solar System, distant stars and neighbouring galaxies, somehow do not feel complete without an accompanying soundtrack. The forms and colours seen in the increasingly brilliant pictures we receive from our telescopes, astronauts and rovers, evoke moods and emotions that are only heightened by the use of music.

Music is a soundtrack to the past, present and future – a means of reliving history but also bringing tomorrow into today. For decades, musicians have composed music inspired by the stars and planets – a theme with infinite creative possibilities.

Although everybody has their own musical tastes and definition of what kind of music is best suited to the visions of space and interstellar travel, instrumental music is often the obvious choice, being universal in language. When looking at NASA's detailed photographs of the surface of Mars or astounding images of the cosmos, the atmosphere and mood of a piece of music can compliment and contrast with what we see, providing an immersive experience of space exploration at a time when only the select, skilled and trained can actually blast off into outer space.

The first half of the twentieth century saw a plethora of science fiction masterworks and the emergence of many pioneering authors of the genre. The first science fiction films also appeared and humankind's ambitions to travel into space edged ever closer to reality. Until this point, any relationship between music and space travel was generally minimal, especially when it came to the records that people bought. All that was soon to change, as the 1970s proved to be a visionary decade, bookended by space travel and science fiction. It is often regarded a golden age for music, film and television as well as SF literature.

The Apollo 11 Moon landing in the summer of 1969 was the obvious catalyst for many artists who would emerge in the following years. Numerous space-influenced albums would appear from artists such as Hawkwind, Jefferson Starship, Gong, Rush and Pink Floyd, to name but a few. There was however, one artist whose music would change the relationship between space and popular music forever.

David Bowie was yet to become a household name when he originally released *Space Oddity* in June 1969. This proved to be quite timely as, despite it not receiving any airplay, the BBC chose to use the song during their coverage of the Moon landing, which undoubtedly helped create Bowie's first hit single. Thanks to *Space Oddity* – a song about a fictional astronaut called Major Tom getting stranded out in space – David Bowie had unintentionally raised awareness of interstellar travel through his music, which would go on to dominate music charts around the world for the next forty years.

It was during the early years of Bowie's career in which he would make a lasting impact, with his iconic album and alien persona of Ziggy Stardust. In 1972, very few people sung about the stars, space or aliens; this was generally left behind in popular music. The combination of Bowie's unusual appearance, distinctive voice and ambitious musical vision caused ripples in popular culture in a country on the brink of recession. Songs such as *Starman* offered the listener escape from daily life and the glimpse of a possible future with proof of life on other worlds. Captivated by Bowie's unconventional but innovative style, fans were sucked into his world, perhaps reconsidering any opinions about the possibility of an alien visiting Earth – somehow Bowie's music at that time, made it all the more plausible.

David Bowie's most famous album, *The Rise and Fall of Ziggy Stardust and the Spiders from Mars*, was a concept album roughly based on Bowie's Ziggy

Stardust character bringing a message of peace to the Earth during its final five years of existence. The impact of both his music and on-stage persona propelled Bowie to superstardom and, even after 'killing' the character off during the final show of his 1973 tour, Ziggy's ghost would continue to haunt Bowie.

In 1975, David Bowie was cast in his first starring role as Thomas Jerome Newton in Nicholas Roeg's adaptation of Walter Tevis' *The Man Who Fell to Earth*, a story about an extraterrestrial who crash lands on Earth while on a mission to find a means of transporting water to his home planet, which is suffering from a devastating drought. Bowie was also ready to produce the film's soundtrack, but contractual complications resulted in him withdrawing from that part of the project. However, aspects of this work would later be used on the album *Low*, in 1977.

Bowie's fascination with space would continue throughout his career, with hit singles such as *Starman* and *Life On Mars?* becoming synonymous with space travel, even in instances where the lyrics themselves bear little relevance. Even *Ashes to Ashes* – Bowie's first number one single of the 1980s – re-introduced the Major Tom character from *Space Oddity*. Such interstellar motifs, though minor, would be recurrent in his work, as evident in the recent *Dancing Out In Space* from his 2013 album, *The Next Day*, his first new music for a decade.

From every aspect of his career, Bowie's work continues to inspire; his music has both a timeless quality and something to make you stop, think and ask questions, as opposed to your run-of-the-mill disposable pop. Through his ever-changing appearance and musical styles, David Bowie has made himself musically and artistically relevant in each coming decade. Yet despite his vast and varied back catalogue, *Space Oddity* still continues to prove its lasting appeal, evident as recently as May 2013, when astronaut Chris Hadfield performed the track in microgravity on the International Space Station. Hadfield's performance soon became an Internet sensation as well as the world's first music video shot in space.

David Bowie probably never even dreamed of this old tune actually being performed live from space, but out of all the classic twentieth century pop songs surely *Space Oddity* is an obvious choice. While an experienced astronaut,

Canadian Chris Hadfield's interstellar performance also served as a reminder that he was still an ordinary guy with a love of music. This was Chris simply showing us that it can be done and that, one day, we'll be able to do the same.

After Bowie had left Ziggy behind and the flurry of space-themed rock albums in the mid-1970s waned, musical visions of the future soon felt like a passing trend as disco and punk arrived. Yet in a decade of shifting musical styles, things were about to change yet again, as a new wave of musicians emerged, armed with synthesisers!

SYNTH PIONEERS

German artists Kraftwerk and Tangerine Dream had already been producing electronic music for some time in their home country. Towards the end of the decade, their influence would make its way to Britain, with the emergence of cutting edge artists such as John Foxx, Gary Numan and The Human League,

Floating in the most peculiar way: *Big Dog*, inspired by the Brian Aldiss novel *Non-Stop*. © Alex Storer.

each fashioning their own futuristic sound and style through the use of synthesisers. Pioneering French artist Jean Michel Jarre would lead the new synth revolution, with one of the first commercially successful all-electronic albums, *Oxygène*.

Jean Michel Jarre's *Oxygène* is one of many classic twentieth century albums often associated with space and interstellar travel, thanks to its otherworldly sounds and atmospheres. Yet, the underlying theme of the album is actually a serious message of pollution and climate change, evident in both its title and the haunting cover painting by French artist, Michel Granger, which was part of Jarre's original inspiration for the album.

Released in France in 1976 and internationally the following year, *Oxygène* presents a musical soundscape like no other, ahead of its time back then and just as relevant today. Originally recorded at Jarre's home studio in Paris, Oxygène is entirely instrumental, purely electronic and structured like a classical movement with no song titles, just simply parts I–VI. The fact that the album contains no lyrics or vocals posed a problem for Jarre when approaching record labels with his original demos – they felt it was too big a risk to market such music. French publisher Francis Dreyfus eventually saw potential in Jarre's experimental work and tentatively pressed 50,000 copies. However, *Oxygène* would go on to sell in excess of 15 million copies, with the success of the album propelling Jarre from being a relatively unknown musician to one of the genre's most successful and celebrated international artists.

Oxygène dispels any theory that electronic music might be devoid of soul or emotion. It is laden with catchy riffs and memorable hooks, alongside bubbling analogue sequences and effects that bounce playfully from speaker to speaker. Yet despite being entirely electronic, the music has an organic quality to it, somehow making at one with the environment. The almost alien nature of the evolving atmospheres and textures of the album make it ideally suited to accompany interstellar visions. Our brilliant senses and imaginations are naturally inclined to make associations with particular sounds and images. With that in mind, *Oxygène* really is a feast for the imagination – every sound that shimmers and glitters evokes images of the stars; every icy blast of white noise could be a meteor in passing; each sonic swirl, the formation of a new galaxy; and every gliding string the drifting of spacecraft in orbit. Maybe this explains why the album still remains frequently associated with space today and is one of the reasons why future generations will be playing this album and making similar connections for decades to come.

Jarre re-visited familiar territory in 1997 with *Oxygène 7–13*, picking up where the original had left off, sounding as if two decades hadn't passed. As the album's title suggests, this is a continuation of the theme rather than a sequel or attempt to create a modern day version. While the album fuses modern digital synths with vintage analogue, the soundscape it creates instantly transports you back to the vivid musical landscape of the groundbreaking original.

A UNESCO Goodwill Ambassador, Jarre would actually go on to have several associations with space, including a collaboration with Arthur C Clarke for a concert staged in 2001 and, most notably, a spectacular outdoor concert illuminating the Houston skyline in January 1986, in collaboration with NASA and launching Jarre's album, *Rendez-Vous*.

Rendez-Vous is a dramatic electronic overture with all the power of a symphony orchestra. Space travel is one of the themes driving the album, making the Houston concert the perfect setting for its first live outing. Jarre composed the album's final track, *Dernier Rendez-Vous*, for musician and astronaut Ronald McNair, who was due to make history by performing his moving saxophone solo live from space, aboard the space shuttle Challenger during the concert. Sadly, this was never to be.

Both the album and concert remain tainted with the tragedy of Challenger. Yet despite this devastating loss, NASA urged Jarre to continue with the concert, which had an estimated live audience of around 1.3 million people. The concert was dedicated to Challenger and her crew. Even to those discovering *Rendez-Vous* and its history over twenty-five years later, the sombre saxophone of the album's closing track never fails to bring a lump to the throat.

The main-belt asteroid, (4422) Jarre (discovered on 17 October 1942 by Louis Boyer), is named after both Jarre and his musician father, Maurice.

FOREVER AUTUMN

Towards the dawn of the 1980s, more and more artists were experimenting with electronic music, although one album in particular would prove there was still a place for rock music in science fiction.

Released in 1978, *Jeff Wayne's Musical Version The War of the Worlds* remains a best seller around the world with timeless appeal. This prog-rock and narrative extravaganza is also the closest adaptation of *The War of the Worlds* to date, with a cast including David Essex, Justin Hayward and Richard Burton, whose instantly recognisable voice always sends a shiver down the spine as he opens the LP with that famous introduction.

The War of the Worlds still sounds surprisingly modern today thanks to brilliant song writing, production and instrumentation from Jeff Wayne, blending solid, catchy rock music with electronic sound effects and exciting narrative. It's a dynamic listen that instantly sweeps you away with the frenetic pace of the story, told in such a way you can almost see what's going on.

Even if the chances of anything coming from Mars are a million to one, *Jeff Wayne's Musical Version of The War of the Worlds* is told with such brilliant conviction that if you heard it on the radio or booming out of the stereo as a child, you might just for a moment think it was actually happening.

In recent years, *Jeff Wayne's Musical Version of The War of the Worlds* has been adapted into an all-star, high-tech arena concert tour, reinforcing its longevity, as if any proof were needed. A new version of the album recorded with a modern-day cast has introduced it to a new generation, although you simply cannot replicate or better the unique sound and atmosphere of the original. Very few SF novels could have been adapted this way so successfully – perhaps because the basic premise represents a believable threat that could seem possible in any generation. Despite the fact that we now know the surface of Mars is not populated with roaming Tripod machines, H G Wells' original 1898 novel still reads like a contemporary work, proving that this is simply one masterwork that in all its forms will continue to go on exciting, scaring and inspiring.

MUSIC FOR SPACE

Another artist who shot to fame in the 1970s, whose work has been widely used in relation with all things interstellar, is British multi-instrumentalist, Mike Oldfield.

The year 1971 saw the release of Oldfield's groundbreaking album *Tubular Bells*, the first instrumental rock album and famously used as the title music to the horror film, *The Exorcist*. Oldfield's influences ranged from rock to classical and folk, and all of this and more would go into his music – epic movements of diverse styles, with Oldfield usually playing every instrument himself.

However Oldfield's music would not be directly directly linked with space travel until 1979, when NASA commissioned director Tony Palmer to make a film celebrating the tenth anniversary of the Apollo 11 Moon landing. *The Space Movie* used music from most of Mike Oldfield's discography to that point, including *Tubular Bells*, *Hergest Ridge*, *Ommadawn* and *Incantations*, alongside orchestrated versions.

In 1994, Mike Oldfield – a self-confessed *Star Trek* fan – released a concept album, *The Songs of Distant Earth*, which was both named after and inspired by the Arthur C Clarke novel. Oldfield took the themes of Clarke's novel and transformed them into a beautiful and evocative collection of interlinked tracks, following the basics of the book but also playing out like a musical evolution of the Earth, beginning with the voice of an astronaut reading the opening lines of the Book of Genesis.

The Songs of Distant Earth is a rich and textured album, with Oldfield's signature guitar taking centre stage, alongside an arsenal of synthesisers and worldly instruments to form a truly global soundscape. Cleverly structured transitions guide you through space and time, culminating in the birth – or discovery – of a new civilisation. *The Songs of Distant Earth* makes a perfect companion piece to the book, which itself ends with a musical concert.

In addition to the futuristic nature of the music (and indeed Clarke's novel), the original CD version of *The Songs of Distant Earth* was one of the first albums to include enhanced interactive content. When the disc was inserted into an Apple Macintosh computer, you entered an interactive three-

Rendez-vous. © Alex Storer.

dimensional world, set to the music of the album. This short but impressive audio-visual experience was rendered on Silicon Graphics computers, (which back then were known as computer workstations; high-end image manipulation, state-of-the-art technology). After navigating your way into the Hibernaculum, you were greeted with a message from Oldfield himself, having emerged from a stasis pod. The album's liner notes contained an introduction written by Arthur C Clarke, which neatly ties the album and book together.

The work of Arthur C Clarke had previously inspired Oldfield's music, one such example being the opening track on his 1992 album, *Tubular Bells II*, entitled *Sentinel*. Like his musical contemporaries Jean Michel Jarre and Vangelis, Mike Oldfield also has an asteroid named after him – (5656) Oldfield (A920 TA), originally discovered in October 1920).

Mike Oldfield's high profile relationship with space continues to this day. In April 2014, the European Space Agency used *Sentinel* as the official soundtrack to the launch of Sentinel-1A, the first satellite for Europe's environmental monitoring programme, Copernicus. Arguably one of the artist's most evocative tracks, *Sentinel* was a perfect choice of music for the event and Oldfield was filmed watching the live launch, alongside his young sons.

In 2008, Oldfield released his first classical recording, *Music of the Spheres*. The title was translated from 'Musica Universalis' – an ancient philosophical concept proposing that vibrations caused by the movements of celestial bodies produce a form of music. Greek philosopher Pythagoras had identified that the pitch of a musical note is proportional to the length of the string that produces it, with simple numerical ratios formed by the intervals between sound frequencies. In comparison, Pythagoras proposed in his 'Harmony of the Spheres' theory that stars and planets each move according to mathematical equations and each emit their own unique orbital resonance – or hum – although this celestial concerto is inaudible to the human ear.

TELEVISION SOUNDTRACKS

Prior to the frequent use of synthesisers in film and television, most science fiction film scores were orchestral, with Holst's *The Planets* being a benchmark in the genre. However, the first signs of a change in direction appeared back in 1956, when Louis and Bebe Barron composed the all-electronic soundtrack to *Forbidden Planet*.

Electronic music was still very much in its infancy when the Barrons wrote their innovative soundtrack and it was perhaps not until 1963, when the BBC's Radiophonic Workshop scored the haunting title music to fledgling science fiction show *Doctor Who*, that electronic music really made its mark. Delia Derbyshire's arrangement of Ron Grainer's *Doctor Who* theme remains one of the most timeless, unique and distinctive television and SF theme tunes to date. It sounded unlike anything else at the time, compared to other shows that had fairly standard chirpy music by the BBC orchestra.

Although the title music sounded as mysterious and alien as the show itself, the incidental music in *Doctor Who* was equally important in creating the right atmosphere. One particular example where the clever use of sounds had more impact is the 1964 story *The Dalek Invasion of Earth*. The ghostly soundtrack by Francis Chagrin generally comprised low drones and ominous throbs. Filmed around London's famous landmarks as well as derelict industrial backstreets and war-torn parts of the city that were yet to be rebuilt, this unsettling sonic climate gave the story a tense and disturbing ambience as Terry Nation's iconic Daleks dominated the landscape.

The following years would also give us interstellar programmes such as *Star Trek, Space: 1999, UFO, Battlestar Galactica* and *Blake's Seven*. Previously, the exploration of space was reserved for the big screen, but by the mid-1970s onwards science fiction was commonplace on television and each series would have music as memorable as the shows themselves. One particularly chilling theme tune was Ken Freeman's music to the BBC's 1984 adaptation of *The Tripods* – a tense and haunting electro-symphonic piece, carrying all the suspense and adventure of John Christopher's original trilogy of books.

Famous existing compositions were often adopted as signature tunes, with which they would become forever associated, such as Sibelius' *At the Castle Gate*, which featured as the title music for every episode of the BBC's *The Sky At Night* from 1957 to the present day. This choice of music was actually Sir Patrick Moore's after he was unhappy with the BBC's original proposal – and it was undoubtedly down to the late Sir Patrick that this title music was never changed. This particular piece evokes a certain academic mood, encompassing the science, technology and ambition involved in astronomy and, above all, taking the subject very seriously at the same time as bringing it to a wider audience.

It's impossible to talk about music and space without mentioning Stanley Kubrick's *2001: A Space Odyssey*. Richard Strauss' *Also Sprach Zarathustra* has become forever linked to the film, thanks to its use in three distinctive scenes, in particular the opening titles during which the camera gradually overlooks the Earth, the Sun and the Moon, and the classic scene where apes discover the monolith (or Sentinel, should you wish to follow Arthur C Clarke's original description of the mysterious black object). The slow and gradual build up of the music culminating in the thunderous timpani never fails to make the hairs on the back of the neck stand on end. The sound and vision go together perfectly, yet listen to the piece in isolation and, while there's no denying it's impact, it doesn't quite have the same effect. It is almost as if it needs Kubrick's stunning visuals to be complete.

The film still looks fantastic even by today's standards but one of the key factors in its maintained appeal is the use of music. Johann Strauss II's best-known waltz, *The Blue Danube* is the other piece of music used frequently throughout the film – a classical piece juxtaposed with the futuristic images of space stations and a crew in zero gravity; their floating movements almost as poetic as the waltz when performed by dancers gliding silently across the stage. It is a combination that shouldn't work, yet it's just perfect.

Into the Unknown – a starship voyages close to the swirling maelstrom of an accretion disc around a black hole, in an image straight out of science fiction. © Alex Storer.

An orchestral score has the power and grandeur required to portray the astounding visions of interstellar travel. It serves as a glorious fanfare to the splendour of space cruisers in the same way it had been to majestic battleships in times past. Symphonic music doesn't date easily, giving timeless appeal across generations.

Electronic music is perhaps more subjective and easier to date in certain instances. Yet because of the ever-increasing technology behind it, whatever period you choose, it still remains incredibly futuristic. You only have to listen to soundtracks such as Wendy Carlos' *A Clockwork Orange* or Vangelis' epic music for *Blade Runner* to realise that you wouldn't want it any other way. In recent years, a common approach has been the combining of symphonic and electronic sounds or the distortion and treatment of acoustic instruments, evident on Clint Mansell's soundtrack to the 2009 film *Moon* (directed by David Bowie's son, Duncan Jones), or Cliff Martinez's haunting music to the 2002 remake of *Solaris*. Electronic music in all its forms automatically lends itself as an ideal style to depict visions of a technologically driven future, both on Earth and in space.

MUSICAL EMISSARY TO THE STARS

Music is one of humankind's greatest art forms and whether it is something unique to us on Earth or not, it remains one of our key means of communicating with life outside of our own planet.

On several occasions we have put our music into time capsules or sent it on a journey into outer space. The Voyager spacecraft, launched in 1977, carry with them 'Golden Records' that comprise a wide range of traditional music from around the world, with artists ranging from Beethoven to Chuck Berry. In 1997, original compositions by French artists Julien Civange and Louis Haéri were placed on board NASA's and the European Space Agency's Cassini–Huygens probe – the intent of this musical project being to leave traces of humanity on these unexplored worlds. The music did indeed reach Titan in January 2005, after travelling seven years and four billion kilometres!

Our music will continue into the approaching centuries. Music of all trends and genres will still be out there in various forms, ready for discovery by the coming generations or launched into space, delivering a message to other civilisations, and even returning to a very different Earth in the future. Another example, although its content is our messages rather than music, is the KEO project, a time capsule satellite, conceived in 1994 by the late French artist/scientist, Jean-Marc Philippe. This ambitious project has sadly seen several false starts, although at present KEO has a tentative launch date of 2015.

What does the future of music hold? In the last decade we've seen an alarming transition from physical product to virtual (although our natural want for something tangible will never disappear). The Internet age has seen music become a fast market product; immediate and disposable. On the other hand, technology has advanced to the point that musicians don't necessarily need to spend a fortune in studio time and face rejection by record labels. It is now possible to record a professional-sounding album from a home studio and distribute it via the net and that, in turn, has given greater freedom of speech and expression to artists.

Music will undoubtedly accompany the progress of our interstellar endeavours, through creative application and at events that bring together visionary thinkers and pioneers. Music creates a vital unity between people, becoming a soundtrack to success. In the past concerts and events have drawn attention to environmental or humanitarian issues; why can we not harness that same musical energy to raise the profile or funding of interstellar projects, looking to the day when we take our music and musicians with us, beyond the stars?

Music remains the great communicator, bridging times, cultures and languages. Where words on a page or images on a screen help us to visualise and inform, there is something about the sonic experience that cuts right through to the core of our imaginations. Music works in ways other media cannot – it reaches a part of our emotions in a unique way, stimulating the senses and transporting you to that other place. An epic and thrilling piece of music can leave you feeling aspirational and inspired – you want to be a part of it and to go with the music on its journey.

Music triggers our creativity and, with it, our ambitions. While even the best painting or photograph remains an image, with music you are surrounded by and immersed in it. With this in mind, interesting and creative music remains a crucial companion to the interstellar vision, providing that drive to keep us moving towards our galactic goals.

AFTERWORD

When I was three or four years old, on the wall at home hung a large print of a painting, showing a vast, rocky alien landscape set below a burning star. To the eyes of a child, this was like a window into another world – a scene of fascination and wonder, and my first taste of life on another planet.

I fondly remember listening to my father playing Jean Michel Jarre's *Oxygène*, while I'd be staring out through this portal into space. The two things seemed to go together perfectly.

The painting in question was *Stellar Radiance* by British space artist David A Hardy. Little did I realise at the time that this was the starting point for my lifelong obsession with electronic music, space art and science fiction, along with the connections that would form between them.

As a musician and artist with a huge interest in science fiction and the prospect of discovering the secrets of the stars, the idea of creating soundscapes to accompany an interstellar vision actually poses a great number of challenges and is something I continue to explore in my work. This interest stems from the work of the artists and composers detailed in the above chapter. Their music continues to inspire me and it is also by coincidence that they are among those whose works are regularly connected to space exploration.

I have been composing and recording my own brand of instrumental electronic music since 2006, under the name of 'The Light Dreams'. An artist at heart and graphic designer by day, I have no traditional musical training. I am what you might consider to be self-taught, producing music from the comfort of my home computer set-up. For me, creating a piece of music is the same as creating a piece of artwork, only working with layers of sounds, rather than colours (although being a synesthete, everything has colours in my mind's eye!).

My lifelong interest in science fiction provides regular inspiration to my music and that influence is at its strongest in my latest releases for the Initiative for Interstellar Studies: *Future Worlds*, which takes the listener through various science fiction scenarios, and *Beyond the Boundary*, a specially composed interstellar voyage.

With the creation of music or art comes a tremendous feeling of accomplishment, knowing you have made something that may be discovered by people in years to come. There is also the satisfaction in the culminating result of an artistic outpouring. I make the kind of music I like and, as an artist, if you're doing something that really excites you, it's bound to excite and inspire others.

About I4IS

The mission of the Initiative for Interstellar Studies (I4IS) is to foster and promote education, knowledge and technical capabilities that lead to designs, technologies or enterprise that will enable the construction and launch of interstellar spacecraft. The vision of the Initiative for Interstellar Studies is to aspire towards an optimistic future for humans on Earth and in space, and for I4IS to be central to catalysing the conditions in society over the next century to enable robotic and human exploration of the frontier beyond our Solar System. Our values are to demonstrate inspiring leadership and ethical governance, to initiate visionary and bold programmes co-operating with partners inclusively, to be objective in our assessments yet keeping an open mind to alternative solutions, acting with honesty, integrity and scientific rigour. Our motto is 'Scientia ad sidera' or 'knowledge to the stars'.

I4IS is built around five core committees that include: an Educational Academy Committee for the purpose of fostering educational abilities to conduct research relating to a broad set of subjects pertaining to interstellar studies; a Technical Research Committee for the purpose of conducting innovative theoretical and experimental research and development, including Dragonfly, a flagship laser-sail project; an Enterprise Committee to encourage entrepreneurship and business innovation initiatives related to the objects of the organisation; a Sustainability and Development Committee to seek space-based technological solutions to solving problems on Earth and in space; and an Advisory Committee to provide guidance on decisions and strategic directions.

To find out more about I4IS' various projects and initiatives, events and interstellar news, visit www.i4is.org. You can also follow I4IS on Twitter @ I4Interstellar, on Facebook at www.facebook.com/InterstellarInstitute and by joining our LinkedIn group.

Meet the Authors

Rachel Armstrong

Rachel Armstrong is Professor of Experimental Architecture at the Department of Architecture, Planning and Landscape, University of Newcastle. She is Project Leader for Persephone, which is part of the Icarus Interstellar group's work to construct a starship within 100 years, and Director of Sustainability and Development for the Initiative for Interstellar Studies. She is also a 2010 Senior TED Fellow that is establishing an alternative approach to sustainability that couples with the computational properties of the natural world to develop a twenty-first-century production platform for the built environment, which she calls 'living' architecture.

Stephen Ashworth

Stephen Ashworth is an occasional astronautical writer whose day job with an academic publisher within the University of Oxford is focused on the eighteenth-century Enlightenment. He has been convinced of the importance of space exploration and settlement ever since watching the first televised moonwalks as a teenager, and favours a gradual, step-by-step approach that relies on the complementary strengths of both government and commerce. He is the author of a science-fiction novel, *The Moonstormers*, available online in electronic formats. He writes an online blog called 'Astronautical Evolution', and plays jazz saxophone.

Jonathan Brooks

Jonathan Brooks is a teacher of science and electrical engineering at schools and technical colleges in the United Kingdom. He is also an industrialist, having previously founded numerous companies associated with shipping, road haulage and energy supplies. He is a keen advocate of helping the next generation achieve their personal aspirations. He is an experienced scuba diver with over half a century of ocean-exploring experience behind him.

MARTIN CIUPA

Martin Ciupa studied Physics as an undergraduate and Cybernetics as a postgraduate at Kings College University of London. His career has been spent in the telecommunications and IT sectors, with long spells in finance. Currently he specialises in data-science systems, in particular for operational support systems for mission critical markets, predictive analytics and big data. His current business (Cybernetic Analytics Ltd) has offices in both the UK and Singapore. Martin's research areas in interstellar studies include the ethical implications and management of interstellar missions, techno-sociological examinations of the Drake Equation and Fermi Paradox, applications of AI systems in space science and the building of mission operational support systems (and the subsequent analysis of the data produced by space probes). Martin has an ongoing research interest in existential risks and mitigation strategic planning.

JEREMY CLARK

After completing a Bachelor's Degree in Physics at The City University, Jeremy went on to work for some of the world's leading financial institutions. The technological skills acquired from studying science took him to Asia, Europe and America. The statistical and logical discipline of physics provided the foundations for the postgraduate study of psychology at The Open University. However, only as a postgraduate student of Astrophysics at Queen Mary, London, was there an opportunity to glimpse the really big picture. Jeremy's academic interests were combined when returning to City University to complete a master's degree in business with a dissertation on using neural networks to identify social clusters in the population. Since childhood Jeremy has been enthralled by space exploration and being a part of I4IS has allowed Jeremy to join his professional experience with his academic knowledge to achieve those childhood dreams.

IAN CRAWFORD

Ian Crawford is an astronomer turned planetary scientist, and is currently Professor of Planetary Science and Astrobiology at Birkbeck College, University of London (www.bbk.ac.uk/es/). The main focus of his research is in the area of lunar exploration, including the remote sensing of the lunar surface and the laboratory analysis of lunar samples. He is a strong advocate for the renewed human exploration of the Moon, the eventual human exploration of Mars, and the development of a spacefaring infrastructure within the Solar System that will one day make interstellar travel a practical undertaking. A more detailed summary of interests, and a list of publications, can be found on his personal website (www.homepages.ucl.ac.uk/~ucfbiac/).

BILL CRESS

Bill Cress is an entrepreneur located in Wayne, New Jersey, USA. He is often referred to as a 'renaissance man' for his many diverse endeavors. As a founder and president of Breakthrough Technologies Corporation, he currently heads a company engaged in disruptive innovation systems for space travel and terrestrial power sources. He is a Director of Icarus Interstellar and a corporate member and consultant for I4IS. He serves as President of William Cress Corporation, a national building and development firm in the USA as well as Cress Photo, which is a supplier of re-invented old technology to the film industry, the military and scientific testing agencies worldwide. Bill chairs the Passaic County NJ Film Commission, heads Global Realty Associates Corp and is a broker member of The Real Estate Board of New York City.

ADAM CROWL

Adam was born in Bendigo, Victoria, Australia in 1970. His first memory of TV is watching the BBC documentary on the Mars Viking landings and (black and white) episodes of *Space 1999*. At age nine he learnt of a star-probe named 'Daedalus' and was given a little book, *The Road to the Stars*, written by Iain Nicolson. A teenage dream career in space-probe design suffered a few hiccups along the way – the Challenger disaster, going to university hoping to do physics but finishing with a BSc in Psychology. Since then Adam has re-taught himself mathematics and physics, web-published polemics against creationism, written an essay on SETI for the late Chris Boyce and semi-completed an engineering/computing degree. Currently he is writing on a broad variety of interstellar and SETI topics while changing day-jobs. Adam is on the board of Directors for Icarus Interstellar.

DAVID FIELDS

David E Fields is a physicist and currently Director of Tamke-Allan Observatory at Roane State Community College. He received his PhD in Experimental Solid State Physics from the University of Wisconsin and worked at Oak Ridge National Laboratory (ORNL), specialising in environmental transport and human risk from chemicals and radionuclides. He subsequently consulted with NASA Marshall Space Flight Center, designing radiation shields to protect astronauts from excessive radiation exposure associated with solar coronal mass ejections, galactic cosmic radiation, van Allan trapped radiation and nuclear propulsion reactors. David has taught at Murray State University, the Federal University of Brazil (UFMG), Pellissippi State and Roane State. Current active research interests include robot antennas, radio interferometry, software-defined radio and developing a novel detection system for radio astronomy research.

TIFFANY FRIERSON

Tiffany Frierson is a senior physics major currently at North Georgia College. She plans to do a PhD in Theoretical Physics, and study breakthrough propulsion physics, especially the more exotic methods including wormholes and warp drives. With a life-long interest in space travel, she gave a speech at Boston University's 50-Year Space Vision conference in 2007, whose attendants included Freeman Dyson, John Mather and Russell Schweickart. Tiffany programs in C++, C#, Java and Python. She currently resides in Georgia, USA.

REMO GARATTINI

Remo Garattini completed his first degree in Theoretical Physics at the University of Milan and later completed his PhD at the Mons-Hainaut

University in Belgium with a thesis on Space-Time Foam. He has a permanent research position at the University of Bergamo, in Faculty of Engineering. He has served on the Editorial Board for the MDPI of Zurich's journal <I>Entropy<I> (a special issue, 'Entropy in Quantum Gravity', ISSN 1099-4300). He is also leader of I4IS' Project Casimir. His research activities are in quantum gravity and quantum cosmology (inflation, dark energy, space-time foam) and the Casimir effect used for MEMS and graphene.

ANGELO GENOVESE

Angelo Genovese received a Master's Degree in Aerospace Engineering (specialising in Space Propulsion) at the University of Pisa, Italy, in 1992. He started to work as an Electric Propulsion Engineer in the research center 'Centrospazio' in Pisa, developing FEEP ion thrusters for ultra-precise positioning of scientific spacecraft. In 2000 he moved to the Austrian Research Centers in Vienna, Austria, where he contributed to an Indium FEEP Micro-propulsion System for ESA's LISA Pathfinder mission. He is currently working at Thales Electronic Systems in Ulm, Germany, on the innovative ion thruster HEMPT, suitable for telecom satellites and interplanetary missions. Angelo has published more than 50 papers in conferences and scientific journals, has contributed to two patents on Indium FEEP ion thrusters and he is a visiting lecturer at the International Space University in Strasbourg, France. He has joined the Initiative for Interstellar Studies as Senior Researcher. He is also a member of the Mars Society, Icarus Interstellar and the Planetary Society.

DAVID HARDY

David A Hardy has been a space artist since 1950, with his first published work appearing in a book by the British astronomer Patrick Moore in 1954. He has provided artwork for covers and interiors to many science fiction and non-fiction books and magazines and has written several books himself. Aside from Sir Patrick Moore, with whom David was a frequent collaborator, he has worked with luminaries such as Arthur C Clarke, Carl Sagan and Isaac Asimov. David is

European Vice President (and former President) of the International Association of Astronomical Artists and Vice President of the Association of Science Fiction and Fantasy Artists. David has been nominated for two Hugo awards, one for 'Best Professional Artist' in 1979 and the other for his 2005 book with Sir Patrick Moore, *Futures: 50 Years in Space*. He also won the best cover art readers award from *Analog Science Fiction and Science Fact* in 2003, 2004, 2005 and 2007. He also jointly received, with Sir Patrick Moore, the Sir Arthur C Clarke Award in 2005. His website is www.astroart.org.

ANDREAS HEIN

Andreas received his master's degree in aerospace engineering at the Technische Universitaet Muenchen. He is working towards a PhD degree at the same university in the area of space systems engineering. Andreas is currently Deputy Director of the Initiative for Interstellar Studies as well as its Director of Technical Programmes. He also founded and leads Icarus Interstellar's Project Hyperion: a design study on manned interstellar flight. At his home university he founded the space elevator group, the interstellar flight group, as well as a research group focusing on the applications of 3D-printing in space. Andreas leads teams that have won several awards in international design competitions, such as the Project Icarus Concept Design Competition in 2013.

ROBERT KENNEDY

Robert G Kennedy III, PE, was educated in classics and foreign languages (Latin, Greek, Arabic, and Russian) before earning a BSc in Mechanical Engineering at Cal Poly. He designed industrial robotics at Douglas Aircraft Company (1987–1991) and pursued research in artificial intelligence at Oak Ridge National Laboratory (1987). He founded Ultimax Group, Inc. (1992-present, www.ultimax.com), a Russian–American company in Oak Ridge, Tennessee. He has written about space-based solar power, shell worlds, climate change, and green energy. He co-founded and sponsored the non-profit Tennessee Valley Interstellar Workshop (www.tviw.us). His work has appeared in the *Journal of the*

British Interplanetary Society, Acta Astronautica and a story on 'Soviet Star Wars' in *Smithsonian Air & Space*. As American Society of Mechanical Engineer's 1994 Congressional Fellow, he spent a year working for the Subcommittee on Space in the US House of Representatives and was a technical consultant on the movie *Deep Impact*. Currently, Robert is a senior systems engineer at Tetra Tech.

JON LOMBERG

Artist Jon Lomberg is best known for his many collaborations with astronomer Carl Sagan. He was the Emmy Award-winning Chief Artist of the original *Cosmos* series, designed the opening animation for the movie *Contact* and was Design Director of the Voyager Golden Record, creating a montage of images describing the Earth for extraterrestrial life. He has also landed message artifacts on Mars aboard NASA spacecraft Phoenix, Spirit, Opportunity and Curiosity. He is currently Project Director for the One Earth Message, a digital successor to the Voyager Record to be uploaded to NASA's New Horizons Pluto mission. He created the Galaxy Garden, the world's first large-scale, accurate and explorable model of the Milky Way done as a flower garden in Hawaii where the artist resides.

KELVIN F LONG

Kelvin F Long is an aerospace engineer, physicist and author. He is the Chief Editor of the *Journal of the British Interplanetary Society* and the co-founder and Executive Director of the Initiative for Interstellar Studies. He is also the founder and Managing Director of the scientific and aerospace consultancy company Stellar Engines Ltd. He has published numerous articles, technical papers and books on the subject of interstellar studies.

RICHARD OSBORNE

Richard Osborne is a consultant rocket scientist as well as a strategic foresight consultant and IT architecture consultant. He chairs the Technical Committee of the British Interplanetary Society (BIS) and is the I4IS Director of Technology and Strategic Foresight. Trained as a physicist, he works as a consultant for companies including Reaction Engines Ltd, Airborne Engineering Ltd and Commercial Space Technologies Ltd. He has also worked on systems engineering for a number of space industry projects ranging from payloads for the Mir space station to Mars missions. He specialises in launcher technology, rocket propulsion and technological and strategic forecasting.

KENNETH ROY

Kenneth Roy is an engineer living and working amidst the relics of the Manhattan Project in Oak Ridge, Tennessee. He invented the 'Shell Worlds' concept that might be useful for future terraforming efforts. In 1997 he made the cover of the prestigious *Proceedings of the US Naval Institute* for his forecast of anti-ship, space based, kinetic energy weapons. With his co-authors Robert Kennedy and David Fields, he has appeared multiple times in the *Journal of the British Interplanetary Society* and *Acta Astronautica* with papers on terraforming and space colonisation. He is an engineering graduate of the Illinois Institute of Technology and the University of Tennessee at Knoxville.

DIVYA SHANKAR

Divya Shankar is a space enthusiast from India. She is currently pursuing an MSc in Space Science and Technology at the Skolkovo Institute of Science and Technology in Russia, which is in collaboration with the Massachusetts Institute of Technology. She has her bachelor's degree in Electronics and Communication Engineering from Nitte Meenakshi Institute of Technology in India. She has worked on university small satellite missions in India that were in collaboration with the Indian Space Research Organisation. She has worked on STUDSAT-1, India's first picosatellite that was launched in 2010, for which she was awarded the Limca Book of National Records. She has also worked on the initial design phase of STUDSAT-2, a twin nanosatellite mission. Divya has worked as a Spacecraft Engineer for Dhruva Space Private Limited and participated in stratospheric ballooning experiments with the Indian Institute of Astrophysics. She is associated with Icarus Interstellar as a student designer and the Initiative for Interstellar Studies as a researcher to execute interstellar spacecraft design research activities.

ALEX STORER

Alex has been working under the name of 'The Light Dreams' since 2006, producing electronic instrumental music. What started out as a personal hobby gradually evolved into something he was confident in releasing independently and Alex now has a growing discography of atmospheric and thought-provoking albums. Alex is also an artist, illustrator and graphic designer. A lifelong passion for science-fiction art came to fruition in 2010, when he finally started working on his own SF-themed digital paintings. Examples of Alex's cartoon illustrations were featured in two internationally published books on comic art in 2003. In 2011, his H G Wells-inspired painting entitled 'Awakening' was included in the exhibition 'Brave New Worlds', in Richmond-Upon-Thames. In 2012, he composed an original soundtrack to space artist David A Hardy's English edit of the 1957 Russian film, *Road to the Stars*. Visit his website at www.thelightdream.net.

GIOVANNI VULPETTI

Giovanni Vulpetti received his PhD in Plasma Physics in 1973. He has spent much of his professional life studying the concepts underlying interstellar flight, astrodynamics and space propulsion, in particular matter-antimatter annihilation propulsion. Since 1992 he has been involved in solar-photon sailing. In 2001, he served as a consultant at NASA/MSFC for the Interstellar Probe project. He has published over 115 scientific papers, and co-authored one popular book, *Solar Sails: A Novel Approach to Interplanetary Travel* in 2008. He is also the author of the scientific book *Fast Solar Sailing: Astrodynamics of Special Sailcraft Trajectories* in 2012. Since 1994 he has been a full member of the International Academy of Astronautics.

GLOSSARY

ASTRONOMICAL UNIT (AU)
The mean distance between Earth and the Sun, which is 149,597,871 kilometres.

BLACK HOLE
A point in space so dense and massive that not even light can escape its gravitational grasp.

COSMIC RAY
A highly energetic particle moving at almost the speed of light and coming from deep in interstellar space. Supernova remnants have been implicated in their creation.

EINSTEIN'S FIELD EQUATIONS
The equations set out in Albert Einstein's 1915 General Theory of Relativity that describe how gravity is a fundamental result of space-time being curved by matter and energy.

HELIOSPHERE
The Sun's magnetic field is carried out into space on its solar wind of charged particles. The magnetic bubble that the solar wind creates is called the heliosphere. Voyager 1 crossed its outer boundary, the heliopause, at a distance of 121 AU from the Sun, but this distance will vary depending upon factors including location, the outward strength of the solar wind and the inward pressure of the interstellar medium and its magnetic field.

INTERSTELLAR MEDIUM
Space between the stars is not empty; it is filled with various gasses, particles of dust, cosmic rays and bathed in radiation and magnetic fields, collectively called the interstellar medium.

ION ENGINE
An ion drive uses a fuel such as xenon, which is ionised by injecting electrons into a chamber with the xenon atoms, before magnetic fields generated by an electric current accelerate the ions and force them out through the exhaust, producing thrust.

KUIPER BELT
A disc of icy, cometary bodies and frozen dwarf planets that lurks beyond the orbit of Neptune, out to a distance of about 50 AU.

LIGHT YEAR

The distance that light, moving at 300,000,000 metres per second, travels in one year (which is 9.46 trillion kilometres).

NUCLEAR FUSION

The generation of energy by fusing atoms together under extreme temperatures and pressures. No fusion reaction in the lab has yet been able to produce more energy than is put into the reaction on a consistent and efficient basis. Fusion produces no nuclear waste other than neutrons that irradiate the reaction chamber.

OORT CLOUD

Beyond the Kuiper Belt is a roughly spherical cloud of comets that could extend up to a light year from the Sun. Although the Oort Cloud has never been directly observed, it is thought that long-period comets originate there.

RAMSCOOP

A starship engine that works by using huge electromagnetic fields to scoop up hydrogen from the interstellar medium for use in a fusion engine. As the starship accelerates, it scoops up more and more hydrogen, thus allowing the ship to go faster and faster.

SAILS

A spacecraft employing a sail can turn the momentum imparted by photons of light incident on the sail, be they from the Sun or from focused lasers or microwave beams, to push the spacecraft towards its destination.

SPECIFIC IMPULSE

A measure of how efficiently a thruster consumes propellant and is calculated by dividing the propellant exhaust velocity with the gravitational acceleration constant.

WARP DRIVE

A hypothetical means of interstellar travel made possible by bending space-time so that a spacecraft in effect brings its destination closer to it. Warp drive requires colossal amounts of energy far beyond our current means, and unidentified 'exotic matter'.

WORLD-SHIP

A large interstellar starship moving at a slow pace such that it takes many generations to reach its destination. The vessel is so large that it can potentially house whole environments – parks, lakes, habitats – inside its confines.

INDEX

Abiogenesis 380–381
AC+79 3888 (star) 159
A Clockwork Orange 417
Active SETI (see also METI) 377–378, 388, 390–392
Adaptive Control of Thought (ACT) 290–292, 295
Adaptive Resonance Theory (ART) 289
AEGIS software 277
'Age of Discovery' 146
Alcubierre, Miguel 229–230, 246
Aldebaran 154
Algae bioreactor 87, 97–99
Alien (movie) 359
Allen, Paul 132
Alpha Centauri 28, 30, 153, 199, 212, 228, 229, 257, 313, 315, 318, 328, 359
    exoplanet around 30, 302, 304, 313, 316–317
    flight time 50, 163, 328
    fly-by of 223
    minimum power requirement to reach 216
    priority target for a mission 44
    proper motion 200
Also Sprach Zarathustra 415
Analog Science Fiction and Fact 145, 402
Andrews, Dana 218
Andromeda Galaxy 309, 321
Angara rocket 128, 130
Antares rocket 125, 131
Antikythera mechanism 278
Antimatter propulsion 44–47, 50, 227, 229, 246–247, 402
Apollo Programme 19, 54, 82–83, 187, 330

Architecture of human habitation in space 89
    sustainable designs 95
Arcjets 183
Arcologies 91
Arecibo signal 378
Ares V rocket 194
Ariane 5 125, 128, 197
Ariane 6 128–129
Art 395
Artificial intelligence (AI) 141, 275–277, 282, 285–288, 290–291, 293–294
Artsutanov, Yuri 138
Ashes to Ashes 407
Asteroid 237, 350–361, 395, 398, 413
Asteroid Belt 183, 186, 191, 223, 255, 307, 317
Astrobiology 21–22
Astron, Graeme 210–211
Astrosphere 320
Atlas rockets 105–106, 113–114, 117–119, 121, 125, 130
At The Castle Gate 415
A Trip to the Moon 2
Australian Centre for Astrobiology 381
Autonomous cars 141
Autopilot 294–295
Avalanche photodiode (APD) 261
Avatar 359

Babbage, Charles 278
Babylon 5 317
Baikal launcher 128, 131, 141
Barnard's Star 68, 313, 399
Barron, Louis and Bebe 414

Battlestar Galactica 3, 415
BBC 385, 414–415
Beamed Electric Propulsion 190
Beam lightsail mission 166
Beesley, Philip 97
Beethoven 417
Bekenstein Bound 366–367
Bell's Theorem 268–269
Belousov Zhabotinsky reaction 92
Berend, Nicolas 207
Bernal, John Desmond 352
Bernal Sphere 353
Berry, Chuck 417
Beyond the Boundary (music) 420
Beyond the Solar System 397
Bezos, Jeff 132
Big Bang 230
Bigelow Aerospace inflatable space station 127
Billingham, John 391
Binary code 390–391
Biology merging with engineering 6
BIOS-3 87
Biosignatures 22
Biospheres 87, 89, 93
Biosphere II 87–88, 90
Biotechnology 93, 95
Black hole 162, 242–245, 321, 398
Blade Runner 417
Blake's 7 415
Blue Origin 132
Bond, Alan 68, 399
Bonestell, Chesley 397
Boron 332
Bowie, David 406–408, 417
Bradyons 241
Brain emulation 364–368
British Interplanetary Society 24, 45, 168, 399
Brown dwarf 69, 301, 328
    binary system 28

proximity to Solar System 153, 318
Brown, Mike 333
Brown, William 190
Buck Roger 275
Burroughs, Edgar Rice 1
Busch, Michael W 391

California Institute of Technology 333
Cambrian Explosion 382
Cambridge University 303
Cape Canaveral 106, 110, 130
Carlos, Wendy 417
Carbon dioxide 344
Carbon planets 308
Carbon–silicate cycle 308, 316
Cardiff University 229
Carter, President Jimmy 222
Casimir effect 232, 234–235, 247, 249–250
Cassini–Huygens 19, 72, 184, 187, 417
Cassiopeia 398
Cathedrals 82–84
Caughran, Dan 107
Centauri Dreams 67
Ceres 186, 337
Charon 336–337
Chaos theory 279, 282
Challenge of the Stars 397, 402
Challenger (space shuttle) 410
Chemical propulsion 170
ChipSat 169
Chorost, Michael 318
Christopher, John 415
Circumstellar disc 18, 34
Civange, Julien 417
Clarke, Arthur C 2, 71, 96, 138, 299, 395, 397, 410, 412–413, 415
Cleaver, Gerald 231
Close Encounters of the Third Kind 3
Cluster-II 201

Cochrane, Zephram 228–229
Cole, Dandridge 351
Colliers 397
Colonisation 14, 58–59, 146, 301, 321,
        327–328, 338–340, 346, 349–351,
        356, 361, 366, 373
Colony 338–341, 343, 345, 351–355,
        361–363, 367
Columbus, Christopher 78, 148–149, 172,
        212
Communications 169, 253–266, 288, 327
Conquest of Space, The 397
Conscious Life Expansion Principle
        (CLEP) 58–60
Convolutional coding 254
Cook, Thomas (travel agents) 400
Cosmic rays 163, 192–193, 199, 255, 320,
331–332
Cosmos (TV series) 5–6
Crawford, Ian 151
Crick, Francis 25, 381
Cryogenics 360–361
Cubesat 200
Curiosity rover 113
Cybernetics 92
Cyborg 364

Daedalus, Project 20, 45, 50, 68, 126, 142,
        168, 302, 399
Dalek 415
Dalek Invasion of Earth, The 415
Dancing Out in Space 407
Dark energy 153
Dark matter 153, 162
Davies, Paul 57, 350
Davis, Tamara 381
Dawn mission 183, 185–186, 191
Da Vinci, Leonardon 10
Day the Earth Stood Still, The 2
Death Star 398
Decatur (see also ULA) 108–109, 115

Deep Space Network 157, 255–256
Deep Space One 185, 188–189
De Grote, Hugh 82
De Jure Belli Ac Pacis 82
Delta rockets 105–106, 110–112, 115,
        118–121, 125, 130
Descartes, Rene 276
Deuterium 164
Dickinson, Richard 190
Digitally uploading human consciousness
        351, 364–371, 374
Directed panspermia 381
DNA 93–94
Doctor Who 414–415
Doppler shift (see also radial velocity) 303
Dragon capsule 134, 293
Dragonfly, Project 169, 171
Drake Equation 377–378, 380, 382–387,
        391–392
Drake, Frank 378
Dreyfus, Francis 409
Dwarf planet 149, 150, 191, 255, 333–334
Dyson sphere 384

Earth
        blue planet 306
        environment 87
        exploration 146–149
        future of life on 314
        origin of life 380
        size 307, 344
Earth-like worlds 312, 316, 328, 338, 350,
397, 402, 404
Earth Versus the Flying Saucers 2
Easter Island 81
Ecological systems 90–92
Ecopoiesis 91
Ecosystem 87, 90, 93–94, 350–351
        human body as an 94
Edgeworth, Kenneth 150
Edison's Conquest of Mars 1

Egypt 80–81
Einstein, Albert 240, 243, 267
Einstein's field equations 240–242, 246, 249
Einstein–Rosen bridge 242, 244
Einstein–Podolsky–Rosen paradox
        267–269
Electric propulsion 180–184, 188, 200,
        202, 212
Electrodynamic tethers 170
Electronic music 408–411, 417
Embryo 351, 361–364, 368, 374
Emulation cities 364–365
End of History, The 26
Energy space 70–71
Enterprise 212
Environmental support systems 91
Enzmann, Robert 399
Enzmann starship 145, 328, 331–332, 342,
        399
Epsilon Eridani 18, 29, 33–35, 218–219,
        221, 222, 313, 317
exoplanets around 33–34, 302, 313
asteroids and comets around 317
Eris 66, 150, 333, 335
ESPRESSO 304
ESTEC 204
Eta Carinae 404
Ethics 389
Europa 316, 380, 386
European Space Agency (ESA) 125, 128,
        137, 197, 414
European Southern Observatory 304, 390
Event horizon 241
Evpatoria radio dish 378, 390
Evolved Expendable Launch Vehicle 106,
        113
Existential threats 23–24, 391
Exomoons 312
Exoplanets 14, 153, 301–302, 398
        atmospheres 306–308, 310, 316
            commonness 14, 36–37, 315, 379

effects of tidal locking 310, 312,
        314
habitability 379, 380
interiors 308
multi-planetary systems 316
number of discoveries 18
reliability of detection 314
smallest 308–309
study of 15, 19
suitability for human life 25, 38,
        311
Exotic matter 231–232, 246
Exploration of Space, The 397
Extraterrestrial life 14, 25, 299, 343, 381
future depends on 26
search for (also see SETI) 57

51 Pegasi 303–305, 307
Falcon 9 rocket 125, 128, 130–132, 134
Faster-than-light travel 327–328, 367–
        368, 400
Fearn, David 169, 179, 203–204, 207, 210
FEEP (field emission electric
        propulsion) 183, 201–202,
        205–206
Fermi Paradox 25, 327, 377–378, 385–387,
        391–392
Flamm, Ludwig 242
Flight time 48–51, 53, 160, 163, 167–170,
        191, 199, 208, 215, 239, 246, 328
Fly-bys 18, 20, 198
        disadvantages of 20
        duration of 68
        benefits of orbiting compared to
        fly-bys 20
        within working life of scientists
        56
FOCAL mission 152
Fomalhaut 404
Forbidden Planet 2, 414
Ford, Larry 235–236

Foreign Policy 391

Forward, Robert 146, 166, 172, 211, 217–219

FOTEC 201

Fountains of Paradise, The 138

Foxx, John 408

Free-floating planets 153

Freeman, Fred 397

Free-space optical communication 258–260

Freitas, Robert 222, 354

Fresnel lenses 186, 218–219

Friedrich Schiller University 315

Fukuyama, Francis 26

Fuller, Richard Buckminster 89–90

Futures: 50 Years in Space 402

Future Worlds 420

Gain 254–256, 263–264

Galactic Tours 399

Galileo mission 187

Gamma-ray burst 237

Ganymede 312

General Fusion Project 140

General Theory of Relativity 51, 152, 166, 228–231, 240–241, 243, 266

Generation ship (see also Worldship) 353, 374

Geneva Observatory 303

Genta, Giancarlo 172

GEOTAIL 201

Gilbreath, William 222

Gilster, Paul 67

GJ 674 (star) 29, 35

GJ 1214b (planet) 308

Glaser, Peter 220

Gliese 445 9, 199, 320

Gliese 581 379, 390, 402

Gliese 667 311–312

Gliese 674b 312

Gliese 849 379

Gliese 876 planetary system 301, 312

Globular cluster 398

Goddard, Robert 1, 180

Gödel, Kurt 279

Golden Record 158

Goldilocks Zone (see also habitable zone) 315

Gong 406

'Grand Tour' (see also Voyager 1 and 2) 154, 157, 253

Grasshopper 1 and 2 133

Gravitational lens 151–152, 266

Gravitational waves 153

Gravity assist 73, 157

Grey, Charles 80

Greenstone device 97–98

Greenwich, University of 97

Grinspoon, David 383

GSLV rocket 125

Guth, Alan 229–230

H II rocket 125

H III rocket 128

Habitable zone 315–316, 351, 379–380

Hadfield, Chris 407–408

Haëri, Louis 417

HAL-2000 177, 276

Hall thrusters 183

Hanson, Robin 365

Haqq-Misra, Jacob 390

Hardy, David A 419

Hatzes, Artie 315

Haumea 333

Hawking–Ellis energy requirements 232

Hawking radiation 246

Hawkwind 406

HD 69830 (star) 317

HD 70642 379

HD 154345 379

HD 189733b (planet) 310

Heavy-lift rockets 105, 110–112, 115, 120

Hebb, Donald 281
Hein, Andreas 51
Heinlein, Robert A 4
Heisenberg Uncertainty Principle 233, 248
Helion Energy Project 140
Heliopause 46, 48, 65–66, 151, 156, 163, 164
Heliosheath 54, 156, 192–193, 320
Heliosphere 9, 56, 64, 73, 158–159, 192–193, 199, 255, 319
Helium 3 50, 164
Hibernation 351, 359–362, 368, 374
High Accuracy Radial velocity Planet Finder (HARPs) 304–305, 311, 314, 390
High Frontier, The 352
Hilbert, David 279
HiPEP (High Power Electric Propulsion) 184
Holst, Gustav 414
Horizon problem 229–230
Hot jupiters 304–305, 307, 310, 316
Hot neptunes 307, 313, 317
HoTOL 126
Hubble Space Telescope 293, 306
'Human job time' 56
Humanity
    approaching end of ideological evolution 26
    civilisation 27
    colonising other environments 88
    expansion into Universe of 24, 38
    post-humanity 368
    survival of 23–24, 87–88, 236–237
    Human League, The 408
    Human reproduction in zero gravity 330
    Hydrogen 332
    Hylozoic Ground Series 97
    Hyperion, Project 359

Icarus Concept Design Competition 359
Icarus Interstellar 359
Icarus Pathfinder 168
Icarus, Project 126, 142, 168, 328, 359
Icarus Starfinder 168, 169
IKAROS sail mission 198
Inflation 229
Initiative for Interstellar Studies 10, 99, 169
Innovative Interstellar Explorer 163–164, 193–196, 199
Intelligent life 25, 153, 382, 387–388, 392
International Academy of Astronautics (IAA) 56, 172, 391
International Astronomical Union 333
International Space Station 71, 185, 329, 407
International Thermonuclear Experimental Reactor (ITER) 140
Interstellar Boundary Explorer (IBEX) 320
Interstellar flight 10, 43, 350
    affordability 141–142
    benefits of 237
    current knowledge insufficient for 47
    global enterprise to achieve 43
    helping to create a stable civilisation 27, 54, 153
    link to intelligent life 58
    objectives of 349, 371
    physiological problems 276
    realistic future 10
    reinvigorated by discovery of exoplanets 14
    research into 43
    scientific requirements of 44
    societal and cultural motivations of 14, 23, 37, 54, 327, 368
    technology required 349
Interstellar Heliopause Probe 198
Interstellar medium 15, 16, 63, 69, 162,

191–192
  dust 167
  entry into 63, 156, 254
  making measurements of 16, 151, 163
Interstellar Precursor Mission 165
Interstellar Probe mission 161
Inertial Confinement Fusion (ICF) 140
Inertial Electrostatic Confinement Project 140
Ion Drive mission 169
Ion engine 164, 165, 181, 183, 186, 196–197, 201, 204, 207, 210, 401, 404
  Four-Grid Ion Thruster 204, 207

James Clerk Maxwell Telescope 317
James, William 276
Japanese Space Agency (JAXA) 128
Jarre, Jean Michel 409–410, 413, 419
Jefferson Starship 406
Jeff Wayne's The War of the Worlds 411
Jet Propulsion Laboratory (JPL) 63–64, 154, 156, 157, 161, 165, 190, 207, 210, 260, 277
Journal of the British Interplanetary Society 49, 51
Jupiter 154–157, 165, 253, 255–256, 301, 307, 317
  Galilean moons 184, 380
  orbit of 305
  starship built in orbit around 399
  Jupiter Icy Moons Orbiter 184–185, 187

Kardashev, Nikolai 222
Kardashev scale 222, 384, 389
Kare, Jordin 220
Kennedy, President John F 212
KEO Project 418
Kepler 37b (planet) 309

Kepler Space Telescope 18, 34–36, 305, 308, 316, 379–380, 402
  number of discoveries 304
  results from 36
Kim, M 332
Kipping, David 312
Klep, Rolf 397
Kohonen, Teuvo 288
Kraftwerk 408
Kuiper Belt 46, 66, 73, 150, 153, 162, 166–167, 191, 254, 317, 333–334, 336
Kuiper Cliff 150–151
Kuiper, Gerard 150
Kurzweil, Ray 365, 368

Laser electric propulsion 45, 190
Last Starfighter, The 3
Launch systems 85
Landis, Geoffrey 166, 211, 219
Large Magellanic Cloud 321
Laser 211, 217, 219–220, 248, 258–260, 264
Launius, Roger 368
Lawrence Livermore National Laboratory 140
Lawrenceville Plasma Fusion Project 140
Lens for sail beaming laser 167
Ley, Willy 397
Life on Mars? 407
Liftport Group 140
Light Dreams, The 419–420
Lineweaver, Charles 381
Link budget 256
Link margin 257, 263–264
Lithium 164, 332
Lobo, Francisco 249
Local Bubble 16
Local Group of galaxies 321
Local Interstellar Cloud 16
Locke, John 27
Loeb, Horst 196

Lohman 16 328
Longevity of a civilisation 383–385
Long March rockets 125, 130
Lone Signal 390
Low Earth orbit (LEO) 105

Maglev 341
Magnetoplasmadynamic thrusters 183
Makemake 333
Manhattan dome 90
Manned missions 51
Mansell, Clint 417
Mantovani, James 223
Man Who Fell to Earth, The 407
Mariner space missions 20, 293
Mariner 10 157
Mars 19, 20, 92, 126, 181, 193, 275–276,
        287, 306, 328, 336, 344, 381–382,
        386, 404–405, 411
Mars Colonial Transporter 134
Mars Direct 354
Mars Exploration Rovers 19, 277
Mars-like world 338, 344
Martinez, Cliff 417
Matloff, Greg 45, 172, 351]
Matrioshka brain 364
Matrix, The 385
Mayor, Michel 303
Mayr, Ernst 382
McCurdy, Howard 368
McNair, Ronald 410
McNutt, Ralph 163, 194, 198
Mega-cities 89
METI (messaging extraterrestrial
        intelligence) 377, 390, 392
Mercury 220–221, 303–304, 306, 309,
        312, 328
Metropolis 275
Metzger, Philip 222
Michaud, Michael 57, 391
Microgravity 228, 321, 329–331

Microlensing 302
Microwave electric propulsion 45, 190
Milky Way 217, 301, 309, 315, 319, 379,
        384–385, 387, 398
Mill, John Stuart 24
Miller, George 289
Minkowski space-time 240–241
Minovitch, Michael 9, 154
Mira 397
Mir space station 71, 201
Moore, John H 342
Moore, Patrick 397, 400, 402, 415
Moravec, Hans 365
Morganthau, Hans J 391
Morris, Michael 243
M-Theory 230–231
Moon
        Apollo astronauts 330
        Apollo seismic experiments 19
        building a base on 404
        industry on 221–223
        stabilising Earth's rotation 382
        tidal locking 309–310
        tourists 400
Mueller, Robert 223
Muscatello, Anthony 222
Music 405–418
Music of the Spheres 414
Musk, Elon 132

N-1 rocket 127
Nano-replicators 363–364
NanoSail-D 198
Nanosat 169
Nanotechnology 140, 219–220, 365
NASA
        Celebrating anniversary of Moon
        landings 412
        Centennial Challenge 140
        collaboration with Jean Michel
        Jarre 410

Eagleworks Labs 233
  funding of 83
  funding private launch
    vehicles 132
  Heliospheric Division 320
  In Space Propulsion Technology
    Programme 170
  Interstellar Probe 161
  Moon landings 82
  NERVA Project 184, 211
  payloads 105
  photographs of Mars 405
  roadmap towards interstellar
    flight 46
  self-replicating facility on the
    Moon 222
  spin offs from 82
  use of popular culture 4
  vision for the Voyager interstellar
    mission 9
  Voyager programme 156
National Radio Astronomy Observatory 378
Near interplanetary space 74
Nebulae 321
Negative energy 232–236, 244, 246, 248
Neptune 66, 68, 75, 150, 156–157, 253,
    307, 333, 397, 399, 404
Neural network 284–285, 287, 292
Neurons 280–282, 284–285, 288
Neutron star 321
New Challenge to the Stars, The 398–399
New Horizons 74, 113, 149, 187, 255–256
Newton's Law of Gravity 282
NEXIS (Nuclear Electric Xenon Ion
    System) 184
Next Day, The 407
NEXT thruster 202
Nitrogen 337–338, 344–345
Niven, Larry 4
Nordley, Gerald 220–221, 223
Norem, P C 217

NSTAR ion engine 186
Nuclear electric propulsion 54, 56, 141,
    188–190
Nuclear propulsion 7, 169, 206–207, 216
  fission 44–45, 217
  fusion 45, 47, 140, 168, 217, 359,
    395, 402
Numan, Gary 408

Oberth, Hermann 181–182
Obousy, Richard 231
Oldfield, Mike 412–413
O'Neill cylinder 93, 96, 339–340, 349, 355
O'Neill, Gerard 349, 351–353
Oort Cloud 64, 66, 68–69, 73, 152–153,
    167–168, 172, 191, 203, 205, 210,
    318, 334, 336
Oort, Jan Hendrik 152
Orbital Science 125, 131
Orbits 70, 316
Orcus 333
Orgel, Leslie 381
Orion, Project 44, 399, 402
100 Year Starship 85
Oxygene 409, 419
Oxygene 7–13 410

Pale Blue Dot 69
Panspermia 25–26, 37
Paradox 240
Parker, Eugene 332
Parkinson, Bob 402
Passive House 91
Pellet propulsion 219–220, 223
Penn State University 390
Penrose, Roger 279
Philippe, Jean-Marc 418
Philips Research Labs 247
Phoenix lander 1
Photodetectors 260–261
Photon rockets 215–217, 398

Pink Floyd 406
Pioneer plaque 158
Pioneer 10 and 11 73–74, 154–156, 158,
        162, 170, 187, 255
Planetary Encounters 67
Planetary migration 307
Planetary Resources Inc 223
Planets, The 414
Plasma drive 168, 183
Plate tectonics 308, 316
Pluto 66, 149–150, 166, 255–256, 334–
        339, 341
Pluto Orbiter 165
Reclassification 333–334
Plutoids 191, 333–339, 341, 345
Posner, Arik 320
Post-Detection Protocol Committee 391
Post-humans 368
Post natural fabrics 95
Power-sats (solar power satellites) 220–223
Powers, Robert M 67–68, 74
Precursor mission 46, 55, 153, 160–170,
        183, 202, 205, 355–358
Procyon 28
Prometheus, Project 184–185
Proton launcher 125
Protoplanetary disc 307
Proxima Centauri 33, 50, 210, 227, 230,
        313, 328, 398
        time to reach at Voyager 1's
        speed 48, 160–161, 179, 227
Pulsar 302, 404
Pulse Position Modulation (PPM) 262
Pythagoras 414

Quantum computing 53
Quantum Electro Dynamics 248
Quantum entanglement 265, 267–271
Quantum Field Theory 51–52, 230,
        232–233
Quantum mechanics 267–268, 270–271

Quaoar 333
Qubits 269
Queloz, Didier 303

Radial velocity 36, 302–303, 314
Radiation 157, 311, 331–332
Radioisotope electric propulsion 187–188,
        193–194, 196–197, 199
Radioisotope thermoelectric generator
        (RTG) 155, 158, 164, 184,
        187–188, 197, 202
Ramjet/ramscoop 50, 395, 400, 402
Rare Earth hypothesis 380
Reaction Engines Ltd 126, 134
Realistic Interstellar Explorer 163
Rectenna 190
Red dwarf (star) 29, 36–37, 301, 308–313,
        320, 344, 379–380, 390
Red supergiant 397
Reed–Solomon coding 254, 262
Rendez-Vous 410
Rendezvous with Rama 96
Resistojets 183
Resources 87, 89–90
*Rise and Fall of Ziggy Stardust and the
        Spiders from Mars, The* 406
Robosloth 172
Robotics 53
Robots 365
Rocket equation 180–181
Rogue planets 318
Roman Empire 383
Roman Law 81–83
Rosen, Nathan 243, 267
Rosetta 201
Ross 248 9, 159, 320
Rotary Rocket Roton 126
RS-25 rocket engine 127
RS-68 rocket engine 111–112, 120
Rush 406
Russian Academy of Science 378

Rutan, Burt 132

SABRE engine 137
Safe Affordable Fission Engine 190
Sagan, Carl 5, 69, 158, 384, 398, 400
Sails 45, 54, 56, 141, 167, 169–170, 172,
      180, 193, 197–199, 202, 211,
      laser sail 217–222
      magnetic sail 218–220
      micro sail 220
      stage sail 219
Salyut 71
Sänger, Eugene 44
Saturn 71–72, 155, 157, 184, 253, 255,
      307, 380
Saturn V 127
S-band 254–255
Scaled Composites 131–132
SCARLET 185
Schrödinger, Erwin 94–95
Schwarzschild solution 240–241
Science fiction
      1970s 406
      adaptations 411
      as test bed for the future 1
      authors 405
      cliche 279
      collaboration with real science 7
      first science fiction films 405
      masterworks 405
      possibilities of interstellar
      colonisation 25
      realistic spacecraft 4
      role in envisioning interstellar
      travel 2
SCUBA sub-mm camera 317
Sedna 65–66, 150, 333
Self-replicating probes 55, 58, 221–223,
      318, 354, 359
Sentinel (music) 413–414
Sentinel (monolith) 415

SERT-1 spacecraft 182
SETI 46, 58–59, 152–153, 265, 270, 288,
      321, 343, 377–378, 383, 390, 398
Shadow biosphere 381
Shaw, Bob 399–400
Shell worlds 344–345
Shepherd, Les 44
Shermer, Michael 383
Singer, Clifford 219
Single-Stage-To-Orbit 125–126, 134,
      137–138, 141
Sirius 28, 159
Sky at Night, The 415
Skylab 71
Skylon 126, 134–137, 141
Sleeper ships 362
Small Magellanic Cloud 321
Smith, Adam 84
Smith, Cameron 353
Smith, R A 397
Snowball Earth 382
Snow line 334
Solar electric propulsion 170, 185, 191,
      193, 196–197, 199
Solaris 417
Solar System
      colonisation of 27, 60, 71, 353
      economy 126, 142
      exploiting resources 224
      exploration of 19–21, 50, 55
      extent of 63–64, 67–68, 71, 73, 151
      life 386
      orbit around galactic centre 382
      origin of 162
      travel across 71–72
      types of planet in 306
Solar wind 64, 67–68, 159, 255, 319
Soleri, Paolo 91
Songs of Distant Earth, The (music) 412
Soyuz 125
Space 1999 3, 415

Spacecraft 2000–2100 AD 400

Space elevator 126, 138–140

Space Launch System 126–127

Space manufacturing 354

Space Oddity 406–407

Space settlements 317, 338–342, 345,
    351–353, 373
      size of a colony 342, 353

Spaceship art 2

'Spaceship Earth' 89

Space telescope 126

Space-Time Alteration Drives, see warp drive

SpaceX 125, 131–134, 141

Special effects 3

Special Theory of Relativity 227, 229,
    239–240, 266

Specific impulse 165, 179, 183, 186, 188,
    197, 203, 205, 210

Specific power 216

Spitzer Space Telescope 317

Standard Model 60

Stanford Torus (see also O'Neill cylinder) 353

Stapledon, Olaf 24–25, 38, 215

Star Maker 215

Star formation 334, 379

Starman 406–407

Starseeds painting 5

Starship
    as a positive image 7
    benefitting science 15
    colonisation targets of 356
    communication system of 253–264
    construction of 126
    crewed starship 372–373
    design of 45–46, 99, 253
    ecology of 95–96
    launch 105
    purpose of 253
    scientific objectives of 15
    shielding 332–333, 338, 353
    to explore exoplanets 14

Starship Cities 99–100

Star Trek 2–5, 179, 212, 228–229, 412, 415

Star Trek: The Next Generation 313

Star Wars 3–5, 398

Starwisp 172

Stellar astrophysics 17–18

Stellar Radiance 419

Stem cells 366

Stine, G Harry 145

Stirling Radioisotope Generators 88, 200,
    202

Stone, Ed 63

Stonehenge 278

Stratolaunch 131

Strauss, Richard 415

String Theory 230–231

Stuhlinger, Ernst 181, 182

Sub-probes 18, 20, 22, 44, 258

Sun 16–17
    closest stars to 28
    evolution of 24
    flares 332
    gravitational influence 65, 67
    intensity of sunlight 220–221
    magnetic field 64, 163

Sun Ship 181, 182

Superconductivity 218

Super-earth 307, 310–311, 315, 317–318

Sustainable design 91

Svaiter, N F 235–236

Synthesisers 408

Tachyon 239–240

Tangerine Dream 408

Tau Boötis 319, 402

Tau Ceti 28, 29, 34–35
    exoplanets around 34

Tau Gruis 404

Taylor Cone 201

Technical University of Munich 359

Technological Singularity 385

Technology Readiness Level 198
Teleportation 366–368
Telescope 259–261
Termination shock 48, 159, 163, 320
Terraforming 329, 343–345, 373
Tin Tin, Project 199–202, 212
Thales 79
Things to Come 2
The Thing From Another World 2
Tholins 336
Thorne, Kip 243
Thousand Astronomical Unit (TAU)
        mission 165, 207–209, 212, 264
3D printers 343, 354
Tidal locking 309, 312, 314, 379
Time travel 52–53
Titan (moon of Saturn) 19, 312, 380, 417
Tom Corbett, Space Cade 2
Tour of the Universe 400
Transhumanism 88
Transit 36, 302, 304–306
Transit Timing Variations (TTV) 312
Trans-Neptunian objects 191
Tri-Alpha Project 140
Triangulum Galaxy 321
Trifid Nebula 398
Tripods, The 415
Triton 66
Tsiolkovsky, Konstantin 138, 180, 352
Tubular Bells 395, 411, 413
Turing, Alan 276–280, 282–283
Turning an interstellar spacecraft 217
Tuomi, Mikko 34
Twilight Zone, The 2
2001: A Space Odyssey 2, 4, 275, 295,
        415–416
2007 OR10 333
2012 VP113 66, 151, 333
Tziolas, Andreas 199–200

UFO (TV show) 415

UFOs 2, 387
Ultraplanetary space 67–68, 74
Ulysses probe 73, 187
United Launch Alliance (ULA) 105–110,
        112–113, 115, 120–122, 125
Universal laws 57–58
University of New South Wales 381
Uranus 156, 157, 253, 307

Vandenburg Air Force Base 106, 110, 130
Vangelis 417
VASIMR 72, 183
Vega rocket 125
Very Large Telescope 304
Venus 306, 344
Vesta 186
Viking Mars missions 20, 22, 187
Volatiles 337
Von Braun, Wernher 181–182, 239, 397
von Däniken, Erich 387
Von Neumann, John 280
Vos Post, Jonathan 218
Voyager 1 9, 48, 63–64, 71, 74, 154,
        158–161, 169–170, 187, 191–192,
        194, 196, 253–255, 319–320, 322
    cease transmitting 163
    cruise speed 48, 160, 179, 212,
        227, 253
    destination 159, 199
    distance from Earth 64, 159, 179
    escape from heliosphere 9, 63, 65,
        73, 151, 191, 254
    Golden Record 417
    robotic ambassador 159
    signal from 158, 254
Voyager 2 9, 48, 64, 74, 154, 158–161, 169,
        170, 187, 191, 196, 253–255,
        319–320, 322

Wallace, Richard 198
War of the Worlds, The 1, 411

Warp drive 51–53, 55, 179, 228–237,
        246–247
Wason, Peter 283
Water 307–308, 312, 315–317, 332, 336,
        338, 344–345, 407
Weak Anthropic Principle 57
Weakly Interacting Massive Particles
        (WIMPs) 162
Wealth of Nations 84
Webb, Stephen 387–388
Wells, H G 1–2, 411
Wentworth, James Jason 67
Wheeler, John 249
White dwarf 321
White, Harold 233
White hole 242, 244–245
Wide-field Infrared Survey Explorer
        (WISE) 318
Wolf 359 313
Woodcock, Gordon 402
World-ship 7, 51, 54, 58–59, 77, 89, 90,
        92–93, 95, 96, 97, 351, 353–359,
        362–363, 368, 374
Wormhole 228, 240, 242–247, 249–250,
        364, 367–368, 372
Wright Flyer 10

X-30 NASP 126
X-33 126
X-band 254–255

Zaitsev, Alexander 378, 390
Zenit launchers 125
Zeta Aurigae 398
Zeta Cancri 398
Zeta Leporis 317
Ziggy Stardust 395, 406–408
Zubrin, Robert 218, 354
Zygotes 351, 361–364, 374